Environmental Regime Effectiveness

Global Environmental Accord: Strategies for Sustainability and Institutional Innovation

Nazli Choucri, editor

Environmental Regime Effectiveness

Confronting Theory with Evidence

Edward L. Miles, Arild Underdal, Steinar Andresen,
Jørgen Wettestad, Jon Birger Skjærseth, and Elaine M. Carlin

The MIT Press
Cambridge, Massachusetts
London, England

This book was set in Sabon by Best-set Typesetter Ltd., Hong Kong.

Printed and bound in the United States of America.

Library of Congress Cataloging-in-Publication Data
Environmental regime effectiveness : confronting theory with evidence / Edward L. Miles
. . . [et al.].
     p.   cm.—(Global environmental accord)
   Includes bibliographical references and index.
   ISBN 0-262-13394-6 (hc. : alk. paper)—ISBN 0-262-63241-1 (pbk. : alk. paper)
   1. Environmental policy—International cooperation—Case studies.   I. Miles,
Edward L.   II. Global environmental accord.
GE170 .E578   2002
363.7′0526—dc21
                                                                                    2001044183

To the memory of Meg O'Hara (1944–1998), who also lived this book

Contents

Appendixes

Preface

The story of this book begins in 1987,* and there have been so many twists and turns to that story that it is worthwhile setting them down. In that year, I was finishing a large research project and wrestling with the question of what to tackle next. Working in Washington, D.C., in October, I was handed a proposal by the chair of an interagency committee on oceans for a major U.S. initiative on the Law of the Atmosphere that addressed the global climate-change problem, an effort akin to the Third United Nations Conference on the Law of the Sea that had begun in 1973 and concluded in 1982. "Let us know what you think of this idea," he said with a smile, thereby signaling that I was the chosen hatchet man. Indeed, having spent more than ten years of my life in the Law of the Sea Conference, my instinctive reaction to a Law of the Atmosphere initiative was intensely negative for a variety of reasons.

In the first place, such large-scale and difficult issues—even without the liability issues—would require at least a decade of negotiations. Moreover, global effort on this scale would be certain to trigger a North-South confrontation on another massive scale, thereby adding to the delays. Finally, during all of this activity and conflict we would lose much time because no actions would be taken in the interim: everyone would have an incentive to await the final outcome of the conference.

I told the chair I would take the weekend to think over his request, returned to Seattle, and communed with myself on Alki Beach. Firmly convinced that a U.S. initiative on a Law of the Atmosphere was a bad idea, I struggled with what I would propose as an alternative, since it was clear that some of the Europeans would in fact propose something like a U.N. Conference on Global Climate Change. I remember that the conversation with myself went something like this:

---

* The text in this preface has been reprinted with permission from *Policy Sciences*, vol. 31, no. 1 (1998), where this article first appeared.

Why do most of the processes you study not work?

Well, what do you mean by *work*?

I mean, why don't they resolve the problems for which they were designed?

Ahh! Then you're saying that they are not effective?

That's right!

Then what do you mean by *regime effectiveness*, and how would you measure it? And once you've answered those questions, how would you design a regime that would solve the problems it was intended to solve?

These musings led me to argue against a major U.S. initiative on climate change and to propose that the Department of State investigate an alternate approach based on a study I wanted to undertake. My argument was that global environmental change, as a set of policy problems, falls into the class of collective- or public-good problems, where a well-developed theory (Olson 1965, 1982) predicts that unless the number of individuals in the group is small or unless coercion or some other special device ("selective incentives") makes individuals act collectively in their common interest, rational self-interested individuals will not act to achieve common or group interests. In this sense, the problem of global change is not a new type of policy issue in the international system (Brooks 1977).

Thus I argued that states seek to collaborate in the international system either to gain benefits they define as important but cannot achieve through their own individual actions or to minimize or avoid negative consequences of their own or others' individual actions. But the mere *attempt* to collaborate does not guarantee any results, let alone optimal outcomes. Collaboration must be negotiated, state behavior must be coordinated, and costs must be shared. Furthermore, states normally insist that the benefits from collaboration must exceed the cost of collaboration, and if they do not, the incentive to collaborate declines and the amount of the collective good produced becomes suboptimal.

Given this perspective on cooperation, to presume that global regulation by treaty negotiation is the most effective response to the set of policy problems posed by global climate change is questionable. I anticipated, for reasons elaborated below, that despite international cooperative agreements, uncontrolled climate change would occur until the costs of leaving it uncontrolled become visible. At that point, working out a collective solution to better climate conditions might be easier.

Four sets of specific objections can be raised to using treaty negotiation to respond to policy problems requiring international collective action for their solution. First, Underdal (1982) shows that international regulation of common property resources must meet two fundamental requirements: all participants in the regime must agree to the proposed regulation, and all significant states must participate in the regime. These two conditions give rise to "the law of the least ambitious program." This law can be expressed as follows: where international management is established through agreement among all significant parties, and where such regulation is considered only on its merits, collective action will be limited to those measures acceptable to the least enthusiastic party. It is possible to alter these preferences only through what Olson (1982) calls "selective incentives"—such as side payments, political pressure, and education. The dynamics of decision making therefore favor players who oppose collective action.

Second, Underdal argues that even the least ambitious program may be unattainable, in practice, as a result of the following dilemma: unless the parties face the distributional consequences of regulation, there may be no regulation that can attract the degree of collective support necessary to have the desired effect; however, when the parties face the distributional consequences of regulation, the risks of aggregation deadlock significantly increase.

Third, unless the treaty contains provisions for monitoring performance of the signatories with effective sanctions to deter cheating, there is in fact a serious risk of defection (Munro 1987).

Fourth, considerable uncertainty surrounds current model predictions of the magnitude and effects of global climate change. While this uncertainty alone provides no adequate argument against regulation, it means that research will be ongoing during negotiations and as regulations are put into place. Since research results could change the nature of the regulations required for an effective response, the regulatory system will have to be sufficiently flexible to change over time, but such an open-ended system implies changing distributional consequences, and herein lies the problem. Miles (1987) found that such flexibility is permitted by states (1) when links are *indirect* between research results and regulatory or management decisions or, (2) when links are direct and the decision rules provide each party protection in the resolution of the interests it is seeking. (The latter usually involves a veto, which induces rigidity, not flexibility.) Where neither of these protective conditions exists, governments have a considerable incentive to constrain the search for and use of scientific research results.

So we appear to be on the horns of a dilemma. To deal effectively with the policy problems posed by global environmental change, global regulatory action is necessary. Since formally independent sovereign states comprise the international system, global regulation will require the negotiation of treaties. However, for the reasons identified above, treaty negotiation, by itself, will be unlikely to provide effective solutions to the problems identified. One key to resolving this dilemma lies in developing and expanding Olson's notion of selective incentives. To do this, we need to evaluate the levels of feasible international collaboration, the relative effectiveness of available joint action on international collective-good problems involving science and technology, and the dynamics inherent in these situations.

Accordingly, the research project I proposed had three parts. The first part sought to answer the following questions based on a set of cases chosen for analysis:

· Under what conditions is international treaty negotiation an effective response to scientific and technological policy problems requiring coordinated action?
· Do these conditions include the use of selective incentives? If so, what are they, and how do they work?

The second part of the project sought to answer the following questions:

· If most international collective responses to collective-good scientific problems are inadequate, what types of incentives and strategies facilitate adequate levels of collaboration and effectiveness of joint action?
· Given the nature of global climate change as an international political issue, which actors can devise and implement them?

The third, applied part of the project sought to answer the following questions:

· Is hard global regulation—clear standards, precise goals, and firm targets—a necessary point of departure for facilitating international cooperation on the global climate-change problem?
· If hard global regulation is not politically feasible and if selective incentives are required in the creation and maintenance of international regimes, how should states be induced to establish effective regulations to cope with the global climate-change problem?

After considering the results of the investigation in the first and second parts of this project, I proposed to assess the possibility of pursuing soft regulation in a decentralized fashion as a strategy for buying time in the face of uncertainty. Soft regulation—avoiding clear standards, precise goals, and firm targets—would seek to initiate a process combining diffuse and specific reciprocity strategies initially in Organization for Economic Cooperation and Development (OECD) member states and then in only the USSR, Central Europe, and China. Significant emission reductions in the OECD group would leave room for emission growth in developing countries without triggering a major confrontation in a global setting. Side payments (in technology transfers and financing) would play a prominent role in this process, as would epistemic communities and the building of consensual knowledge. The decision process would be designed to maximize learning potential over time and to allow participants to build their confidence to negotiate specific protocols sequentially. For regime-design purposes, we would need to answer the following questions:

· On what scale does regulation have to proceed? Since group size is an important criterion, what can be done effectively on a regional rather than global basis?
· Is it possible and useful to mute the potential North-South dimension to conflict by focusing most regulatory action initially on the North?
· What strategies are available for ensuring maximum participation with the least potential for conflict?
· How can states most effectively appraise the performance of decentralized actions so that they can have confidence that the standards are being met?

The initial response of the Department of State to my ideas was enthusiastic, and I was encouraged to pursue a formal proposal. Little did I realize the size, shape, and speed of the rollercoaster I was about to board.

During 1988, a significant divergence of view on global climate change emerged between the Western Europeans, led by Prime Minister Margaret Thatcher of the United Kingdom and the Bush administration in the United States. Thatcher led a very high-profile initiative culminating in a major meeting held in Noordwyk, The Netherlands, in November 1989 and in the same month delivered a major speech calling for strong action at the U.N. General Assembly.

Secretary of State James Baker was the leader of the U.S. delegation at Noordwyk and made a speech before the Intergovernmental Panel on Climate Change (IPCC) outlining three principles that were to underlie U.S. policy: (1) action should not be delayed until scientific uncertainties are resolved, (2) the immediate focus should be on steps that are justified on other grounds (that is, a no-regrets policy), and (3) proposed solutions should be specific, cost-effective, and fair to all concerned. These words, however, were the last on the subject ever to pass Baker's lips in public.

As the movement for a Framework Convention on Climate Change gathered steam, pushed enthusiastically by Thatcher and her cohorts at Noordwyk, and this issue was identified as one of the highest priorities to be included in the Rio meeting on sustainable development then set for June 1992, a new architect of U.S. policy emerged in the person of John Sununu, chief of staff of the Bush White House. Sununu asked the assistant director for geosciences of the National Science Foundation for the computer codes for the general circulation models (GCMs) that the U.S. team used for the IPCC assessments and had these codes run on the super-computer in the White House Situation Room. When he saw that the data points were 500 kilometers apart in the first-generation models, he decided that GCMs were not credible and could not be used to make policy.

The U.S. policy that emerged after this point and until the Clinton administration was therefore based on the following operational precepts:

· Oppose and seek to stall all attempts at a law-making conference on climate change.
· Increase U.S. investments in research concerning the physical dynamics of climate change, and reduce levels of uncertainty in the GCMs.
· Prohibit all major research into the policy aspects of global climate change.

To highlight this contrast between the United States and its partners in the OECD on the global climate-change issue, I have gotten ahead of my personal story. Back in 1988, when the Department of State was initially interested in my argument against a U.S. initiative on a Law of the Atmosphere, I could not develop a full-scale proposal because I was involved in other commitments. I also thought that the research credibility among the Group of 77 would be enhanced if this were not solely an American effort. Based on my observations from the Law of the Sea Conference, I thought the output would be more acceptable if it were either a joint U.S.-Canadian effort or a U.S.-Norwegian effort. Since the Canadians did not then have a university-based group working on these issues and the Norwegians did, I chose Norway and the occasion of the thirtieth anniversary of the Fridtjof Nansen Institute to make my proposal. Although the first draft of a joint proposal was ready by late 1988, and the Norwegian team was funded by early 1989, I ran into problems on the U.S. side.

My proposal was later submitted to the Office of Ocean, Environment, and International Scientific Affairs (OES) of the U.S. Department of State. It went in through the Ocean side, but Environment was expected to have a major say since it would be in charge of climate-change negotiations. At the time, Ambassador William Nitze was Deputy Assistant Secretary of State for Environment, but my proposal picked up significant collateral support from the Policy Planning Staff, Secretary Baker's Office, and the Office of the Permanent Representative of the United States to the United Nations in New York. At the time, the U.S. Perm. Rep. was Ambassador Thomas Pickering.

In spite of this powerful support, my proposal was delayed because Nitze was out of the country for an extended period of time, and some of his senior staff thought it inappropriate for the United States to lend its support to a joint project with Norwegians when Prime Minister Gro Harlem Brundtland (the Green Queen) was giving President Bush a hard time publicly on environmental issues. My other supporters in State were unable to override such concerns, and eventually the issue was sent up the ladder for Assistant Secretary Fred Bernthal to resolve. Bernthal split the difference and proposed chopping off the questions that did not involve the oceans, thereby getting around the need for approval by Environment within OES.

In November 1989, the Department telephoned me in Sweden, where I was then lecturing, to see if I was agreeable to such a compromise. In that conversation, I asked if the United States fully understood that the people who would be collaborating with me would also have access to Prime Minister Brundtland. The reply was that some people understood and that some people chose not to. I said that I could accept the Bernthal compromise but that the questions that remained were far less useful to the Department than the questions that were deleted. The official pointed out that once I had answered the questions that were specified in the contract, I could answer any additional questions that appeared relevant to me. On that basis, I agreed to the Bernthal compromise.

That compromise was never implemented because Nitze returned to Washington, D.C., and when we met on December 9, 1989, he asked me what exactly I wished to do. My response was that I wished to assist the Department in its preparations for negotiating the Climate Convention and that were I czar of U.S. climate-change policy, I would insist on getting answers to the following questions:

· Given the jobs to be done at global, regional, and national levels, what approaches to institutional design facilitate effective action?

· How do the political dynamics of the global climate-change issue affect the feasibility of the recommended approach to institutional design? What design adjustments enhance the feasibility of acceptance without compromising effectiveness too greatly?

· Since rapid action is essential for stabilizing and eventually reducing greenhouse-gas emissions and therefore atmospheric concentrations, should protocols to a framework convention adopt hard or soft reduction targets?

· How should these protocols be phased?

· To what extent should actions be based on governmental regulations? On market incentives? On some combination thereof?

· How would a combination of soft phased targets, market incentives, and decentralized state action work within the framework convention?

Nitze responded that if I would immediately put my questions in no more than a two-page letter to him, he would sign off on the project. I sent him such a letter on December 10, 1989. A short time later, Nitze was fired on orders of the White House, and my contract was killed because it had not then been formally approved by the comptroller-general of the Department of State.

The error Nitze had committed in Sununu's eyes was to take the Climate Convention negotiations, and the role of the United States within them, too seriously for a White House that was then beginning an about-face in U.S. policy. The bureaucratic decapitation of Nitze was an object lesson that was not lost on the rest of the Department of State and others in the U.S. government whose responsibilities included the climate-change issue. I quickly moved to other bases, but they had received orders from Sununu not to fund any "MARS-type stuff"—that is, mitigation and adaptation research studies were unwelcome.

I do not suggest that Sununu even knew I existed. Clearly, once these decisions had been taken, the U.S. government took the position that it did not wish to support any policy research on global climate change because policy-research results would stimulate agitation for policy action. Therefore, for the government it was better not to know the options and therefore to avoid evaluating the options by claiming that the level of uncertainty in the GCMs was simply too great.

With no hope of funding from the U.S. government, I then set out in pursuit of the foundations. Support for my research was eventually patched together with small grants from the Hewlett Foundation, the University of Washington, and the North Atlantic Treaty Organization (NATO). Meanwhile, my Norwegian colleagues had lost no time in jumping on their steeds and charging the problem while I, the initiator of the project, was left at the starting gate lacking the wherewithal even to purchase a nag, much less a steed.

The mismatch in the timing and scale of funding caused further delays because each side became enmeshed in other commitments. While the NATO money enabled the two research teams to meet periodically in Seattle or Oslo, progress was slow. The pace picked up, in the first half of 1994 when, with funding from my then dean, G. Ross Heath of the College of Ocean and Fishery Sciences at the University of Washington, I was able to spend six months of a sabbatical leave in Oslo.

Still bearing the simultaneous burden of other large commitments, we were in sight of the end in 1996 when my wife became seriously ill. Her deterioration over the course of the next two and a half years led to further substantial delays in completing the book. In this connection, I am enormously grateful to my Norwegian colleagues for their patience. My wife eventually died on August 16, 1998, and the book is dedicated to her because she lived it from the time of the 1994 sabbatical in Norway. She never wavered in her conviction that these were important problems and that these particular Norwegian colleagues were exceptionally able. She would share in our satisfaction that the book has finally seen the light of day.

Edward L. Miles
Seattle

References

Brooks, H. 1977. "Potentials and Limitations of Societal Response to Long-Term Environmental Threats." In W. Stumm, ed., *Global Chemical Cycles and Their Alteration by Man* (pp. 241–252). Berlin: Dahlem Konferenzen.

Miles, Edward L. 1987. *Science, Politics, and International Ocean Management*. Berkeley: Institute of International Studies, Policy Papers in International Affairs No. 33.

Munro, G.R. 1987. "The Management of Shared Fishery Resources Under Extended Jurisdiction." *Marine Resource Economics* 3(4): 271–296.

Olson, Mancur. 1965. *The Logic of Collective Action*. Cambridge, MA: Harvard University Press.

Underdal, Arild. 1982. *The Politics of International Fisheries Management: The Case of the Northwest Atlantic*. Oslo: Universitetsforlaget.

Acknowledgments

Edward Miles wishes to thank the Hewlett Foundation for grants in 1990 and 1991 that enabled him to build a team in Seattle at the University of Washington to begin this research project. He also wishes to thank then dean G. Ross Heath for critical support provided during a sabbatical year in 1993 to 1994 and the University of Washington for the award of the Virginia and Prentice Bloedel Chair of Marine Studies and Public Affairs. Without this support between July 1994 and the summer of 2000 the book could not have been completed.

Thanks are also due to members of the Seattle team, who made significant contributions to the joint work but who left the project before the work was completed. In particular, Kai N. Lee was a vigorous member of the group working on a study of the regime for controlling the international trade in tropical timber until he left the University of Washington in 1992. Similarly, Susan Gaertner worked as Miles's research assistant from 1991 to 1992 on an evaluation of the London Dumping Convention regime but left the university before the study was completed. Maaria Solin Curlier, coauthor of chapter 14, prepared several drafts on the CITES regime as her master's thesis for the School of Marine Affairs at the University of Washington. Because preparation of this book took so long, the press of family and professional commitments prevented her from doing the final revisions. Steinar Andresen graciously consented to complete the work for Maaria.

The Seattle and Oslo teams wish jointly to acknowledge the support of the NATO Collaborative Research Grants Program for grant No. CRG 93073 in 1993 to 1994, which was critical in facilitating meetings of the teams in Seattle and Oslo during a difficult period in the life of this project. We jointly acknowledge also the valuable comments made on an earlier draft by the four anonymous reviewers of MIT Press.

Elaine Carlin wishes to acknowledge the extraordinary hospitality of Willy Østreng, director, and the entire Fridtjof Nansen Institute (FNI) and of Arild Underdal and his good offices during her time in residence in Oslo.

Arild Underdal has benefited from the supportive environments of the University of Oslo Department of Political Science, the Center for International Climate and Environment Research (CICERO), and the Centre for Advanced Study, Oslo. A grant from the Research Council of Norway to CICERO and FNI enabled him to set aside extra time at a critical juncture. He is grateful to Lynn P. Nygaard for editing assistance that not only improved his English but also helped clarify the argument in several instances.

The FNI team wishes to acknowledge support from the Norwegian Research Council that made possible its early work on this project from 1990 to 1992 in close cooperation with Arild Underdal. Additional support from the Research Council was obtained in 1997 through a joint grant to FNI and CICERO to complete the work. Because, for reasons that Miles has explained in the preface to this book, completion was not possible within the time frame of the grant, support for completing the project was then provided by FNI. Since the FNI team would not have been able to finish the project without this support, its members extend their sincere gratitude to FNI.

Finally, all the authors would like to acknowledge a large debt of gratitude to Suanty Kaghan, secretary to Edward Miles in the School of Marine Affairs at the University of Washington. Kaghan bore the brunt of moving the manuscript into final production from the last two drafts and typed all of Miles's chapters in the book through several versions. We also thank Jeannette Waddell for her valuable assistance in preparing the index.

I

Introduction

1

One Question, Two Answers

Arild Underdal

The Research Question

Many of the major policy challenges facing governments today are in some sense collective problems calling for joint solutions. However, even when effective solutions can be developed and implemented only through joint efforts, voluntary cooperation can be hard to establish and maintain, making it all the more important to understand the conditions for success and the causes of failure. This study addresses the question of why some efforts at developing and implementing joint solutions to international problems succeed while others fail.

Leaving aside sheer luck, at the most general level this question seems to have two possible answers. The first lies in the character of the *problem* itself: some problems are intellectually less complicated or politically more benign than others and hence are easier to solve. The second possibility focuses on *problem-solving capacity*: some efforts are more successful than others because more powerful (institutional) tools are used or because greater skill or energy is used to attack the problem.

This study attempts to explore the contents and merits of these two general propositions. We started out with an equal interest in both, but along the way we became increasingly preoccupied with the latter, for two main reasons. First, it points us toward factors that can—at least in principle—be deliberately manipulated by decision makers and hence used as *tools* for problem-solving. If the study of regime formation and implementation is ever to contribute to the development of praxis itself, it will have to do so primarily by providing insights into the functioning of accessible tools or instruments. Second, we think it is fair to say that—thanks largely to developments in the application of game-theory constructs to the analysis of international cooperation—more is known about what makes a problem politically malign than about what determines our capacity to solve it. In particular, we became puzzled (and also encouraged) by the fact that some of our malign problems were indeed solved or alleviated rather effectively, and we wanted to understand better what made such successes possible.

After having formulated the research problem, we must define our key concepts. What precisely do we mean by *regime effectiveness*, and how do we go about measuring it? What distinguishes a benign problem from one that is malign? What are the critical components of problem-solving capacity? Once we have constructed this conceptual platform, we can move on to try to translate the two basic propositions formulated above into specific form. The general statement that some problems are harder to solve than others becomes interesting only if we can specify *which* kinds of problems are malign and what makes them so. Likewise, the proposition that the problem-solving capacity of some institutions or systems is greater than that of others becomes interesting only to the extent that we can specify *which* kinds of institutions and systems have the greatest capacity and what determines this capacity. The remainder of this introductory chapter is devoted to these questions. In a final section we introduce our empirical testing ground—fourteen cases, all but one focusing on problems of environmental protection or resource management.

The Dependent Variable: The Concept of Regime Effectiveness

In a common-sense understanding, a regime can be considered effective to the extent that it successfully performs a certain (set of) function(s) or solves the problem(s) that motivated its establishment.[1] Although useful as a point of departure, it soon becomes obvious that this definition of *effectiveness* is not sufficiently precise to be useful as an analytical tool for systematic empirical research. Thus before venturing into comparative research, some conceptual groundwork is needed to clarify *what* precisely our dependent variable is. This is all the more important since analysts studying the impact of international institutions have construed their dependent variables somewhat differently.[2] Even though an encouraging trend toward convergence can be seen over the last decade, it can still be difficult to distinguish substantive differences from those that are merely terminological.

From a methodological perspective, evaluating the effectiveness of a cooperative arrangement means *comparing* something—let us provisionally refer to this object simply as *the regime*—against some standard of success or accomplishment. Any attempt at designing a conceptual framework for the study of regime effectiveness must, then, cope with at least three (sets of) questions: (1) what precisely constitutes *the object* to be evaluated? (2) against which *standard* is this object to be evaluated? and (3) *how* do we go about comparing the object to this standard—in other words, what kind of measurement operations do we have to perform to attribute a certain score of effectiveness to a certain regime? Let us briefly consider these three questions.

What Constitutes the Object to Be Evaluated?

This may at first glance appear a trivial question; the object clearly must be the cooperative arrangement (regime) in focus. A second look reveals that identifying the object to be evaluated may not be this simple.

First, we must determine whether we are interested only in the impact of the co-operative arrangement itself or also in the costs incurred and positive side effects generated in the efforts to establish and maintain it. The former is the appropriate basis for evaluating the *regime* itself, while the latter provides a basis for evaluating (also) problem-solving *efforts* or processes. The distinction is not merely one of academic hair splitting. Establishing and operating a regime usually entail various kinds of costs—some of which are significant—and a rational actor presumably makes his choices on the basis of some estimate of *net* rather than gross benefits.[3] Problem-solving efforts can also have significant positive side effects. Thus, international negotiation processes are often large-scale exercises in *learning*, through which at least some parties modify their perceptions of the problem and of alternative policy options and perhaps see their incentives change as well. As a consequence, the process itself may lead governments as well as nongovernmental actors to make *unilateral* adjustments in behavior—even in the absence of any legal obligation to do so (see Underdal 1994). The aggregate impact of such side effects may well be more important than the impact of any formal convention or declaration signed in the end. Since the costs of tracing the various process-generated effects systematically and in depth can be very high, we nonetheless in this project have to confine ourselves mainly to studying the effects of the regimes themselves. But the overall implication of what we have said above should not be missed: problem-solving *efforts* usually generate their own consequences, over and beyond those that can be attributed to the cooperative arrangement they may establish. And some of these process-generated costs and benefits may, indeed, be far from trivial (see, e.g., Miles 1989; Underdal 1994).

Second, as Easton (1965, 351–352) and others remind us, a distinction should be made between the formal *output* of a decision-making or regime-formation process (that is, the norms, principles, and rules constituting the regime itself) and the set of consequences flowing from the implementation of and adaptation to that regime. In the context of environmental policy, the latter may be further specified by drawing a distinction between consequences in the form of changes in human behavior (here referred to as *outcome*) and consequences that materialize as changes in the state of the biophysical environment itself (*impact*). Environmental regimes are, at least officially, established to protect some environmental values. The ultimate interest thus most often pertains to biophysical impact. In all cases, however, this goal is to be accomplished through changes in human behavior (outcome). In the regime-formation stage and its immediate aftermath, the norms and rules of the regime itself are, however, all we know. Actual change in behavior and environmental impact can be determined only at a later stage (usually after several years of operation). Moreover, there is no straightforward method for inferring outcome or impact from information about output. Targets of regulation sometimes respond by making ingenious adjustments that are hard to predict or by more or less flagrant noncompliance. Such adjustments may render even the most stringent rules ineffective in practice. Similarly, since the role of a specific set of human activities in causing environmental change is sometimes not well understood, the change prescribed in human behavior (such as cutbacks in emissions) will not always in fact lead to the change predicted in the state of the environment. Even perfect compliance with a strong regime is therefore not a *sufficient* condition for achieving policy goals defined in terms of biophysical impact.

In general, then, the effectiveness of a particular regime (E_r) can be seen as a function of the stringency and inclusiveness of its provisions (S_r), the level of compliance on the part of its members (C_r), and the side effects it generates (B_r)—that is,

$$E_r = f(S_r C_r) + B_r.$$

Analytically, we treat output, outcome, and impact as three distinctive steps in a causal *chain* of events, where one serves as a starting point for analyzing the subsequent stage(s) (see figure 1.1).

Furthermore, we distinguish between the stage of regime *formation* (the end product of which is a new set of rules and regulations—that is, output) and that of regime *implementation* (the first product of which is behavioral change—outcome), leading—if the diagnosis is correct—to some change in the state of the biophysical environment (impact) further down the road. At the output stage, a regime can be assessed on the basis of criteria such as the stringency of its rules and regulations, the extent to which the system of activities targeted is in fact brought under its jurisdiction or domain (its inclusiveness), and the level of collaboration established.

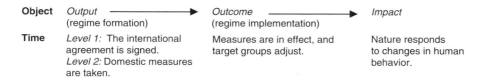

Figure 1.1
Objects of assessment

In this study, we are particularly interested in the relationship between level of collaboration and effectiveness measured in terms of behavioral change. The basic question may be formulated as follows: does more cooperation lead to better substantive results? For the purpose of this part of the analysis, we measure level of collaboration in terms of a six-point ordinal scale:

0   Joint deliberation but no joint action.

1   Coordination of action on the basis of tacit understanding.

2   Coordination of action on the basis of explicitly formulated rules or standards but with implementation fully in the hands of national governments. No centralized appraisal of effectiveness of measures is undertaken.

3   Same as level 2 but including centralized appraisal.

4   Coordinated planning combined with national implementation only. Includes centralized appraisal of effectiveness.

5   Coordination through fully integrated planning *and* implementation, with centralized appraisal of effectiveness.

In line with the logic of the flowchart in figure 1.1, we treat level of collaboration as an *intervening* variable. More specifically, we assume that level of collaboration is affected by problem malignancy and the problem-solving capacity of the system that established the regime[4] but also—in turn—that it makes a positive, albeit modest, contribution to effectiveness.

Against Which Standard Is the Regime to Be Evaluated?
Defining an evaluation standard involves at least two main steps. One is to determine the *point of reference* against which actual achievement is to be compared. The other is to decide on a standard *metric of measurement*.

It seems that basically two points of reference merit serious consideration in this context (see figure 1.2). One is the hypothetical state of affairs that would have come about had the regime not existed. This perspective leads us to conceive of

Distance to Collective Optimum

	Great	Small
High	Important but still imperfect	Important and (almost) perfect
Low	Insignificant and suboptimal	Unimportant yet (almost) optimal

(Left axis label: **Relative Improvement**)

Figure 1.2
Two dimensions of effectiveness

effectiveness in terms of the *relative improvement* caused by the regime.[5] This is clearly the notion we have in mind when considering whether and to what extent a regime matters. The other option is to evaluate a regime against some concept of a good or ideal solution. This is the appropriate perspective if we want to determine to what extent a certain collective problem is in fact solved under present arrangements. The words *good* and *ideal* are chosen to indicate that there are at least two different suboptions that can be adopted—known as *satisficing* and *maximizing*, respectively. Since actors tend to have different standards of satisfaction, we will here refer to the *collective optimum*.[6] We elaborate on this notion below, but at this point we define a collectively optimal solution as one that accomplishes, for the group of members, all that can be accomplished—given the state of knowledge at the time.[7]

These two approaches are clearly complementary. Even a regime leading to a substantial improvement may fall short of being perfect. Conversely, in more fortunate situations a minor adjustment may be sufficient to reach the joint optimum. Moreover, both dimensions are interesting in their own right. International regimes are, it seems, typically evaluated in terms of how well they (can be expected to) perform compared to the state of affairs that would have come about in their absence (the noncooperative solution) *as well as* in terms of their ability to solve the problems they are designed to cope with.[8] This suggests that students of international regimes should be able to play with *both* these notions of effectiveness and perhaps combine them into an integrated measure as suggested by Helm and Sprinz (1999).[9] It is, however, important not to confuse the two.

Each of these approaches calls for further conceptual refinement. Consider, first, the notion of *relative improvement*. Although intuitively meaningful, the provisional definition given above leaves open at least one critical question: what precisely is the *baseline* from which change or improvement should be measured?

In principle, it seems that we most often have a choice between two main options: one is the hypothetical state of nature that would have obtained if, instead of the present regime, we were left in a no-regime condition (a *fully* noncooperative solution).[10] The alternative option is to take as our baseline the hypothetical situation that would have existed had the previous order or rules of the game been left unchanged. The former measures effectiveness in absolute terms, while the latter measures change from one order to another (effectiveness differentials). The former may ultimately be the more interesting, but the notion itself is elusive and thus virtually impossible to measure. For practical reasons, then, we are left with the previous order as our baseline—the underlying assumption being that it would have continued by default in the absence of the present regime.

Conceiving of effectiveness in terms of the distance between what is *actually* accomplished and what *could have been* accomplished immediately puts before us the intriguing question of what constitutes the *maximum* that a particular group of actors can accomplish. A natural scientist would probably answer by referring to environmental sustainability, assimilative capacity, or some other notion of eco-system health. An economist would frame the answer in terms of social welfare, probably measured as net economic benefit to the group of regime members. Whenever we are talking about solutions that have to be shaped through political processes, however, such technically perfect solutions will not always be politically feasible. What is politically feasible depends on the institutional setting, particularly the decision rule. Whenever we are dealing with collective decisions that can be made only through consensus, the appropriate notion of the political optimum is the *Pareto frontier*. This frontier is reached when no further increase in benefits to one party can be obtained without leaving one or more prospective partners worse off. In the most favorable circumstances, a solution maximizing the sum of net benefits to the group will also be Pareto optimal, but in the context of international negotiations there is no guarantee that the two will coincide (see Underdal 1992b). In this study we therefore started with an ambition to assess regimes in terms of a political as well a technical notion of the ideal solution. We were soon forced to conclude, however, that we had no reliable method for determining distance from the Pareto frontier with sufficient confidence. As a consequence (and with some reluctance), we had to drop the notion of political optimum from the comparative analysis.

To define a standard of evaluation we need not only to decide on a point of reference against which actual achievement is to be compared; we also need to agree on some standardized *metric of evaluation*. In most of our cases we seem to face a choice between measuring effectiveness in terms of social welfare (usually translated into some measure of economic costs and benefits) or in terms of ecological sustainability or some other biophysical criterion.[11] For example, the performance of the International Whaling Convention (IWC) may be evaluated in terms of net economic benefits, in terms of sustainable biological yield, or in terms of some preservationist notion of protection. As we demonstrate in chapter 15, the score that we give to the IWC depends critically on which of these values we choose as the basis for our evaluation. The most important lessons to be drawn here seem to be that we should be explicit about the choices we make and should realize that scores obtained by using different evaluation metrics cannot be used interchangeably—at least not without a critical examination of compatibility.

How Do We, in Operational Terms, Attribute a Certain Score to a Regime?

The discussion of the practicalities of operational measurement is left to chapter 2. Here we say just a few words to clarify our ambitions. In this study no attempt is made to go beyond *ordinal*-level measurement. The purpose of the project—to help improve our understanding of why some international problem-solving efforts are more successful than others—does not require a higher level of measurement. Nor do we know how to construct a cardinal scale that would make sense in this context. In fact, in comparing different regimes or components we use scales including only four (in the case of distance to the collective optimum) or five (in the case of relative improvement) levels of effectiveness. Even such a crude coding is by no means a straightforward exercise. The major challenges that we face in scoring cases in terms of effectiveness are (1) to determine empirically our point of reference (whether it is the collective optimum or the hypothetical state of affairs that would have occurred in the absence of the regime) and (2) to distinguish the causal impact of the regime itself from that of other factors affecting human behavior or the environment. We take some comfort in the fact that these challenges are not unique to this particular study. In fact, they are encountered—and somehow apparently solved—by all students and practitioners who dare to assign a certain score of effectiveness to a political institution.

Methodological Challenges to Measuring Regime Effectiveness

It should by now be abundantly clear that scoring cases in terms of effectiveness confronts us with intriguing methodological challenges. This is particularly so when we compare scores *across* different regimes; diachronic comparison *within* one regime or regime component is, fortunately, somewhat less complicated. The reader should understand that however careful and systematic we try to be in examining available evidence, all assessments of effectiveness will inevitably involve some element of subjective judgment or inference on the part of the analyst. We do believe that many of our conclusions are quite robust, but we realize that others are not. We owe it to our readers to try to distinguish the former from the latter, and we attempt to do so in the presentation of the individual cases.

To summarize, we have taken as our point of departure a common-sense notion of effectiveness, saying simply that a regime is effective to the extent that it successfully performs some generic function or solves the problem that motivated its establishment. For most *environmental* regimes the ultimate test will be to what extent they improve the state of the environment itself. Environmental objectives are to be achieved through changes in the human behavior that causes environmental damage (such as pollution and nonsustainable harvesting). Accordingly, human behavior is the immediate target. The relationship between rules and regulations (output), change in human behavior (outcome), and biophysical change in the environment (impact) is a question to be settled through empirical research. In assessing the effectiveness of a regime we use two basic points of reference—the hypothetical situation that would have existed in its absence (the noncooperative situation) and a notion of what would be the ideal solution (the collective optimum). These distinctions are summarized in figure 1.3.

As we have seen, some other studies have used different notions of effectiveness. Combined with the fact that the scores we end up with will sometimes depend critically on which standard we apply, this observation raises two important questions.

First, does it make sense to try to develop some *composite* or *aggregate* score to cover all or at least the main options considered above? Our answer can briefly be summarized as follows: (1) If we are talking about using different operational *indicators* for the same theoretical concept, computing some aggregate score basically means constructing an index. This may certainly be a sensible thing to do, provided that the indicators included are believed to capture different aspects of the phenomenon we are trying to get at. (2) Aggregating scores across *different* basic concepts is a straightforward operation only as long as one case strongly dominates another.[12] For example, referring to figure 1.2, we can see no problem in rating cases in the upper right-hand category as "more effective" than those in the lower left-hand cell; in terms of our two criteria the former strongly dominates the latter. (3) To produce a meaningful aggregate score where one case does *not* strongly dominate another, we would have to combine different notions into one integrated

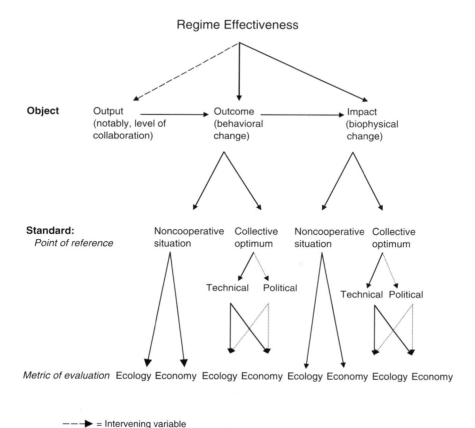

Figure 1.3
Regime effectiveness: Objects and standards of assessment

formula. As we have seen above (note 9), a sensible integrated measure can be created by computing an effectiveness coefficient in which *actual* improvement is expressed as a fraction of *potential* improvement. As far as we can see, this particular kind of combination is the only way to produce aggregate scores in cases where one regime does not strongly dominate another.

As we now move on to define and examine our independent variables, a second question arises: Can the same model, including the same set of independent (and intermediate) variables, be used to account for variations in performance *irrespective* of how we define *effectiveness*? Can, for example, the same model that we would use to explain variance in what Young (1994) labels "effectiveness as problem solving" equally well account for variance in what he calls "process effectiveness" or "constitutive effectiveness"?[13] This question is much too complex to be answered adequately here. The general rule of thumb must be that the greater the substantive difference between the definitions in question and the better specified the independent variables, the less likely that the answer will be an unqualified yes. Note, for example, that Young's definition of *process effectiveness* and some notions of *regime strength* (e.g., Haggard and Simmons 1987) focus only on the level of implementation and compliance. The relationship between effectiveness as defined here and compliance is by no means straightforward; for malign problems, the situation may very well be that the level of compliance tends to be *inversely* related to the amount of behavioral change required by regime rules (see, e.g., Downs, Rocke, and Barsoom 1996). In another study, Young (1998) divides the process of regime formation into three distinct stages—agenda formation, negotiation, and implementation—and argues that the political dynamics unfolding at each stage are sufficiently different that a hypothesis formulated with reference to one of these stages will not necessarily hold for the other two.

Moreover, any attempt at measuring effectiveness will have to refer to the state of affairs *at one particular point in time*. For several reasons—one being that producing effects usually takes time—scores may vary depending on when the assessment is made. Everything else being equal, we would expect the effectiveness of a regime to increase when it has had the time to mature and penetrate the system of activities in question. This is not to suggest that we should expect a linear increase in effectiveness over time; rather, we would expect the typical pattern to be *curvilinear*—increasing as the regime matures but diminishing as it ages into obsolescence. The general point is simply that scores may vary over the lifetime of a regime. One very important implication of these observations is that comparing the outcomes or impacts of two or more regimes is a straightforward exercise only if these regimes are measured at similar stages in their life cycles. Whenever we are unable to synchronize observations in this particular sense, caution is required in comparing scores *across* regimes.

The Independent Variables: What *Determines* Regime Effectiveness?

As indicated above, we see regime effectiveness as a function of two main (sets of) independent variables—namely, the character of the problem and what we have called problem-solving capacity. Before defining more precisely what we mean by these notions and formulating our hypotheses in more specific terms, however, a few words are needed to clarify the relationship between our two (sets of) independent variables. More specifically, at least two questions deserve to be addressed: one is that of relative importance, and the other is that of interplay.

In a survey of empirical cases, we may very well find that one of our independent variables—and in this case, problem type would be our candidate—accounts for most of the variance in effectiveness that we can actually observe. Such a finding would warrant the conclusion that one determinant is basically more important than another if and only if both independent variables are measured by the same yardstick. In this study we will definitely not be in a position to claim that one unit of *benignity* equals one unit of *capacity* (except, perhaps, in the derived statistical terms of standard deviation). When the requirement of standardized measurement cannot be met, the finding that one independent variable has a greater impact than another will be open to diverging interpretations. The amount of variance accounted for by any given independent variable can be seen as a function of its relative basic weight and its range of variation. In the absence of a standardized unit of measurement we have no firm basis for distinguishing the impact of the former from that of the latter.

Even when their effects cannot be separated empirically, basic weight and range of variation should be distinguished *conceptually*. By itself, a constant accounts for *none* of the variance observed. Yet it may be an important determinant of the phenomenon we want to understand. As an illustration, most intergovernmental decisions are made by consensus. In a particular survey of international decision-making processes, we may therefore be justified in considering the decision rule of consensus as basically a constant. If the decision rule remains the same throughout a given sample of cases, it can obviously not by itself help us explain why outcomes *differ*. But this hardly justifies the jump to the conclusion that the decision rule is therefore irrelevant to understanding negotiation behavior or outcomes. Only by keeping the distinction between basic weight and range of variation clearly in mind can we avoid such misinterpretations.

Finally, problem structure and problem-solving capacity cannot be seen as mutually independent factors. Capacity is always the capacity to do *something*. Beyond a certain level of generality, what constitutes problem-solving capacity can therefore be determined only with reference to a certain category of problems or tasks. Thus, it is by now conventional wisdom that the problem-solving skills and the institutional tools required to solve benign problems are somewhat different from those required to solve problems that are malign in character. For example, the relative importance of technical versus political skills tends to differ; the more malign the problem, the higher tends to be the premium on, inter alia, integration and mediation skills. Similarly, malign problems tend to require higher levels and perhaps also more complex arrangements of cooperation—including more attention paid to procedures for monitoring and enforcing compliance. In what game theorists refer to as *pure coordination games*, common norms or ideas may serve as focal points or clues enabling the parties to achieve effective tacit coordination without formal rules and regulations.

The basic implication for our analysis is that *notions of capacity will have to be matched with notions of problem type and task*. This is potentially a substantial complication, indicating that we should ideally be able to play with a complex and differentiated *set* of capacity constructs. We still have far to go before we can claim to have such a set of analytic tools. To make our task manageable, we confine ourselves in this study to one rather crude three-dimensional concept. The fact that our main interest is to better understand what contributes to the successful handling of *malign* problems serves to narrow the scope of the analysis and thereby also to make our simplifications somewhat less distorting than they otherwise might have been.

Problems: Benign and Malign

A policy problem may be difficult to solve in at least two respects. At the *intellectual* level some problems are substantively more intricate or complicated than others, implying that more intellectual capital and energy are needed to arrive at an accurate description and diagnosis and to develop good solutions. It is, for example, by no means obvious what level and rate of resource exploitation can be expected to maximize the long-term global social welfare that can be harvested from Antarctica. Nor is it easy to determine what constitutes a sustainable pattern of energy consumption and what mix and programming of policy instruments can achieve such a pattern at minimum social cost. To answer these and many other technical questions we need powerful theory, large amounts of accurate data, and the creative imagination and perseverance of skilled people. But collective-action problems are also *political* issues, and as such they can vary in their degree of malignancy. The political malignancy of a problem will here be conceived of primarily as a function of the configuration of actor interests and preferences that it generates. According to this conceptualization, a perfectly benign problem would be one characterized by identical preferences. The further we get from that state of harmony, the more malign the problem becomes.

In this study we are concerned primarily with the *political* aspects of policy problems. The intellectual dimension will be considered only as it interacts with political characteristics. This is by no means a trivial aspect; intellectual complexity and political malignancy do in fact often interact—most often with the consequence of making a problem more intractable, but sometimes with the benign consequence of facilitating agreement.[14] For example, descriptive uncertainty about the state of a recipient or stock—and particularly theoretical uncertainty about the causal impact of particular human activities on the environment—may fuel political controversy. In turn, political conflict may contaminate processes of knowledge production and dissemination (see, e.g., Miles 1989) and thereby serve to obstruct the development of *consensual knowledge*. As students of politics, we thus examine the interplay between knowledge and politics from the perspective of policy making rather than knowledge making. Moreover, we conceive of intellectual complexity in terms of the amount of descriptive and theoretical uncertainty pertaining to the knowledge base rather than in terms of some objective measure of the inherent intricacy of a problem. We do recognize that the two will not necessarily correlate perfectly, but as our goal is to understand actor behavior and outcomes of collective decision-making processes, the former is the appropriate focus in this study.[15]

Before we move on to specify what distinguishes politically malign problems from those that are benign, a few words seem in order to position our approach in relation to the extant literature that examines the impact of problem characteristics on outcomes of international cooperation. Grossly simplified, it seems meaningful to distinguish between two main paths of research in this area. One distinguishes *problem structures* according to properties of the issue or issue area in focus. The most well-known case in point is a typology developed and applied by a research team based at the University of Tübingen (see, e.g., Rittberger and Zürn 1990; Efinger and Zürn 1990; Hasenclever, Mayer, and Rittberger 1997).[16] Their typology distinguishes four main *objects of contention* and is used to develop a set of hypotheses about the prospects for regime formation (see figure 1.4).

The other main approach categorizes problems according to the structure of the strategic games they generate. Students pursuing this path typically distinguish between broad categories of games (such as coordination games and collaboration games) (Stein 1982) or analyze situations in terms of a set of specific game models (see, e.g., Zürn 1992). As their bases for classification are different, there is no straightforward method for translating from one of these typologies to the other. Yet the hypotheses derived by the two schools seem at least largely compatible and based on similar lines of reasoning.

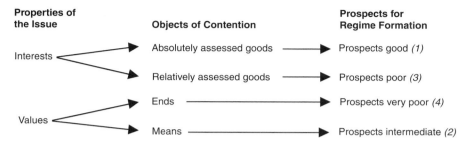

Figure 1.4
Objects of contention: Tübingen team distinctions

Like the Tübingen group, we focus not on specific game structures but on generic *mechanisms* that generate the kinds of problems that international regimes are established to address. Nevertheless, in terms of its basic premises, the scheme that we use here is closer to the *situation-structure* approach. We take as our point of departure the assumption that the official purpose of international regimes is to coordinate behavior in situations where the absence or failure of coordination will or can lead to *suboptimal* outcomes. At least two broad categories of situations fit this description. In one, the cost-benefit calculus of individual actors includes a nonproportional or biased sample (representation) of the actual *universe* of costs and benefits produced by his decisions and actions, so that "the pursuit of self-interest by each leads to a poor outcome for all" (Axelrod 1984, 7). We refer to these situations as problems of *incongruity*. In the other category of situations, (1) the overall result depends on the matching of actions taken by individual parties, (2) more than one route can lead to the collective optimum, and (3) the choice between or among these routes is *nontrivial*, meaning that compatibility cannot be taken for granted—even in the absence of conflict of interests or values. We refer to the latter category as problems of *coordination*. In political terms, the former can be considered more or less malign, while the latter are basically benign. Let us explain why.

Problems of Incongruity   To repeat, the defining characteristic of this category of problems is that the cost-benefit calculus of an individual actor is systematically biased in favor of either the costs or the benefits of a particular course of action. Such a bias may be due to the objective distribution of material consequences, the perspective applied in assigning value to these consequences, or both. Transfrontier pollution and common-pool resources are prominent examples of situations where the former mechanism is at work. To the extent that pollution generated in one country causes damage in other countries, the polluting country typically harvests all the benefits of the activities causing pollution and suffers only a certain fraction of the damage. Similarly, to the extent that the fishing fleet of one country depletes a common-pool resource, it would share the loss in terms of future fishing opportunities with others while reaping all the short-term benefits of higher catches itself. Now, such objective physical impacts do not speak for themselves; they serve as decision premises only as interpreted and evaluated by the actors involved. One important aspect of this evaluation process concerns the criteria used in assigning value to perceived consequences. For a moderately altruistic actor—being equally concerned with his own welfare and the welfare of others—the physical distribution of impact would not matter; his own cost-benefit calculus would positively incorporate also what happens to others. By contrast, a competitively motivated actor would be concerned about relative gains and losses and hence assign negative value to the welfare of others. Finally, an individualistically motivated actor would be concerned only with his own payoff and assign zero value to the welfare of others. The bottom line is that what comes out as problems of incongruity depends on both the objective distribution of material consequences and the perspectives and criteria used by the actors in assigning subjective value or utility to perceived consequences. In the absence of evidence pointing in another direction, we assume that individualistic motivations dominate.

Incongruity can be caused by at least two different mechanisms—externalities and competition.[17] The term *externalities* here denotes external leaks—those effects of an actor's behavior that hit others and (therefore) disappear from the actor's own cost-benefit calculations. Assume that the government of country A is considering the level of pollution abatement to demand from its industries. Half the total damage costs are suffered within A's own borders, and half are suffered by neighboring countries. Assume, furthermore, that all goods produced by A's industries are consumed domestically and are subject to no competition from abroad. If A's government seeks to maximize national economic welfare, and no issue linkages are constructed, it will balance 50 percent of the benefits derived from abatement against 100 percent of abatement costs.

Let us now change one of these assumptions so that A's industries face strong competition in foreign as well as domestic markets. In this case, the real costs of pollution abatement would include not only the direct costs of installing and operating new filters or production equipment, but also—at least if the "polluter-pays" principle is observed—indirect costs in terms of loss of market shares. The latter can be attributed to competition rather than leaks and will quite often be the major cause of concern.

In general, a relationship of *competition* exists whenever one actor's (subjective) welfare depends on how well he performs compared to others. Competition is found in many spheres of social life—from markets to sports and arms races—and, to use the terminology of the Tübingen team, whenever we are dealing with some "relatively assessed good." The difference between externalities and competition as defined here can be described as one between effects that just leak and those that boomerang. In the case of externalities, the worst that can happen to an actor contributing unilaterally to the provision of some collective good is that the benefits he thereby produces are harvested by others (acting as free riders). In a relationship of competition the worst that can happen is virtual extinction from the game in question. Similarly, if effects just leak, an actor causing a net reduction in social welfare will be able to shift some of the costs onto others. In the context of competition, noncooperative behavior may conceivably be rewarded up to the point where the defector's share of overall group benefits from the activities in question approaches 100 percent. This implies that competition can distort actor incentives by *amplifying* the costs of cooperative behavior or by positively *rewarding* defection in ways in which pure leaks can never do. Thus, other things being equal, competition is inherently the more malign problem of the two.[18]

Problems of incongruity can be particularly hard to solve through voluntary cooperation to the extent that they are also characterized by asymmetry or cumulative cleavages. A problem is *asymmetrical* to the extent that the parties involved are (or perceive themselves to be) coupled in such a way that their values are incompatible or their interests negatively correlated. The typical upstream-downstream relationship is a good example of negatively correlated interests. Unless some kind of cost-sharing scheme is established, measures to control unidirectional transfrontier pollution will benefit victims at the expense of polluters. Equally asymmetrical can be situations where one party values a particular species for consumptive uses and another for its beauty or its place in nature. Other things being equal, the more asymmetrical an incongruity problem, the more difficult it will be to find a solution that will be accepted by both or all parties. This is particularly so if the distribution of costs and benefits is determinate—that is, easy to predict for the parties themselves. When we are dealing with multidimensional problems, the presence of cumulative cleavages can be an additional source of complications. Cleavages are *cumulative* to the extent parties find themselves in the same situation on all dimensions or issues, so that those who stand to win (or lose) on one dimension also come out as winners (or losers) on the other dimensions as well. Compromises and package deals are easier to find for problems characterized by crosscutting cleavages.

Problems of Coordination   Contrary to the position taken by Keohane that "Cooperation takes place only in situations in which actors perceive that their policies are actually or potentially in conflict" (Keohane 1984, 54; cf. also Oye 1985, 6, and Stein 1982, 302), we argue that even in situations characterized by perfect harmony of interests coordination of behavior may be required to ensure collectively optimal outcomes (cf. Schelling 1960). In general, coordination can be useful also in situations where (1) the overall result depends on the compatibility of individual choices, (2) more than one route can lead to the collective optimum, and (3) the choice between or among these routes is not a trivial or obvious one, meaning that compatibility cannot be taken for granted even when actor interests are identical. We refer to such situations as problems of coordination. These problems are politically benign in that actor interests are fully compatible. Cooperative solutions will be stable in the sense that once such a solution is established, no actor will have any incentive to defect unilaterally.

Many international regulations, ranging from the allocation of radio frequencies to the designation of shipping lanes, are designed essentially to help actors tackle coordination problems. In some cases, such as regulation of maritime or air traffic, the major purposes of coordination are to avoid accidents or reduce transaction costs. In other cases, particularly in what Young (1996) refers to as "programmatic" activities, there may even be positive synergy to be tapped into. For example, the overall utility of emissions or environmental quality monitoring may be enhanced substantially by such international coordination as standardization of methods and a system for pooling of observations. This is also the case for quality of weather forecasting.

Problem Type: A Summary   Summing up, we conceive of problem malignancy as a function of incongruity, asymmetry, and cumulative cleavages. We consider incongruity as the principal criterion for classification; the other two are properties that tend to augment the political intractability of incongruity problems (see table 1.1). The major differences between problems of incongruity and problems of coordination are summarized in table 1.2.

Table 1.1
Characteristics of malign and benign problems

Malign	Benign
Incongruity (in particular relationships of competition)	Coordination (synergy or contingency relationships)
Asymmetry	Symmetry or indeterminate distribution[a]
Cumulative cleavages	Cross-cutting cleavages[a]

a. As indicated above, these dimensions are relevant primarily for problems of incongruity.

Table 1.2
Characteristics of incongruity and coordination problems

Dimension	Incongruity	Coordination
Essence of problem	Incentive distortion ($q_I \neq k_i$)	Imperfect information, communication failure
Essence of cure	Incentive correction	Information or communication improvement
Consequences of unilateral cooperative moves	Risky, particularly in relationships of competition	No risk (except for transaction costs of own efforts)
Tactical implications for negotiations	Manipulation or coercion likely	Integrative negotiations, persuasion
Postagreement implications	Incentives to unilaterally defect tend to persist; transparency, monitoring, and enforcement mechanisms important	Self-enforcing; no incentives for unilateral defection from an agreed solution

The simple typology developed above can easily be refined—for example, by using game-theory concepts to introduce finer distinctions as well as to derive a more differentiated and precise set of testable hypotheses. We do not pursue that path here for two reasons. First and most important, to get full mileage out of fine-grained distinctions on an *in*dependent variable, we would need equally fine-grained distinctions on the *de*pendent variable. Since we use only a crude, four-level ordinal scale to measure effectiveness, the marginal utility of further specification of game structures declines rapidly. The general implication is that a strategy of specification that would be essential in formal deductive analysis may be rather futile for purposes of comparative empirical research—particularly studies with a small number of cases. Second, in encounters with complex problems, fairly simple and crude distinctions often seem to provide the most robust and useful tools, at least as a first cut of comparative analysis. Moreover, we think a two-step procedure can help us avoid the danger of pounding square pegs into round holes. Game-theory models may be most useful when they are applied to fairly specific decision points rather than to broader problems. There is, for example, no compelling reason to assume that the essence of (for example) transboundary air pollution—let alone externality problems *in general*—can be adequately modeled as a (symmetrical) prisoner's dilemma game, even though many *specific* situations may well have that structure.

The line of reasoning that we have pursued in this section leads to the following main hypotheses:[19]

H₁: The more politically malign the problem, the less likely the parties will achieve an *effective* cooperative solution—particularly in terms of technical optimality.

The reasoning behind the latter part of H₁ is that—up to a certain level—the more malign the problem, the more suboptimal the noncooperative state of affairs is likely to be.[20] The worse the non-cooperative situation, the less capacity and effort it takes to bring about *some* improvement, but the more it will take to achieve an ideal solution.

H₂: Political malignancy and uncertainty in the knowledge base tend to interact to increase the intractability of problems.[21] Uncertainty in the knowledge base tends to slow down the development of effective responses to benign problems as well, but here the effect on the end result will be less detrimental.

So far, we have treated each problem on its own merits only. In real life, issues are often linked either through inherent functional (inter)dependence or through deliberate tactical moves. Moreover, an actor may support or oppose a particular solution for reasons that have little or nothing to do with the official purpose of the regime. Issue linkages or ulterior motives may thus contaminate a benign problem or render a malign problem easy to handle. Taking this broader context into account, we can now try to specify the conditions under which malign problems can be successfully solved:

H₃: Regimes dealing with truly malign problems will achieve a high degree of effectiveness only if they contain one or more of the following: (1) selective incentives for cooperative behavior (see Olson 1965), (2) linkages to more benign (and preferably also more important) issues, or (3) a system with high problem-solving capacity. The presence of at least one of these factors is a *necessary*, but not a sufficient, condition for a high level of effectiveness.

Problem-Solving Capacity   The alternative proposition formulated on the first page of this chapter offered the concept of *problem-solving capacity* as the key to explaining regime effectiveness. The general argument is that some problems are solved more effectively than others because they are dealt with by more powerful institutions or systems or because they are attacked with greater skill or energy. The time has come to look inside the elusive and complex notion of problem-solving capacity and to try to identify at least some of its main elements.

Let us first of all repeat that in this study our interest pertains primarily to the capacity to deal with the *political* rather than the intellectual aspects of a problem and to the interplay between those two dimensions. When solutions are to be shaped through collective decisions, problem-solving capacity (in its political interpretation) can provisionally be conceived of as a function of three main determinants:[22]

· The institutional setting (the rules of the game),
· The distribution of power among the actors involved, and
· The skill and energy available for the political engineering of cooperative solutions.

We try to deal with all three determinants, but our analysis is confined essentially to organizational structures and political capabilities and not to what might be called cognitive or behavioral aspects. An attempt at measuring the impact of behavioral skill or efficacy would have required a kind of in-depth analysis of individual actors and political processes that we had neither the time nor the resources to undertake on a broad scale.[23]

Second, it bears repeating that problem-solving capacity can be determined precisely only with reference to a particular category of problems or tasks. We have indicated some of the implications of that observation above. It also means that the significance of at least two of the determinants listed above clearly differs from one category of problems to another. Most obviously, while coercive power can be an important tool in dealing with malign problems, it is likely to be largely irrelevant— and, if exercised, often counterproductive—in benign situations. The premium on skills in political engineering is higher for malign than for benign problems. And the institutional arrangements conducive to integrative negotiations are to some extent different from those that facilitate distributive bargaining (Walton and McKersie 1965). We do not aim at exploring all these nuances here. Since our main interest pertains to the conditions under which malign problems are successfully dealt with, the remainder of this section is framed primarily with such problems in mind.

The Institutional Setting   As a basic social science concept, *institution* refers to constellations of rights and rules that define social practices, assign roles to participants in those activities, and guide interactions among those who occupy those roles (see Young 1994, 3). In this study, we use the term *institutional setting* broadly as a label for two different notions of institutions—namely, institutions as arenas and organizations as actors. The distinction refers to functions and does not imply a ranking in terms of importance. Institutions can shape outputs and outcomes as much by coupling actors and problems and determining the rules of the game, as by entering the game as more or less independent actors. Arenas are important in their own right and for different reasons.

Institutions as Arenas   When we talk about an institution as a "framework within which politics takes place" (March and Olsen 1989, 16), we conceive of it as an arena. Arenas regulate the access of actors to problems and the access of problems to decision games. Moreover, they specify the official purpose as well as the rules, location, and timing of policy games. Institutions as arenas can be described by answering the following question: *who* deals with *what, how, when,* and *where?*

   Arenas differ in terms of rules of access, decision rules, and rules of procedure, as well as in terms of informal culture. For example, the membership of a regime is in some cases restricted to countries that satisfy certain criteria (the Antarctic Treaty System would be a case in point). Others (such as the International Whaling Commission) are open to any state that cares to submit a formal application and pay its membership fee.[24] Consensus is the decision rule most frequently used in international organizations, but a number of organizations have some provisions for decision making by voting (usually requiring a qualified majority on substantive matters). As we discuss below, the decision rule is an important determinant of the capacity of an institution to aggregate diverging preferences. Other things being equal, aggregation capacity reaches its maximum in strictly hierarchical structures and is at its lowest in systems requiring unanimity. Furthermore, we know that rules of procedure may differ in several respects—for example, in their differentiation into subprocesses (committees) and in the amount of discretion vested in committee or conference chairs to, for example, draft proposals ("negotiating texts"). Finally, in addition to these sets of formal rules, many arenas—particularly those that are in active use over a prolonged period of time—develop their own *in*formal codes of conduct or cultures.

An important research question is generated by these observations: to what extent and how do different rules of access, decision making, and procedure affect the capabilities of arenas to fulfill particular functions in the regime-building process? In other words, how does an institution's design affect its ability to provide actors with incentives to adopt and pursue an integrative, problem-solving approach (Walton and McKersie 1965), provide opportunities for transcending initial constraints (such as by coupling or decoupling issues) (see, e.g., Sebenius 1983), and enhance the institutional capacity to integrate or aggregate actor preferences?

This project cannot cover this comprehensive agenda in full. What we do, however, is focus particularly on *decision rules and procedures*—arguably the most important determinant of institutional capacity to aggregate actor preferences into collective decisions.[25]

Consensus is the default option and also the most commonly used decision rule in international politics.[26] Even though less demanding rules are sometimes adopted in specific contexts, we normally find that consensus is the master principle in one or both of two meanings. First, in a basically anarchical system any other decision rule adopted by a group of states will have to be approved by consensus. Second, in many instances we find that provisions for majority voting are coupled to some kind of right of reservation, meaning that a party who has strong objections to a particular regulation can—by filing a formal reservation by a certain deadline—declare that that decision will not apply to itself.

After unanimity, consensus is the most demanding decision rule there is. If we combine the requirement of consensus with a requirement of *inclusiveness* (meaning that all parties in a given group must join for a solution to be implemented) and assume that each option is evaluated only in terms of its own merits, it leads to "the law of the least ambitious program" (Underdal 1980, 36), meaning that collective action will be limited to those measures that are acceptable to the least enthusiastic party (that is, limited by the Pareto frontier). In fact, even the least ambitious program may prove unattainable in practice. It is easy to see that the amount of collaboration actually achieved may fall short of what even the least enthusiastic party considers desirable. The rule of consensus gives each party a veto not only over the overall amount of (for example) emission reductions; the right of veto also pertains to every conceivable means of achieving that reduction (including, of course, the distribution of reductions or costs). Moreover, inherent in the process of (distributive) bargaining are certain "perversities," providing "incentives to actors to behave in ways that have the effect of hindering mutually beneficial cooperation" (Keohane 1988, 29; see also Johansen 1979 and Underdal 1987). A well-known economist has even formulated what is sometimes referred to as the "law of bargaining inefficiency," saying that "bargaining has an inherent tendency to eliminate the potential gain which is the object of the bargaining" (Johansen 1979, 520).[27]

Even though pervasive and resilient, there is fortunately nothing inevitable about these "perversities." Institutional arrangements as well as behavioral strategies can be designed to counter such risks. Most fundamentally, institutionalization can help by encouraging actors to extend their time horizons beyond those of one-time encounters (see, e.g., Axelrod and Keohane 1985, 232f) and shift from norms of *specific* to *diffuse reciprocity* (see Keohane 1988). Moreover, as Sand (1990) has reminded us, actors can use several strategies to beat the "law of the least ambitious program." These strategies include the creative use of selective incentives, differential obligations (including loopholes), and promotion of voluntary overachievement by pusher countries (leading to two- or multiple-track cooperation schemes). Even though there can be no doubt that the decision rule of consensus is a major constraining factor in international cooperation, there are indeed quite a few things one can do to enhance the possibilities of cooperation.

To summarize, we hypothesize:

H₄: The establishment of (negotiation) arenas as formal *institutions* that exist and are used over an extended period of time tends to facilitate cooperation and enhance the effectiveness of international regimes by encouraging actors to adopt extended time horizons and norms of diffuse rather than specific reciprocity and by reducing the transaction costs of specific projects.

H₅: In dealing with *malign* problems, the decision rules of unanimity and consensus tend, other things being equal, to lead to less effective regimes than rules providing for (qualified) majority voting. More precisely, the use of majority voting tends to lead to more ambitious regulations (output). However, to the extent that this is accomplished at the cost of sacrificing the interests of significant actors, it will do so at the risk of impairing compliance (outcome). Except for strongly malign issues, the former effect will most often be somewhat stronger than the latter.

Organizations as Actors   All organizations can serve as arenas, but only some can also qualify as significant actors in their own right. International organizations can be considered actors to the extent that they provide independent *inputs* into the problem-solving process or somehow amplify *outputs* of these processes. To qualify as actor, an organization must have a minimum of internal coherence (unity), autonomy, resources, and external activity. Without a certain minimum of coherence, an organization cannot be considered *one* actor. Without some autonomy (notably in relation to its members), it would be a mere puppet commanded by its masters. By definition, an actor must somehow in fact *act*. Without a certain minimum of activity within its environment, an organization can hardly be said to play the role of an actor (even though it may have the *capacity* to act). Finally, unless an organization has a certain minimum of resources at its disposal, its own contributions to its activities would tend to be inconsequential.

Organizations, and even specific bodies within international governmental organizations (IGOs), vary considerably in terms of scores on these three dimensions. One obvious candidate for a top score would be the European Union. At the level of specific organizational bodies, the same can be said about the Commission of the European Union. Most of the secretariats serving international regimes seem, however, to find themselves closer to the critical minimum required (particularly, it seems, on the autonomy and resources dimensions) to achieve actor status, and some even fail to meet that threshold. Also other systemic roles, such as those of conference presidents and committee chairs, are most often quite narrowly circumscribed in terms of formal authority.[28] As a consequence, the amount of organizational energy available to pursue "the common good" in international politics is most often only a small fraction of the aggregate amount of energy geared to the pursuit of national interests. We nevertheless believe that variance in the actor capacity of IGOs or subordinate bodies and officials can affect the amount of success obtained in attempts to create and implement international regimes. Thus, we suggest

H_6: Actor capacity on the part of the international organization in charge and its subordinate bodies and officials tends, *ceteris paribus*, to enhance regime effectiveness. The impact tends, other things being equal, to be larger in the case of malign than in the case of benign problems and greater for moderately malign than for strongly malign problems.

The institutional setting itself is not a truly independent variable in the sense of being an exogenous parameter completely beyond the control of the members of the regime. On the contrary, institutions are social constructions, established and modified through joint decisions. A hard-nosed realist might even argue that international organizations are essentially epiphenomena—puppet instruments that powerful states can use to promote their own interests and will transfer power to only when most or all other options are exhausted (see, e.g., Mearsheimer 1995). The fact that institutions are negotiated entities has two important implications. First, in bargaining about the design of arenas or organizations, each actor is likely to evaluate alternative settings primarily as a means to protecting or promoting its *own* substantive interests. Whenever these interests diverge, as they most often do, the setting may itself become an issue of delicate bargaining. This means that the more malign the substantive issues to be dealt with by a particular institution, the more difficult it will be to reach agreement about the shape of that institution and the weaker and more constrained it is likely to be. Where two or more existing institutions present themselves as alternatives, the same can be said about the choice between or among these institutions. As a consequence, *institutional capacity tends to be most difficult to build up or draw on where it is most needed.* The overall causal relationship thus looks something like this:

Problem malignancy $\xrightarrow{\quad-\quad}$ Decision rule/actor capacity $\xrightarrow{\quad+\quad}$ Effectiveness

Second, some regimes are served and managed by an organization or body established with a particular purpose in mind, while others are served by an organization established (primarily) for another purpose. In the former case, the institutional setting in which a regime was formed often is different from the setting in which it is implemented. Even when the setting is constant, its significance may change; it is by no means obvious that the institutional arrangements that contribute to success in the phase of regime formation contribute as much in the phase of implementation. Some functional requirements are specific to one particular phase; for instance, implementation and compliance-review mechanisms can no doubt enhance effectiveness at the implementation stage but can hardly contribute much in the phase of regime formation. Similarly, it is easy to see that some of the behavioral strategies that can be helpful in forging agreement—such as the use of what Henry Kissinger referred to as "creative ambiguity"—may jeopardize implementation. Much work remains to be done, however, before we can differentiate with precision and confidence institutional design principles with reference to the various stages that international regimes typically go through (see, though, Wettestad 1999).

Power The more complex the problem and the more demanding the decision rule, the more critical leadership of some kind becomes. And the less formal authority that is vested in systemic actors (such as conference chairs and secretariats), the more important *inf*ormal sources of leadership become. One such source is *power*, here defined narrowly as the control over important events (Coleman 1973).

Following Coleman (1973), two faces of power may be distinguished. One derives from control over events important to *oneself*; the other from control over events important to *others*.[29] The former provides autonomy—the privilege of being able to pursue one's own interests without having to worry about what others might wish or do. The latter provides an actor with the means to impose its own will on others. The notion of *hegemony* combines the two (see, e.g., Snidal 1985). A *benevolent* hegemon is an actor sufficiently predominant to be able and willing to provide collective goods at its own expense, or—more generally—to establish and maintain unilateral solutions to collective problems.[30] By contrast, the *coercive* hegemon rules by virtue of its control over events important to others, and it uses this control to induce their submissive cooperation or—more bluntly—to impose its own will on them. Either way, power can be a device to bypass or break aggregation deadlock. We may therefore conclude that the existence of a *unipolar* distribution of power tends to enhance the decision-making capacity of a system, and—by implication— also the probability that at least *some* collective decision will be made. In this particular respect, a unipolar distribution of power can be seen as a functional substitute for formal hierarchy or other strong decision rules.

Control over events important to others may be transformed into coercive leadership through at least three different mechanisms. First and most obviously, an actor can use its control over important events as a device to provide selective incentives (rewards) to those who join or comply or to punish anyone who refuses to go along with or defects from an established order. Although perhaps a less predominant mechanism of hegemony than one might think (see, e.g., James and Lake 1989), examples of leadership by (promises of) rewards or (threats of) punishment are not hard to come by, even among friends. One example: in the period of reconstruction after World War II the U.S. government forged economic cooperation in Western Europe by making Marshall aid grants contingent on a commitment to join the Organization for European Economic Cooperation (OEEC) and to accept a modest level of coordination of economic policies. Coercive leadership involves tactical diplomacy; an actor may promise or threaten to do things it would not contemplate except for the purpose of influencing the behavior of others.

Second, unilateral measures taken by a predominant actor for its own benefit may simply set the pace to which other parties will find it in their own interests to adapt. Thus, the industries and governments of the small European Free Trade Association (EFTA) countries have little choice but to adapt to whatever standards the European Union may establish for products or services in their own internal market. They are, to be sure, not automatically bound in a legal sense by EU decisions, and they may even have been able to wield some occasional influence over the substance of EU regulations. But the sheer weight and importance of the European Union as an economic actor is such that when the European Union moves to establish a certain standard for its own area, its small neighbors will probably find it in their own best interests to make sure that they are capable of meeting that requirement— even without any hint from Brussels that such a move would be appreciated.

Coercive leadership may also be exercised through a similar but more indirect and complex mechanism. As described by James and Lake (1989, 6f) it works as follows. A unilateral policy choice made by the strong actor (the hegemon) alters the structure of opportunities facing other societies. The greater the market power of the hegemon, the greater the impact of its actions. As some industries or subgroups begin to take advantage of their new opportunities, these sectors tend to expand relative to others. As a consequence, their economic weight and domestic political influence will tend to increase. Moreover, their private interests tend to become wedded to policies that maintain or reinforce the new order. By causing a restructuring of the *domestic* configuration of interests and by affecting the distribution of political influence *within* other societies, a unilateral policy decision made by the hegemon may generate and strengthen domestic demands within weaker nations for making adaptive adjustments in their own policies.

Since we are interested in the extent to which collective problems are solved, we need to investigate not only whether a system is capable of producing *some* joint decision but also the *substance* of whatever decisions it produces. This implies that we must ask not only whether the distribution of power is unipolar, bipolar, or multipolar but also whether it is skewed in favor of parties advocating strong or weak regulatory measures. We would, in other words, like to know the distribution of power *over the configuration of preferences*.[31] The basic assumption behind this coupling can most simply be stated as follows: the probability that a particular solution will be adopted and successfully implemented is a function of *the extent to which it is perceived to serve the interests of powerful actors*.

At least two mechanisms might disturb this proposition. In dealing with *benign* problems, a benevolent hegemon may well provide collective goods at its own expense but in doing so may inadvertently weaken the incentives of others to contribute. The net balance, then, depends on the relative strength of these two effects. When it comes to malign problems—particularly those that are characterized by severe asymmetries and cumulative cleavages—concentration of power in the hands of *pushers* might generate fear among *laggards* and possibly also other prospective parties that *their* interests will not be accommodated within the regime. If so, they may very well conclude that they had better not get involved at all. This suggests that at least for problems characterized by severe asymmetry, a rough balance between pushers and intermediaries may in fact be more conducive to the development and implementation of cooperative solutions than a high concentration of coercive power in the hands of the former.

The argument outlined above has been framed essentially with reference to what we might call power in the *basic game* itself—that is, in the system of activities that is the subject of regulation. International regimes are, however, established through political processes and will therefore be influenced also by resources that are specific to those *decision games*. Two important types of decision-game resources are *votes* and *arguments*. Numbers obviously count if the decision rule prescribes voting. Also in a system practicing the rule of consensus, coalition size will often be an asset, exerting social pressure on a reluctant minority. Although certainly a soft currency in international politics, there is often some clout also in the better argument—whether based on superior knowledge or moral stature (Risse 2000). The important point to be made here is that the distribution of such decision-game resources will sometimes differ significantly from the distribution of control over important events in the system of activities to be regulated. Such incongruity may be a severe source of strain. The winning coalition may be tempted to use its control over the decision game to establish rules and regulations that are not acceptable to those who control the basic game. Strong polarization of actors in the upper left-hand and the lower right-hand cells of figure 1.5 is likely to be a good indicator of trouble.

Let us now try to summarize the argument:

P_1: Other things being equal, the more unipolar the distribution of power within a system, the greater is its capacity to aggregate actor preferences into collective decisions. This impact is strongest where decision rules are most demanding.

H_7: Concentration of power in the hands of pushers tends by and large to enhance effectiveness, while concentration of power in the hands of laggards has the opposite effect. But

Actor's prospects of promoting own interests *inside* the regime

	Poor	Good
Good	Will resist or withdraw Prefers *weak* regime	Will show moderate support Prefers *moderately strong* regime
Poor	Will be passive or indifferent (or act as revolutionary)	Will show strong support Prefers *strong* regime

Actor's prospects of promoting own interests *outside* the regime

Figure 1.5
Actor stance as a function of prospects of promoting own interests

· In dealing with *benign* problems, unilateral provision of collective goods by a benevolent hegemon tends, although with some exceptions, to weaken the incentives of others to contribute; and

· In dealing with *malign* problems—particularly those characterized by severe asymmetry and cumulative conflict—concentration of power in the hands of pushers tends to generate fear and withdrawal among laggards. For such problems, the most conducive distribution of power is likely to be one in which the aggregate strength of pushers is roughly balanced by the aggregate strength of intermediaries, with laggards in a weaker position but not completely marginalized.

H_{7a}: The prospects for regime effectiveness tend to decline if the distribution of power in the decision game differs substantially from the distribution of power in the basic game.

Previous research has concluded that power-based propositions by and large perform poorly in empirical tests (see, e.g., Efinger, Mayer, and Schwarzer 1993, 269; Young and Osherenko 1993, 228f). We suspect that this conclusion is due to the way power has been conceptualized and measured—mainly as the concentration of capabilities rather than as the distribution of power over the configuration of interests, and as "overall structural power" rather than control over the system of activities to be regulated. Young (1994, 118f) is certainly right that bargaining leverage in a specific negotiation process cannot be inferred in any straightforward way from general capability indices. The conceptualization of power proposed above is intended to bridge that gap. We are not at all prepared to dismiss the distribution of power, as conceptualized in this study, as largely irrelevant to the formation and operation of international regimes.

Skill and Energy The third constitutive element in our conceptualization of problem-solving capacity takes us from the study of structure to the study of behavior. In this project we are not able to undertake the kind of in-depth, comparative analysis of individual behavior that would be required to say much about this component.[32] Some general remarks seem nonetheless appropriate to indicate how we see its place in the overall model.

Two basic questions raised by our conceptualization of problem-solving capacity may be formulated as follows: How well can outcomes of regime-formation and -implementation processes be predicted and explained on the basis of knowledge about the structure of problems and systems? To what extent do skill and effort make a significant difference?

Our model is premised on a set of assumptions about what determines problem-solving capacity. Two of these assumptions may be summarized as follows. First, the structure of a decision situation circumscribes, with some elasticity, the range of politically feasible solutions. More precisely, if we know the configuration of actor preferences, the institutional setting (notably "the rules of the game"), and the distribution of power, we can get a fairly good idea of what will be politically feasible and—although with less accuracy and confidence—indicate a more or less narrow range of likely outcomes. Second, for reasons to be indicated below, the *structural logic* of most situations will to some extent be indeterminate and be so perceived by the actors involved. (This is, of course, the implicit rationale for spending time and energy on negotiation efforts.) Some situations, notably those characterized by high complexity or instability, are likely to be seen as more indeterminate than others.

The logic of a negotiation process may be indeterminate for at least three reasons. First, the problem or situation itself may be ambiguous and be perceived differently by the actors involved. The cause-and-effect relationships that we need to understand to cope with environmental and other types of problems are quite often poorly understood. The present state of knowledge about anthropogenic causes of global climate change is a case in point.

Second, even if actors agree completely on how to describe a situation, its policy implications may be anything but obvious. A closer look at formal bargaining theory itself provides instructive evidence. Research in the axiomatic-static tradition of Nash (1950) has produced several solutions to the same bargaining problem, none of which has so far been conclusively authorized as uniquely compelling by rational choice criteria and even less so by normative standards. Even if one of these general formulas were to emerge as *the* unique solution, we should recognize that the more complex the game to which it is applied, the less conclusive the specific, operational advice it can produce tends to become. A cursory glance at Axelrod's (1984) computer tournaments suffices to demonstrate (1) that once we iterate even the most simple two-actor prisoner's dilemma game for an unspecified number of rounds, it is by no means obvious what the optimal strategy is and (2) that the choice of strategy can make a significant difference with regard to outcomes. Decision makers will probably often find clues and be able to converge on focal points that are alien to formal game theory (Schelling 1960). But it seems likely that just as often they will bring to the problem a wider range of policy interpretations, rendering its structural logic more rather than less ambiguous.

Third (and more important), international problem solving is a different kind of game. While the players portrayed in formal bargaining-theory models face a problem of making a collective choice of one solution from a predefined set of options, actors in international negotiations typically enter negotiations with incomplete and imperfect information and perhaps also with tentative and vague preferences (Iklé 1964, 166f). This implies that *discovering*, *inventing*, and *exploring* possible solutions may be important elements of the process (see, e.g., Walton and McKersie 1965; Winham 1977; Zartman and Berman 1982). Search, innovation, and learning are processes that can hardly be captured in deterministic models. Moreover, while formal rational-choice theory pictures actors as being "prisoners" of a game that is exogenously defined, actors in real-life negotiations can redefine their game. Together they can determine the institutional setting and redefine the nature of the problem. And even though the basic interests and values of states are likely to prove quite nonsusceptible to modification in the short run, an actor may be able to influence the specific positions of its prospective partners through various kinds of techniques—including persuasion, manipulation, and coercion—or by more or less covert intervention in foreign games of domestic politics (see, e.g., Putnam 1988).

Without further specification, the argument that skill and effort *can* make a difference is not in itself a useful proposition. It becomes interesting only to the extent that we can specify, with a fair amount of precision, what constitutes *skill* and *energy* in this context and how these factors influence cooperation processes. This is a complex challenge that we do not take on here. All we can do here is to explore briefly just one dimension that is central to our interest in explaining regime effectiveness—namely, the political engineering of effective solutions.

The political engineering of international cooperation includes at least three major tasks. One is the design of *substantive solutions* that are politically feasible. The most restrictive approach to solution design would take the institutional setting, actor preferences, and the distribution of power as exogenously given parameters and ask three main questions: (1) What are the *minimal* requirements that a solution must meet to be adopted and implemented? (2) What is the *maximum* (in terms of efficiency or fairness) that we can hope to accomplish? And (3) How would we design a solution if our only concern were to maximize its political feasibility? Another major task is to design *institutional arrangements* that are conducive to the development, adoption, and implementation of effective solutions. Here, too, the configuration of preferences and the distribution of power are often taken as givens, but institutional arrangements can themselves have a significant impact on actors' willingness to enter into cooperative arrangements. The third design target is *actor strategies*—notably, strategies that can be effective in inducing the constructive cooperation of prospective partners.

This is not the place to examine the anatomy of leadership or try to formulate relevant design principles.[33] All we can hope to do in the empirical analysis is to explore to what extent one or more of these functions of *instrumental leadership* was in fact performed in the various regime-formation and -implementation processes. The basic hypotheses guiding this part of our analysis can be stated simply as follows:

H_8: Instrumental leadership tends to facilitate regime formation and implementation. And the more skill and energy that are available for and actually invested in instrumental leadership, the more likely an effective regime will emerge.

H_{8a}: The *need* for instrumental leadership tends to increase with problem malignancy. However, *supplying* such leadership tends to become increasingly difficult as malignancy increases. Instrumental leadership thus tends to make the most difference in dealing with problems that are *moderately malign*.

Instrumental leadership may come from several sources: officers of intergovernmental organizations, conference or working group chairs, national delegates, and transnational organizations or informal networks. The former two will be subsumed under the notion of IGO actor capacity. With regard to informal transnational networks, we focus on those referred to as *epistemic communities* (Haas 1990, 55;

1992, 3).[34] According to Haas and his colleagues, such communities can play an important role in the establishment of and, at least in some cases, in the operation of international environmental regimes:

H_9: Informal networks of experts—*epistemic communities*—contribute to regime effectiveness by strengthening the base of consensual knowledge on which regimes can be designed and operate. The more integrated an epistemic community and the deeper it penetrates the relevant national decision-making processes, the more effective—*ceteris paribus*—the regime it serves tends to be.

Finally, let us point out that skill should not be considered a *constant*. On the contrary, there are many good reasons to believe that actors learn from their own experience as well as from knowledge supplied by others. Thus, we hypothesize that

H_{10}: Skills in designing effective regimes tend to improve over time as actors learn. This implies that regimes tend to become more effective over the first decade or two of their lifetime and that the latest generation of regimes tends to be—by and large—more effective than regimes of previous generations.[35]

Summary

The essence of the line of reasoning that we have outlined above can now be summarized in a rather simple core model (see figure 1.6). As this graphical representation indicates, our core model has important limitations, two of which remain to be spelt out clearly. First and most important, it considers each regime as a stand-alone arrangement and assumes that actors evaluate it on its own merits only. In the real world, of course, context matters. Regimes may be embedded, nested, or in some other way linked to other institutions, and these links may be sources of strength or weakness. Moreover, a government has multiple concerns and objectives, and some of these may well influence its negotiation and implementation behavior. We try to determine whether such links and ulterior motives significantly affect regime formation and operation, but we do not study the impact of such contextual elements in depth. Second, figure 1.6 offers a *static* picture. A dynamic representation would also include feedback lines. For example, an effective regime may well, over time, transform the structure of the problem it addresses or enhance the problem-solving capacity of the system of which it is a part. Again, these are effects we recognize and explore, but only summarily. The comparative analysis focuses essentially on the factors included in the core model.

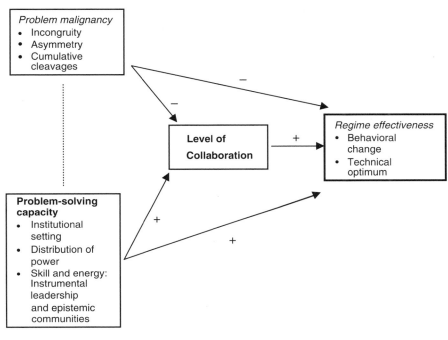

Figure 1.6
The core model

Empirical Evidence

In part II we examine these hypotheses against evidence from fourteen international regimes (see table 1.3). All but one of these (nuclear nonproliferation) deal with environmental protection or management of natural resources.

This set of cases has been selected on the basis of two main criteria. Our principal concern has been to make sure that we had a sufficient range of variance on our dependent variable (regime effectiveness) as well as in terms of problem structure, institutional setting, and distribution of power. Moreover, we wanted to ensure variance not only across cases but also *within* regimes, across components or phases. A second—and also a secondary—criterion has been practical feasibility. To be able to cover such a relatively large number of cases in some depth, we decided to draw heavily on previous work in which we have ourselves been involved over the years and—wherever relevant—to utilize available data and material to which we already had access. Even though we consider such a pragmatic approach to be

Table 1.3
Overview of cases

Regimes

1. Satellite telecommunications
2. Dumping of low-level radioactive waste at sea
3. Ship-generated oil pollution
4. Oslo Convention
5. Paris Convention
6. Barcelona Convention
7. Long-range transport of air pollutants (LRTAP)
8. Stratospheric-ozone regime
9. International trade in endangered species (CITES)
10. International Whaling Convention (IWC)
11. High-seas salmon in the North Pacific
12. Antarctic resources (CCAMLR)
13. South Pacific tuna
14. Nuclear nonproliferation

perfectly legitimate, we should like to emphasize that we make no claim whatsoever that our set of cases is in some sense a *representative* sample of the universe of international regimes. In fact, we know that it is *not*. For one thing, it is clearly biased in terms of substantive problem areas; as pointed out above, all but one of the regimes included deal with environmental protection or resource management.[36] Less obvious but perhaps more important is the fact that our set includes only cases in which at least some level of cooperation has been achieved. Research on collective problems that have integrative potential but have *not* generated cooperation may well have shed important new light on the causes of failure and thus enabled us to transcend conventional perspectives in the study of international regimes. We recognize that this category of problems remains a neglected field of study but have to leave it as a challenge for future research.

In part II, we organize our fourteen cases into three broad categories according to scores on our *dependent* variable. We begin with those regimes that we have scored as effective (in comparative terms) and end with those we have scored as relatively ineffective. For each of these categories we provide a brief introduction, summarizing what we expect to be common, distinctive characteristics of each category. We hope that these summaries help the reader keep the essence of our theoretical argument fresh in mind when examining the details of each of our case stories.

Notes

1. The discussion in this section is based on Underdal (1992a).

2. One option that has been frequently used in previous research is to focus on the formal properties of the regime—such as level or scope of cooperation, usually defined in terms of the kind of functions fulfilled (information exchange, rule making, rule enforcement, and so on) (see, e.g., Kay and Jacobson 1983, 14–18). Others (including Aggarwal 1985, 20; Zacker 1987; and Haggard and Simmons 1987) have focused on notions of regime strength, meaning—for Aggarwal—"the stringency with which rules regulate the behavior of countries." For more recent contributions that are closer to the concept used in this study, see, e.g., Young (1994), Levy, Young, and Zürn (1995), Bernauer (1995), and Helm and Sprinz (1999).

3. We are, though, struck by the fact that much of the political discussion seems to focus mainly on notions of gross benefits.

4. The life of a regime should be seen as a *sequence* of events. At the stage of regime *formation*, level of collaboration is a feature of the arrangement being developed (a *product*). However, once a regime is established and it enters the implementation stage, level of collaboration may be seen as a component of its (institutional) capacity.

5. This formulation does not imply an assumption that a new regime will necessarily improve the present state of affairs. As conceived of here, relative improvement can be negative as well as positive.

6. Although analytically distinct, the two perspectives are often hard to distinguish in practice. For example, faced with a piece of advice from a scientific advisory body, decision makers may find it hard to determine whether the solution prescribed is considered optimal or merely satisfactory.

7. The latter is an important proviso. If a group of actors succeeds in accomplishing all that could be accomplished given the best knowledge available by the time, any distance remaining to the objective optimum would be a failure of *knowledge making* rather than *policy making*. As students of politics, we are more concerned with the latter than with the former.

8. One might also interpret relative improvement as the standard that better fits a *remedial* orientation or an *incrementalist* perspective and distance to the collective optimum as the standard that makes more sense in a *synoptic* approach to decision making (see Braybrooke and Lindblom 1963).

9. They suggest conceiving of effectiveness in terms of *actual* improvement expressed as a fraction of *potential* or maximum improvement. In game-theory terms, the formula can be written as follows:

$$\frac{(\text{Actual regime solution}) - (\text{Noncooperative solution})}{(\text{Fully cooperative solution}) - (\text{Noncooperative solution})}.$$

Applied to their case (LRTAP), the *actual* regime solution is the actual change in emissions of a particular pollutant (such as SO_2) under the regime; the *noncooperative* solution is the change in emissions that would have occurred in the absence of the regulation in question; and the *fully cooperative* solution is the change in emission levels required to maximize social welfare in the region.

10. We do realize that this formulation leads into intriguing conceptual problems if we accept the claim made by Puchala and Hopkins (1982, 247) that "a regime exists in every substantive issue-area in international relations where there is discernibly patterned behavior." The notion of a *no-regime* condition seems to require a stricter definition of *regime*, notably one where the existence of *explicit* norms, rules, and procedures is considered a defining characteristic.

11. The choice between human welfare or ecological sustainability may be seen as a choice of *metric* but also as a choice of evaluation *standard*. Thus, there is a close link between evaluation standard and metric of measurement.

12. One regime dominates another if and only if it has a higher score on one criterion and at the same time is inferior on none of the other relevant criteria. Strong dominance requires one regime to have a higher score on *all* relevant dimensions.

13. *Process effectiveness* is defined by Young (1994, 146) as a matter of "the extent to which the provisions of an international regime are implemented in the domestic legal and political systems of the member states as well as the extent to which those subject to a regime's prescriptions comply with their requirements." A regime is effective in *constitutive* terms if "its formation gives rise to a social practice involving the expenditure of time, energy, and resources on the part of its members" (1994, 148).

14. The latter occurs mainly where imperfect knowledge leaves a "veil of uncertainty" around the *distribution* of costs and benefits (see, e.g., Brennan and Buchanan 1985; Young 1989). As a simple rule of thumb, we might say that uncertainty tends to interact negatively with political conflict to the extent that it can be interpreted as undermining the overall rationale for action but can potentially facilitate agreement to the extent that it makes it hard for each actor to predict the *distribution* of costs or benefits that would flow from the choice of a certain policy option. In conflict-of-interest situations, *in*determinate distributions tend to generate politically more benign games than those that are determinate (everything else being constant).

15. Besides, we know of no straightforward method for determining the intrinsic intellectual complexity of a problem.

16. Substantive-issue typologies, such as the one suggested by Czempiel (1981), also belong to this category.

17. A more technical description is provided in appendix A.

18. One of the points of contention between realist and liberal approaches to the study of international cooperation is that the former see competition as a more pervasive characteristic of interstate relations than the latter. One argument is that even goods that on their own might have been considered absolutely assessed are in many cases indirectly linked to military capabilities—the ultimate relatively assessed good in an anarchical system (see, e.g., Grieco 1990; Hasenclever, Mayer, and Rittberger 1997).

19. All hypotheses are subject to the *ceteris paribus* ("other things being equal") condition.

20. I am grateful to Jon Hovi for reminding me that this relationship is curvilinear. As we approach a zero-sum conflict, the potential gains from cooperation approach zero.

21. As indicated above, the interaction goes both ways: not only does uncertainty that can be interpreted as questioning the rationale for action add to the political intractability of malign problems, but political malignancy tends in turn to contaminate and thereby impede the development of (consensual) knowledge.

22. This is not intended to be an exhaustive list. It is easy to think of other elements. Some of these are captured in the notion of the social capital of a system. At a very general level, *social capital* can be defined as "the arrangement of human resources to improve flows of future income" (see, e.g., Ostrom 1995, 125–126). This definition includes institutions (in a broad sense), networks, and even common beliefs.

23. Moreover, as explained above, our interest in some dimensions of problem-solving capacity grew along the way. We started out with a somewhat narrower focus on problem structure and a small set of regime-design questions.

24. In fact, the latter seems sometimes not to be a strictly necessary condition; failure to pay one's membership fee seems in some international organizations *not* to lead to automatic exclusion.

25. Although decision rules are relevant for both benign and malign problems, they gain particular importance when preferences diverge, as they do when issues are politically malign.

26. In multilateral conferences, a distinction is often made between *unanimity* (requiring the positive approval of all) and the somewhat less demanding rule of *consensus* (requiring "only" the absence of any formal, substantive objections).

27. He then ominously and correctly adds: "It may in fact eliminate *more* than the potential gain" (Johansen 1979, 520).

28. Nonetheless, incumbents often succeed in using such roles as an important basis for exercising influence—sometimes to promote private or national rather than common interests.

29. In formal terms, actor A's direct power over actor B with regard to a specific issue (i) can be conceived of as a function of B's relative interest in i (U_{iB}) and A's share of control over i (K_{iA})—that is, ($U_{iB} \cdot K_{iA}$). As pointed out by Bacharach and Lawler (1981), *interest* in a particular event works two ways: it increases willingness to pay (and hence serves as a source of bargaining weakness), but it also motivates greater tactical effort (which can be a source of strength). It is not obvious which of these effects is the greater.

30. This is equivalent to saying that its own cost-benefit calculations correspond fairly well to that of the group or system.

31. In other words, rather than following the conventional path of treating interest-based and power-based approaches as competing rivals (see Levy, Young, and Zürn 1995; Hasenclever, Mayer, and Rittberger 1997), we *combine* them.

32. To get an idea of what would have been required, see, e.g., Miles (1998).

33. For an attempt to explore what will qualify as politically feasible solutions, see Underdal (1992b).

34. Haas (1992, 3) defines an *epistemic community* as a "network of professionals with recognized expertise and competence in a particular domain and an authoritative claim to policy-relevant knowledge within that domain or issue-area. These communities are, furthermore, characterized by (1) a shared set of normative and principled beliefs, (2) shared

causal beliefs, (3) shared notions of validity, and (4) a common policy enterprise. Sebenius (1992, 354) suggest that epistemic communities can be seen as "de facto, cross-cutting natural coalitions of 'believers.' "

35. The *ceteris paribus* condition is critical in this case. A plausible hypothesis would be that international cooperation tends to start with problems that are relatively easy to solve and then perhaps to move on to deal with more malign problems. Such a pattern would lead us to expect that new steps will tend to become increasingly *difficult*, requiring greater problem-solving capacity to achieve the same level of effectiveness.

36. The one exception—nuclear nonproliferation—has been included mainly to give us an opportunity to explore whether the dynamics of regime formation and implementation in a field closely linked to national security differ substantially from what we can observe in a low-politics area such as environmental protection and resource management. We fully realize, of course, that one single case cannot provide an adequate control sample, and we certainly make no claim of having penetrated the field of national security. At the same time, we do believe that the general line of reasoning outlined in this chapter is valid also beyond the field of environmental regimes. If national security is, as some claim, a very different ballgame, then even one case study may suffice to show that we had better abstain from generalizing our propositions and findings to other issue areas.

References

Aggarwal, V. 1985. *Liberal Protectionism: The International Politics of Organized Textile Trade.* Berkeley: University of California Press.

Andresen, S., and J. Wettestad. 1995. "International Problem-Solving Effectiveness: The Oslo Project Story So Far." *International Environmental Affairs* 7: 127–149.

Axelrod, R. 1984. *The Evolution of Cooperation.* New York: Basic Books.

Axelrod, R., and R. O. Keohane. 1985. "Achieving Cooperation Under Anarchy: Strategy and Institutions." *World Politics* 38: 226–254.

Bacharach, S. B., and E. J. Lawler. 1981. *Bargaining: Power, Tactics, and Outcomes.* San Francisco: Jossey-Bass.

Bernauer, T. 1995. "International Institutions and the Environment." *International Organization* 49: 351–377.

Braybrooke, D., and C. E. Lindblom. 1963. *A Strategy for Decision.* New York: Free Press.

Brennan, B., and J. M. Buchanan. 1985. *The Reason of Rules: Constitutional Political Economy.* Cambridge: Cambridge University Press.

Coleman, J. 1973. *The Mathematics of Collective Action.* London: Heineman.

Czempiel, E. O. 1981. *Internationale Politik.* Paderborn: Schöningh.

Downs, G. W., D. M. Rocke, and P. N. Barsoom. 1996. "Is the Good News About Compliance Good News About Cooperation?" *International Organization* 50: 379–406.

Easton, D. 1965. *A Systems Analysis of Political Life.* New York: J Wiley.

Efinger, M., P. Mayer, and G. Schwarzer. 1993. "Integrating and Contextualizing Hypotheses: Alternative Paths to Better Explanations of Regime Formation?" In V. Rittberger, ed., *Regime Theory and International Relations.* Oxford: Clarendon Press.

Efinger, M., and M. Zürn. 1990. "Explaining Conflict Management in East-West Relations: A Quantitative Test of Problem-Structural Typologies." In V. Rittberger, ed., *International Regimes in East-West Politics*. London: Pinter.

Grieco, J. M. 1990. *Cooperation Among Nations: Europe, America, and Non-Tariff Barriers to Trade*. Ithaca: Cornell University Press.

Haas, P. M. 1990. *Saving the Mediterranean*. New York: Columbia University Press.

Haas, P. M. 1992. "Introduction: Epistemic Community and International Policy Coordination." *International Organization* 46: 1–35.

Haggard, S., and B. A. Simmons. 1987. "Theories of International Regimes." *International Organization* 41: 491–517.

Hasenclever, A., P. Mayer, and V. Rittberger. 1997. *Theories of International Regimes*. Cambridge: Cambridge University Press.

Helm, C., and D. Sprinz. 1999. "Measuring the Effectiveness of International Environmental Regimes." PIK Report No. 52. Potsdam: Potsdam Institute for Climate Impact Research.

Iklé, F. C. 1964. *How Nations Negotiate*. New York: Harper & Row.

James, S. C., and D. A. Lake. 1989. "The Second Face of Hegemony: Britain's Repeal of the Corn Laws and the American Walker Tariff of 1846." *International Organization* 43: 1–31.

Johansen, L. 1979. "The Bargaining Society and the Inefficiency of Bargaining." *Kyklos* 32: 497–522.

Kay, D. A., and H. K. Jacobson. 1983. *Environmental Protection: The International Dimension*. Totowa, NJ: Allanheld, Osmun.

Keohane, R. O. 1984. *After Hegemony*. Princeton: Princeton University Press.

Keohane, R. O. 1986. "Reciprocity in International Relations." *International Organization* 40: 1–27.

Keohane, R. O. 1988. "Bargaining Perversities, Institutions and International Economic Relations." In P. Guerrieri and P. C. Padoan, eds., *The Political Economy of International Co-operation* (pp. 28–50). London: Croom Helm.

Lang, W. (1989): "Multilateral Negotiations: The Role of Presiding Officers." In F. M. Markhof, ed., *Processes of International Negotiations* (pp. 23–42). Boulder: Westview (for IIASA).

Levy, M., O. R. Young, and M. Zürn. 1995. "The Study of International Regimes." *European Journal of International Relations* 1: 267–330.

March, J., and J. P. Olsen. 1989. *Rediscovering Institutions*. New York: Free Press.

Mearsheimer, J. J. 1995. "The False Promise of International Institutions." *International Security* 19: 5–49.

Miles, E. L. 1989. "Scientific and Technological Knowledge and International Cooperation in Resource Management." In S. Andresen and W. Østreng, eds., *International Resource Management*. London: Belhaven Press.

Miles, E. L. 1998. *Global Ocean Politics*. The Hague: Martinus Nijhoff.

Nash, J. F. 1950. "The Bargaining Problem." *Econometrica* 18: 155–162.

Olson, M. 1965. *The Logic of Collective Action*. Cambridge, MA: Harvard University Press.

Olson, M. 1982. *The Rise and Decline of Nations* (New Haven: Yale University Press).

Ostrom, E. 1995. "Constituting Social Capital and Collective Action." In R. O. Keohane and E. Ostrom, eds., *Local Commons and Global Interdependence.* London: Sage.

Oye, K. A. 1985. "Explaining Cooperation Under Anarchy: Hypotheses and Strategies." *World Politics* 38: 1–24.

Puchala, D. J., and R. F. Hopkins. 1982. "International Regimes: Lessons from Inductive Analysis." *International Organization* 36: 245–275.

Putnam, R. D. 1988. "Diplomacy and Domestic Politics: The Logic of Two-Level Games." *International Organization* 42: 427–460.

Risse, T. 2000. " 'Let's Argue!' Communicative Action in World Politics." *International Organization* 54: 1–39.

Rittberger, V., and M. Zürn. 1990. "Towards Regulated Anarchy in East-West Relations: Causes and Consequences of East-West Regimes." In V. Rittberger, ed., *International Regimes in East-West Politics*. London: Pinter.

Sand, P. H. 1990. *Lessons Learned in Global Environmental Governance.* Washington, DC: World Resources Institute.

Schelling, T. C. 1960. *The Strategy of Conflict.* Cambridge, MA: Harvard University Press.

Sebenius, J. K. 1983. "Negotiation Arithmetic: Adding and Subtracting Issues and Parties." *International Organization* 37: 281–316.

Sebenius, J. K. 1992. "Challenging Conventional Explanations of International Cooperation: Negotiation Analysis and the Case of Epistemic Communities." *International Organization* 46: 323–365.

Snidal, D. C. 1985. "The Limits of Hegemonic Stability Theory." *International Organization* 39: 579–614.

Stein, A. A. 1982. "Coordination and Collaboration: Regimes in an Anarchic World." *International Organization* 36: 299–324.

Underdal, A. 1980. *The Politics of International Fisheries Management: The Case of the Northeast Atlantic.* Oslo: Universitetsforlaget.

Underdal, A. 1987. "International Cooperation: Transforming 'Needs' into 'Deeds.' " *Jorunal of Peace Research* 24: 167–183.

Underdal, A. 1992a. "The Concept of Regime 'Effectiveness.' " *Cooperation and Conflict* 27: 227–240.

Underdal, A. 1992b. "Designing Politically Feasible Solutions." In R. Malnes and A. Underdal, eds., *Rationality and Institutions.* Oslo: Scandinavian University Press.

Underdal, A. 1994. "Progress in the Absence of Substantive Joint Decisions?" In T. Hanisch, ed., *Climate Change and the Agenda for Research.* Boulder, CO: Westview.

Walton, R. E., and R. B. McKersie. 1965. *A Behavioral Theory of Labor Negotiations.* New York: Free Press.

Wettestad, J. 1999. *Designing Effective International Regimes.* Cheltenham: Elgar.

Winham, G. R. 1977. "Negotiation as a Management Process." *World Politics* 30: 87–114.

Young, Oran R. 1989. *International Cooperation.* Ithaca: Cornell University Press.

Young, O. R. 1994. *International Governance*. Ithaca: Cornell University Press.

Young, O. R. 1996. "Institutional Linkages in International Society: Polar Perspectives." *Global Governance* 2: 1–23.

Young, O. R. 1998. *Creating Regimes*. Ithaca: Cornell University Press.

Young, O., and G. Osherenko. 1993. "Testing Theories of Regime Formation: Findings from a Large Collaborative Research Project." In V. Rittberger, ed., *Regime Theory and International Relations*. Oxford: Clarendon Press.

Zacker, M. W. 1987. "Trade Gaps, Analytical Gaps: Regime Analysis and International Commodity Trade Regulation." *International Organization* 41(2): 173–202.

Zartman, I. W., and M. R. Berman. 1982. *The Practical Negotiator*. New Haven: Yale University Press.

Zürn, M. 1992. *Interessen und Institutionen in der internationalen Politik: Grundlegung und Anwendung des situationsstrukturellen Ansatzes*. Opladen: Leske & Budrich.

2

Methods of Analysis

Arild Underdal

Any empirical study builds on a number of decisions about research strategy and method that some may find questionable or misguided. To help the reader determine how much confidence to place in the findings reported in this book, in this chapter we briefly review and discuss some of the major methodological problems that we have struggled with and explain the solutions we have opted for. We proceed in two steps. First, we consider the challenge of making causal inferences, arguing for a combination of intensive and extensive approaches. Then we move on to discuss problems pertaining to the measurement of single variables.

Causal Inference: The Case for Combining Methodological Approaches

Our empirical analysis will be a two-track exercise in which we combine a set of in-depth qualitative case studies (to be reported in part II) with a more extensive and rigorous comparative analysis relying on formal logic and statistical techniques (part III). Despite the long and sometimes heated controversy over the relative merits of qualitative versus quantitative methods in political science research, we take the ecumenical view that these two sets of techniques are largely complementary, each providing opportunities that the other does not offer, at least not to the same degree. We also see them as building on the same basic logic of scientific inquiry, meaning that there are no fundamental incompatibilities preventing us from combining them in one integrated study (King, Keohane, and Verba 1994, 4).

Through exploratory case studies we can trace policy-making and policy-implementation processes as they unfold and reconstruct the story behind each individual regime. Although heuristically guided by a more or less well-specified causal model, qualitative case-study research typically starts out with relatively open templates inviting exploration of whatever causal factors or mechanisms seem to be important in shaping developments. This makes the approach well suited for identifying causal mechanisms and pathways, for generating new hypotheses, and for exploring complex and unfamiliar puzzles. In this project, we agreed from the start that case-study authors should pursue causal paths wherever they might lead—also when they point to factors or mechanisms that are neither included in nor consistent with the model outlined in chapter 1.[1] Moreover, by virtue of their diachronic perspective, case studies are uniquely suited for studying *intra*regime dynamics—including specific phenomena such as learning and momentum, as well as more general but elusive mechanisms such as path dependency. Finally, they also provide unique opportunities for what might be called *holistic* analysis, which emphasizes the significance of interplay and context.

Our comparative analysis builds directly on this set of case studies. More precisely, it takes the qualitative description and interpretation provided in each case study, translates them into numerical values for the (main) variables included in the model, and builds two integrated data files containing standardized information about all cases. These two files have important common features: both contain a finite set of variables, and both have a fixed set of response categories for each variable. Moreover, in this project they serve basically the same purpose, which is to enable us to examine systematically patterns of variance across cases. More specifically, we use these files to determine whether and to what extent the patterns that emerge when we examine all cases together are consistent with the hypotheses formulated in chapter 1.

At a technical level, however, the two files differ in important respects. One has the format of a dichotomized "truth table" and is analyzed in accordance with principles of Boolean logic (see Ragin 1987, 2000). The other file contains interval- and ordinal-scale data analyzed by simple statistical techniques. The former approach—known as *qualitative comparative analysis* (QCA)—is particularly suitable for bringing out the *full range* of causal conditions associated with a particular outcome and for identifying *conjunctures* of such conditions. Moreover, it is geared toward identifying necessary and sufficient conditions (Ragin 1987, chs. 6–7). At the same time, however, the fact that all variables are dichotomized makes it a fairly blunt instrument. Moreover, it is not well suited to distinguish the effect of one independent variable from that of another. On both these scores, the other file has a distinct advantage. For most of the main variables, it offers a larger set of categories. And conventional statistical techniques give us more mileage when it comes to determining *how much* of the variance observed can be attributed to *each* independent variable or cluster of variables. Since the files have different strengths, we use them for different purposes—the QCA file for identifying combinations of factors and pathways leading to particular outcomes and the statistical file for separating effects and measuring their (relative) strength.

By combining in-depth case studies with more extensive and formal comparative analysis we can to some extent use the strengths of one to compensate for limitations inherent in the other. Thus, a more rigorous analysis of a larger set of observations can help us determine whether inferences made on the basis of data for one single case are supported by evidence from a larger sample of cases. It can also help us discover patterns that do not emerge when we consider each case on its own, particularly similarities and differences across regimes. Conversely, in-depth qualitative research can help us identify and describe the *causal mechanisms* and stories behind patterns of covariance. It thereby provides a more solid basis for distinguishing correlations that lend themselves to a substantively plausible interpretation from those that do not. In addition, it serves a useful heuristic function in helping us discover pathways that may guide or inspire further statistical analysis.

Measuring Single Variables

One major challenge for empirical research on international politics is that many of the key variables identified in models and theories call for what might be called *judgmental assessment* rather than straightforward observation and measurement. A variable such as the number of houses built in a certain region during a particular period of time can be determined by simple counting. By comparison, measuring the effectiveness of a particular regime or the political malignancy of an environmental problem leaves a much larger role for subjective judgment.[2] Since it is usually very difficult to formulate an exhaustive set of precise rules and operational criteria to guide such judgmental assessment, the process tends to become less transparent, and the results less reliable. This makes it all the more appropriate to try to explain the guidelines that we have followed in determining *regime effectiveness* and *problem malignancy* and in assigning scores on other variables that do not lend themselves to straightforward measurement.

The general procedure may be summarized as follows:

1. As part of the development of the research design, a set of general guidelines for assigning scores to problematic variables—in particular those of regime effectiveness and problem malignancy—were formulated by the team (see below). These guidelines were later refined but not fundamentally changed in response to

problems encountered in their application as well as in response to comments and suggestions from others.

2. Equipped with these guidelines, each case-study author assigned scores for his or her case(s). Wherever feasible, outside experts were consulted—in a few cases through a formal review process. We have benefited substantially from their comments and suggestions but in some cases decided not to follow their advice.[3] Thus, the standard phrase whereby the authors accept the full responsibility for what is reported is in this case more than a mere ritual acknowledgment of the obvious.

3. To ensure consistency, the team as a whole reviewed scores on key variables for all cases. Since further analysis and external comments resulted in some adjustments for a couple of cases, this internal review process was repeated twice.

Normally, consistency in interpretation and measurement is more easily obtained in *intra-* than in *inter*regime analysis.[4] It is, for example, by and large easier to determine whether the effectiveness of a particular regime has increased or decreased over time than to determine whether one regime is more or less effective than another. Furthermore, the difficulties involved in comparing scores across regimes tend to increase as the variation in substantive issues that the regimes are intended to address increases. Thus, it is easier to compare two regimes dealing with pollution control than to compare a pollution regime with, for example, one established to manage marine living resources. We have tried to minimize these problems by keeping ambitions at a modest level: in no instance do we attempt to move beyond *ordinal*-level measurement, and for all our key variables we limit the number of categories to three or four. These measures do not solve the basic problem, however. As a general rule of thumb, it therefore seems prudent to place more confidence in conclusions from intraregime analysis than in comparisons across regimes. However, the former cannot be a general substitute for the latter; some important questions can be answered only by interregime comparison.

From Verbal Description to Numerical Scores

Before describing in more specific terms how we have dealt with some of the key variables included in our core model, we would like to pause for a minute to reflect on an interesting experience that we have had in translating qualitative description into numerical scores. With a few notable exceptions, colleagues and practitioners alike have been significantly more critical of the numerical case accounts than they were of the verbal or qualitative descriptions—even when the two contain the same pieces of information. In this study, the main difference between the two lies in format, not in content; the numerical file contains processed rather than new data.[5] This observation leads to an interesting question: what in the numerical format tends to generate skepticism and criticism not provoked by a purely verbal presentation of the same information?

We suspect that part of the answer may be found in a misconception of the level of measurement involved—more precisely, a belief that numerical scores must refer to at least interval-scale data. This is not correct; numerical scores can, of course, be used to represent even nominal-scale distinctions. The substantive interpretation of scores, however, varies profoundly from one level of analysis to another. On a cardinal scale, 0 means *zero units*, 1 means *one unit*, and so on. On an ordinal scale, numerals are used merely to distinguish *more* from *less*, so when we see a range of scores [0,1,2], all we can infer is that 0 < 1 < 2. In this study *all* our numerical codes are at the *same* level of measurement as the corresponding verbal description. And in no instance do we try to go beyond ordinal-scale measurement.

Second, the numerical format itself is probably seen as more constraining and demanding. It is perceived as more constraining in that it offers not only a finite set of variables but also a finite set of predefined categories for each variable. In the purely verbal mode of description, students probably feel more free to invent or modify categories on an ad hoc basis. We see this freedom as largely illusory, however: any category that you can define in prose can be represented by a numerical code. Moreover, unless the new ad hoc categories are explicitly defined, the substantive significance of subtle nuances in wording will probably be lost on the reader. There is even a real risk that by the time the study is published, the author will no longer remember exactly what those distinctions and nuances mean. Finally, freedom from the constraints of a fixed format may be bought at a high cost; a plethora of ad hoc and ill-defined categories provides a shaky basis for any systematic comparative analysis.

Moreover, the numerical format is probably also seen as more demanding in that it requires an *unambiguous choice*. In prose, for example, you may say that a particular problem is in some (unspecified) respects strongly malign and in other (equally unspecified) respects quite benign and then leave it up to readers to draw their own conclusions. In a fixed format *we* have to decide which of the existing categories provides the better fit—or define a new intermediate category that captures that particular combination of characteristics.[6]

Third, precisely because it can be more constraining and demanding, the numerical format will normally be more transparent. Transparency greatly facilitates criticism. For the individual scholar it may well be tempting to seek protection behind a shield of vagueness. For the research enterprise at large, however, minimizing risk is hardly a good recipe for progress. From that perspective, we concur with King,

Keohane, and Verba (1994, 112) that—as a rule of thumb—it is "better to be wrong than vague." As will become evident when we present the various case studies, we do recognize that in some cases good arguments can be made for alternative assessments and interpretations. This makes it all the more important to give the reader an opportunity to check the scores that we have assigned to the various regimes. To provide such an opportunity, we have included the (main) codebook as well as the corresponding data matrix as appendices B and C to this volume. Moreover, at the end of each chapter we provide a brief summary of scores on the main variables for that particular case.

Let us now add a few words to explain how we have dealt with the three key components of our core model: regime effectiveness, problem type, and problem-solving capacity.

The Dependent Variable: Measuring Regime Effectiveness

As we point out in chapter 1, determining regime effectiveness is not merely a matter of descriptive measurement; it is as much an exercise in causal inference. In addressing the question of whether or to what extent a regime made a difference, we compare the state of affairs that obtains with the regime in place with the hypothetical situation that would have occurred in its absence. In doing so, we not only try to measure difference; we also attribute difference in human behavior or the health of the environment to the existence or operation of a regime. In procedural terms, it often makes sense to perform these operations in sequential order—first to compare the situation that existed immediately before the regime was established (t_0) with the situation that exists when it is in place (t_r) and then to move on to explore how we can explain the change or stability observed. But the comparison of the state of affairs at times t_0 and t_r does not in and of itself tell us anything about the role of the regime. Moreover, there are other procedures—relying on some kind of business-as-usual scenario—that may offer a more direct route in some cases.[7] Similarly, when we ask whether a regime solved or ameliorated the problem it was designed to deal with, we want to know not only how far the current state of affairs is from what can be considered good or optimal but also whether or to what extent the regime brought us any closer to that goal.[8] The latter requires that we can distinguish contributions made by the regime itself from those of other sources. Needless to say, this can be a tall order, requiring a comprehensive understanding of what drives developments within the system of activities to be regulated. Particularly when it comes to assessing effectiveness in terms of impact

Table 2.1
Guidelines for assessing effectiveness

Regime versus No-Regime Counterfactual	Regime versus Collective Optimum
1. Look for available estimates or calculations from recognized experts in the field. Consensual estimates are to be preferred over contested estimates, estimates from nonpartisan sources over those that come from partisan sources, and estimates actually used as a basis for the negotiations over other estimates.	1. Look for available recommendations or advice from recognized experts in the field. Consensual advice is to be preferred over contested advice, advice from nonpartisan sources over advice from partisan sources, and recommendations actually used as a basis for the negotiations over other pieces of advice.
2. If no such estimates can be found in documents to which you have access, see if you can yourself obtain estimates from independent (nonpartisan) and competent experts.	2. If no such advice can be found in documents to which you have access, see if you can yourself obtain assessments from independent (nonpartisan) and competent experts.
3. If neither of these two strategies works, try to make your own estimate. Business-as-usual scenarios derived from a theory-based model or empirical mapping of causal pathways are to be preferred to estimates based on linear extension of (previous) trends. For systems characterized by high stability, you may—as a last resort—use the assumption that the *status quo ante* (at time t_0) would have continued.	3. If neither of the preceding options is available, use the *official purpose* of the regime, as stated in its "constitution" (or the declared objective of the specific protocol or regulation), as your standard of reference. If the official purpose is contested, specify also competing views.
4. If neither of the options above is available, report *missing data*.	4. If neither of the above provides adequate guidance, report *missing data*.

on the biophysical environment, it also requires natural science expertise that most political scientists do not have. There is, though, no easy way out. The concept of effect is inextricably linked to the notion of causation. And since there can be no effect without a cause, there can be no attribution of effectiveness without causal inference.

At an early stage, we therefore adopted a few rather simple rules of thumb to help us navigate these difficult waters. The gist of these guidelines is summarized in table 2.1 (the formulations above refer to the outcome and impact foci only).[9]

To be sure, several problems arise even with our preferred solution. One obvious constraint is that conclusive and consensual expert estimates or advice will be available only in a few of the cases that we want to study. Moreover, even where it is available, it may be hard to translate into a standardized yardstick for measuring effectiveness, particularly when defined in terms of optimal solutions. There are at least two reasons for this: (1) advice may be framed with reference to *different* levels of ambition (such as preventing further deterioration or restoring a stock or ecosystem in *x* years) rather than some unequivocal notion of what constitutes the best solution, and (2) the advice that can be found will probably most often refer to ecological or some other technical criteria, not to social welfare (which, presumably, is the principal concern of governments). Despite these and other pitfalls, expert assessment or advice is clearly one important source of data to be utilized and should normally be preferred over homemade assumptions. When expert advice is not available or appears too inconclusive to determine what would be an optimal solution, a fallback strategy would be to look for some official declaration of a joint goal or purpose. Some, but not all, conferences provide such a declaration (of the format "eradicate hunger in ten years"). It should be recognized, though, that such declared goals are often quite vague—serving hortatory rather than operational purposes—and that the relationship between official goals and the notion of collective optimum is by no means clear. This indicates that this strategy can be used only with caution and also that scores based on different operational measurements cannot always be used interchangeably. The bottom line is that analysts will sometimes have to rely on their own best judgments, modified or supported by whatever judgments by other and better-informed observers they can find or collect.

We use a five-point ordinal scale to measure behavioral change (the scale itself is described on page 483, Var62) and a four-point scale to describe distance from the collective optimum (page 484, Var66). We emphasize that both of these scales invite exercises in *soft* assessment, and we realize that our conclusions may well in some instances be challenged. We take some comfort in the fact that—except in one case—the instances of dissent that we have encountered during the external and internal review processes raise questions about the choice between *adjacent* categories, leading into discussions about, for example, whether the improvement observed qualifies as significant rather than small.[10] In most cases our reviewers have argued that we tend to *under*estimate effectiveness. Recognizing that in-depth case studies designed to trace regime effects may be prone to underestimating the impact of *other* factors, it has indeed been our policy to offer conservative assessments whenever in doubt. To examine the sensitivity of our main conclusions to contested scores, we rerun some of our main tests with the alternative assessments of effectiveness proposed by external reviewers.

Independent Variables: Type of Problem

In chapter 1, we say that a problem may be difficult to solve in two different respects: it may be intellectually complex or poorly understood, and it may be politically malign. Concentrating on the latter aspect, we moved on to explore what distinguishes a politically malign problem from one that is more benign. We ended up with a multidimensional construct and concluded that most of the problems that generate efforts to establish international regimes combine malign and benign features. As a consequence, the distinction is by no means crisp and clear-cut, and measurement becomes a matter of determining which of these features is (are) the most salient in each particular case.[11]

The question that remains to be addressed is whether we nonetheless can construct a crude ordinal scale that provides a substantively meaningful integrated assessment of political malignancy. It is abundantly clear that there are many pitfalls in such an operation. We nevertheless believe that it is a worthwhile exercise; integrated scales or indexes that can be used to reduce complexity are important tools of systematic comparative analysis with a small number of observations. In this case, we decided to proceed in two main steps. First, we ask whether the basic structure of the problem is one of incongruity, coordination, or a balanced mix. The answer to this question tells us whether the problem is predominantly malign or benign. For problems that are predominantly malign, we then proceed to assess the *strength* of incongruity relationships (distinguishing between moderate and strong). This gives us a four-level ordinal scale, summarized in table 2.2. In using this scale, we consider each problem on its own merits only.[12] As a third and final step, however, we proceed to explore links to other issues and ulterior motives that seem to affect actor behavior. Analytically, we treat the latter dimensions as contextual elements, not as characteristics of the problem(s) addressed by the regime.

Independent Variables: Problem-Solving Capacity and Its Components

As we explain in chapter 1, problem-solving capacity is an even more complex construct than malignancy. We have introduced it for the analytic purpose of capturing the aggregate impact of three different factors—the institutional setting, the distribution of power, and the supply of instrumental leadership in engineering cooperative solutions. Much of the current debate among students of international

Table 2.2
Scale used in assessing problem malignancy and benignity

Score	Problem features
Benign	The problem is predominantly one of coordination, characterized by synergy and contingency relationships. Incongruity features are absent or weak.
Mixed	Elements of incongruity are combined with elements of synergy and contingencies, and neither clearly dominates the other.
Moderately malign	Substantial elements of incongruity core limited mainly to externalities. Neither competition nor asymmetry is a salient feature.
Strongly malign	Severe incongruity also includes strong elements of competition and asymmetry. Cumulative cleavages may further aggravate the situation.

regimes centers on the role that each of these factors plays in regime formation and implementation. Our study does not deviate radically from that path. What we want to add to the agenda is a concern also with interplay and aggregate effects. And we offer the concept of problem-solving capacity as a crude analytic tool that may help us frame research on these complex questions in terms of a coherent, instrumental perspective.

As defined here, *institutional setting* can be described in relatively straightforward terms, while *leadership* is a more elusive variable inviting judgmental assessment. The distribution-of-power variable falls somewhere in the middle. A precise, interval-scale measure *can* be constructed from a set of capability indices. For some analytic purposes—for instance, predicting the outcome of an international negotiation process—such a tool is required (see, e.g., Underdal 1998). In this study, however, all we need is a crude, ordinal-scale assessment of whether the distribution of power is skewed in favor of one particular group of parties. This kind of assessment will be based on soft judgment rather than hard measurement.

The data protocol specifies the questions and scales that we have used as a basis for the comparative analysis.[13] Since we are dealing with a complex construct where the various constitutive elements to some extent have different meanings in different contexts, we have to combine these elements in different ways for different analytical purposes. We can best explain how this is done as we go through the various steps of the comparative analysis. We thus refer to the concluding chapter for more detailed information.

Units of Analysis

This volume includes case studies of fourteen international regimes. Most of these regimes have two or more distinct components, and some have been through several distinct stages of development. For some comparative purposes, each of these components or stages is considered a separate unit of analysis. A few words may therefore be appropriate to explain *how* we have distinguished components and phases.

A *component* is a separate protocol or other document specifying a distinct set of commitments. For example, the LRTAP regime has spawned separate protocols for different polluting substances, such as sulfur dioxide and nitrogen oxide. Other regimes—including that pertaining to ship-generated oil pollution—have been through substantial changes in the rules and regulations pertaining to the same activity or substance during their lifetime.[14] In both cases, we are dealing with significantly different sets of commitments, established at different points in time. It therefore makes sense to consider each of these components as a separate unit of analysis.

While distinguishing between components is a fairly straightforward exercise, distinguishing one *phase* from another often involves a substantial amount of discretion on our part. Our guiding principle has been to make a distinction between phases if and only if we observe a significant change in regime effectiveness, in problem malignancy, or in problem-solving capacity over time (with components held constant). The authors of the various case studies explain their reasons for making such distinctions in each specific case.

It goes without saying that components or phases of one regime are not fully independent units; they are parts of and embedded in a larger superstructure.[15] This is duly recognized in the case studies as well as in the comparative analysis of *intra*regime variance. It does, however, introduce an element of autocorrelation that may disturb the broader statistical analysis.

Concluding Remarks

The study of regime effectiveness can be pursued through a wide range of methodological approaches—from formal game theory and computer simulations to in-depth qualitative case studies.[16] We feel no urge to engage in a dispute over which of these approaches is generally best. On the contrary, our point of departure is that different methodological approaches to a large extent *serve different purposes* and that the field at large would be well advised to work with research designs that *combine* complementary approaches. Whatever the merits of the particular methodological solutions that we have opted for in this study, we believe that the combination of theory-guided intensive case studies with more extensive, rigorous, and transparent comparative analysis is a promising strategy (cf. Mitchell and Bernauer 1998).

Notes

1. One important implication of this liberal strategy, though, is that many interesting suggestions may be made in the various case studies that we cannot follow up in the comparative analysis—simply because we lack comparable data for other cases.

2. The difference, though, is a matter of *degree* rather than one of principle. Few if any of the variables used in social science research leave *no* role for judgment. Thus, in counting the number of houses one may well have to make judgmental decisions about, for example, whether a certain building satisfies the defining characteristics of a house or whether it should be counted as a single unit or more.

3. When we did not heed the good advice that we got from competent experts, the reason was in most cases that we discovered that the general standards by which they had scored a case differed somewhat from those that we had agreed to apply throughout this study.

4. One reason, of course, is that in this study each regime is analyzed by a single author. Consistency is normally easier to achieve at the *intra*personal than at the *inter*personal level.

5. Some variables referred to in one or more of case studies are not included in the comparative data files, and some of the variables or indicators included in the latter are not explicitly dealt with in all case studies. These differences, however, are not important in this context.

6. In addition, researchers may have to conclude that they do not have sufficient data to determine the appropriate score. In contemplating that solution, however, they face a real dilemma. Poor data may substantially weaken any database, but so will frequent resort to the "don't know" or "missing data" category. Researchers will have to decide in each specific case which is the lesser evil and provide the information required for readers to make up their own minds as to the implications of that choice.

7. For a good overview and evaluation of different approaches to counterfactual analysis in the study of international politics, see Tetlock and Belkin (1996).

8. This implies that assessment of effectiveness in terms of some notion of good or optimal solutions requires that we can determine whether or to what extent the regime induced behavioral change.

9. Assessing effectiveness in terms of *output* (formal rules, regulations, and so on) is normally a somewhat easier task. Essentially, it is matter of determining whether "the rules on paper" *demand* a significant change in target group behavior and whether or to what degree regime demands are sufficient—assuming compliance—to "solve the problem."

Some tentative suggestions for assessing effectiveness-as-output can be found in Underdal, Andresen, Ringius, and Wettestad (1999).

10. The one noticeable exception is the case of whaling, but here the main object of controversy is the *standard* of evaluation rather than facts about change in human behavior or in the state of whale stocks.

11. Recognizing that the intermediate category of mixed offers an easy way out, we agreed to resort to this score only when we could find no sound basis for using one of the adjacent categories.

12. Moreover, we ask how the problem was *perceived* by the parties involved.

13. Included at the end of the volume as appendix B. In this appendix we also explain how our indexes are constructed.

14. In some cases, changes may be so fundamental that they raise the question of whether it makes sense to talk about the *same* regime.

15. The same observation can in many cases be made about *regimes* as well.

16. For an overview and evaluation of various methodological approaches to the study of regime effectiveness, see Underdal and Young (forthcoming).

References

King, G., R. O. Keohane, and S. Verba. 1994. *Designing Social Inquiry.* Princeton, NJ: Princeton University Press.

Mitchell, R., and T. Bernauer. 1998. "Empirical Research on International Environmental Policy: Designing Qualitative Case Studies." *Journal of Environment and Development* 7(1): 4–31.

Ragin, C. C. 1987. *The Comparative Method.* Berkeley: University of California Press.

Ragin, C. C. 2000. *Fuzzy-Set Social Science.* Chicago: University of Chicago Press.

Tetlock, P. E., and A. Belkin. 1996. *Counterfactual Thought Experiments in World Politics.* Princeton, NJ: Princeton University Press.

Underdal, A. 1998. "Modelling International Negotiations: The Case of Global Climate Change." Paper prepared for the Third Pan-European Conference on International Relations, Vienna, September 16–19.

Underdal, A., S. Andresen, L. Ringius, and J. Wettestad. 1999. "Evaluating Regime Effectiveness: Developing Valid and Usable Tools." Agenda-setting paper prepared for the Second-Year Workshop of the E.U. Concerted Action Project on the Effectiveness of International Environmental Agreements. Fridtjof Nansen Institute, Lysaker.

Underdal, A., and O. R. Young, eds. Forthcoming. *Regime Consequences: Methodological Challenges and Research Strategies.*

II

Effective Regimes

Introduction: Common Features of Effective Regimes

As indicated in chapter 1, we present our case studies in three main sections, based on their scores on our *dependent* variables. In this part we examine regimes that we consider by and large *effective* in comparison with the other regimes in our sample. This classification is based on crude, overall assessments. In some cases *intra*regime variance across stages or components is as great as variance across regimes that we have placed in different categories. We pay due attention, of course, to intraregime variance in each case study as well as in the comparative analysis. However, we also need to deal with each regime in more holistic terms. Moreover, we hope that the reader will have an easier task of digesting the large amounts of information provided in the various case studies if we order our presentations in terms of a simple organizing principle.

The argument that we outlined in chapter 1 suggests that effective regimes will be characterized by particular configurations of scores on our set of independent variables, as indicated in table II.1. More specifically, we suggest that there are two main paths to effectiveness. One goes through type of problems; benign problems are easier to solve than those that are malign. Moreover, problems that are well understood are easier to deal with than those that are clouded in uncertainty about

Table II.1
Hypothesized configuration of scores for *effective* regimes

Independent Variable	Hypothesized Score
Type of problem	• Predominantly *benign* or at least *mixed* • State of knowledge: *good*
Problem-solving capacity	*High*, as indicated by • Decision rules providing for adoption of rules by (qualified) majority • An IGO with significant actor capacity serving the regime • A well-integrated epistemic community • Distribution of power in favor of pushers or pushers + intermediaries • Instrumental leadership by one or a few parties or by individual delegates or coalitions of delegates
Political context	*Favorable*, as indicated by • Linkages to other, benign problems • Ulterior motives or selective incentives for cooperation

cause-and-effect relationships. The other path goes through problem-solving capacity; other things being equal, the greater the problem-solving capacity of a system, the more effective the solutions it produces. High problem-solving capacity is likely to be a *necessary*, but not sufficient, condition for developing effective solutions to truly *malign* problems. *Benign* problems can, however, be solved effectively even with modest capacity.

These propositions apply as long as we consider each problem in isolation and on its own merits only. In real life, regime-formation and -implementation processes always take place within a broader political context that may enhance or impede success. A favorable political context can to some extent reduce the demands on problem-solving capacity but is probably in and of itself neither a necessary nor a sufficient condition for regime effectiveness.

3

Toward the End of Dumping in the North Sea: The Case of the Oslo Commission

Jon Birger Skjærseth

The Success of the Oslo Commission

The twenty-year history of the Oslo Convention—from the early 1970s to the early 1990s—represents a development from dumping anarchy toward international and national regulation and governance.[1] When the coaster *Stella Maris* left Rotterdam in 1971 with 650 tons of toxic waste on board destined to be dumped in the northern part of the North Sea, no international guidelines had yet been drawn up that could give formal weight to the protests that arose. Moreover, national legislation on dumping and permitting procedures were, in most cases, completely absent. In short, anyone could dump anything without interference from national authorities or international bodies. Partly as a consequence of the *Stella Maris* incident, the Convention for the Prevention of Marine Pollution by Dumping from Ships and Aircraft was established. The Oslo Convention covers the entire Northeast Atlantic as far as the North Pole.

The Oslo Convention was implemented domestically by the parties in the form of new legislation covering previously uncontrolled activities. For example, the United Kingdom implemented the Convention by passing a Dumping at Sea Act in 1974, while the Netherlands formally implemented the Convention by the Pollution of the Sea Act in 1977. These acts established permit procedures as required by the Oslo Convention. Thus, when the decisions were taken in the latter part of the 1980s to phase out dumping and incineration at sea, these activities were under full control of the respective governments, and the decisions could be implemented by simply withholding permits. Today, most of the dumping and incineration of hazardous waste have ceased, thanks to alternative disposal methods and the implementation of no- and low-waste technology. The only remaining problem of some significance is dredging operations, which unlike the others leads to no *new* inputs to the marine environment. Thus, the Oslo Convention is probably the only example of an international environmental agreement that has been rendered super-fluous by the ending of the activities the Convention was established to control. In 1992, the Oslo Convention was merged with the Paris Convention on land-based sources, and the Oslo Commission—the convention's executive body—was swal-lowed by the Paris Commission. In 1998, the Oslo and Paris Conventions were replaced by the OSPAR Convention for the protection of the marine environment of the Northeast Atlantic.

A Moderately Malign Problem Seeking a Solution

Increasing economic activity in the years following World War II led to a rapid growth in waste production. Many West European states looked for new ways of disposal, including dumping at sea, and by the late 1960s people became aware of the ever-increasing world pollution. Rachel Carson had already warned us in 1962 in her book *Silent Spring* that we were in danger of poisoning our environment by the use of hazardous substances. The International Council for the Exploration of the Sea (ICES) reported in 1968 that huge amounts of waste were being discharged into the North Sea.[2] In April 1971, a British company announced its intention to collect chemical effluents from West European ports for the purpose of dumping them at sea. According to the London *Observer* newspaper, big business was moving into sea dumping, and the Norwegian government informed the U.K. government of its concerns. The Scandinavian states had already decided to ban such dumping, and the Scandinavian concern was shared by other countries. Moreover, the prepa-rations for the global U.N. Conference on Environmental Protection took place in Stockholm in 1972, and ocean dumping was one of the subjects to be discussed within the framework of this conference.

In July 1971, the *Stella Maris* incident caused immense public alarm and was met with protests from a number of European governments, revealing the need for international guidelines for regulation of dumping. This need was transformed into action when the Oslo Convention (OSCOM) was signed by all thirteen West European maritime states in 1972. However, dumping at sea was only one minor problem among many causes of marine pollution. With the densely populated areas surrounding it, the North Sea has some of the busiest marine traffic in the world, and many of Europe's major ports are located around it. Offshore oil and gas exploitation has also made the North Sea one of today's most important oceans for petroleum production. In addition to dumping, river input, direct discharge, and atmospheric fallout are among the major sources of contaminant inputs to the North Sea.

Dumping at sea has always been a malign incentive problem as compared to more benign coordination problems. Consequently, joint and individual action witnessed cannot be explained merely by the proposition that actors had common interests in coping with the problem. On the contrary, the key actors had significantly different interests and preferences. On the other hand, the degree of malignancy may be characterized as moderate due to the magnitude of the problem. As Kay and Jacobsen (1983) have reminded us, no nation believes that the ability to engage in unrestrained dumping has strategic, economic, or political importance.

Various features made dumping a malign problem in political terms. First, the system of activities was largely asymmetrical. The United Kingdom has been responsible for about half of all waste dumped in the North Sea. Moreover, dumping at sea has also been an attractive option for densely populated continental countries such as Germany, Belgium, and the Netherlands. On the other hand, the Nordic countries have never viewed dumping of hazardous substances at sea as a viable disposal option, and these countries have not been engaged in dumping activities during the 1980s (except Norway, which until 1989 delivered waste for incineration at sea). Second, dumping at sea is mainly an externality problem, which in some instances also carries a potential to distort competition. For example, major investments were necessitated by the titanium dioxide industry to cease dumping of wastes both in Germany and Belgium. In Germany, the cost of installing alternative technology was 230 million DM, and this has increased the cost of producing titanium dioxide by 30 percent (Oslo Commission 1990). In Belgium, the costs of installing similar processes amounted to about 2 billion BFR for one firm alone (Oslo Commission 1990). In the Netherlands, a factory had even to halt production pending a recycling facility becoming operational (Oslo Commission 1990). The phasing out of industrial waste dumping in the United Kingdom led to additional costs for twenty-five companies licensed to dump liquid and solid industrial waste in 1989. The Imperial Chemical Industries Ltd. was the largest dumper and was licensed to dump 165,000 tons of liquid waste. The costs of implementing no-waste technology for this company alone amounted to about £25 million.[3] The phasing out of contaminated sewage sludge in the United Kingdom would imply net costs in the order of billions rather than millions. According to the U.K. government in 1989 (Oslo Commission 1989, 33):

U.K. authorities have concluded that the coastal waters around the U.K. have been little affected and that there has nowhere been a significant or unacceptable detriment to the marine resources which these waters support. In contrast, the social and economic consequences of radically changing disposal policy in each case for which dumping at sea has been used were considerable.

Third, externalities were asymmetrically distributed. The North Sea currents take a long-term anticlockwise circulation from the U.K. east coast throughout the central North Sea. Water discharged from continental estuaries is more confined to the eastern region. Water from these sources eventually leaves the North Sea via the Skagerak along the Norwegian coast. Thus, the United Kingdom in particular may be categorized as an exporter of marine pollution not significantly affected by either its own or others' activities. It is said that (Clark 1990, 279)

Italy and Spain have a climate which allows them to grow oranges; Britain does not. That is not a reason to prohibit those countries from growing oranges. In the same way, the argument went, Britain should be able to benefit from its geographical advantage of turbulent seas for the disposal of waste.

Fourth, the dumping problem was also linked to the dispute on whether regulations should be based on uniform emission standards (UESs) or environmental-quality standards (EQSs). This dispute, which was framed as a matter of principle concerning the sea's assimilative capacity, made the problem more difficult even though the UES-EQS controversy was linked to different economic interests.

In relative terms, however, the dumping problem was less malign as compared to land-based sources for example. Even though these problems share some basic similarities, the order of magnitude was different. Land-based emissions account for approximately 70 percent of contaminated inputs to the sea. Consequently, the size of externalities is much greater. Moreover, the costs of dealing with such effluents are much higher as compared to dumping. For example, although the Norwegian government has prioritized land-based discharges since the early 1970s, the costs of implementing the 1987 and 1990 North Sea Declarations on land-based sources amounted to NOK 10 billion (Norway 1991–1992, 84). Land-based inputs originate from a wide range of sectors ranging from industry to agriculture, and regulation would have a larger potential to distort competition. The dumping and incineration problems were also highly visible compared to diffuse land-based emissions. Since the *Stella Maris* incident, dumping at sea has attracted high media attention, public concern, and attention from the green movement. Greenpeace has been especially active in pushing for a ban on dumping since the 1970s. Dumping and incineration vessels served as perfect targets for campaigns, and these problems were also framed in terms of values and symbols.

However, the positions taken by the parties have to a large extent reflected their material interests even though traces of values and norms can be found in the arguments of the respective governments. The most active parties have been the United Kingdom, the Netherlands, and the Nordic countries. These countries can be characterized, respectively, as the main "laggard," the "intermediate," and the "pushers" in terms of positions adopted concerning a stringent dumping policy. An agreement among these states would in many ways imply an agreement for the whole North Sea. The Nordic countries have preferred uniform emission standards based on the assumption that pollution can best be avoided by limiting discharges at source as far as possible. Since 1985, the Nordic countries have explicitly proposed phasing out of dumping as their preferred option within the Oslo Commission. At the Nordic ministers' meeting in Reykjavik in 1985 it was agreed that "The actual aim is to work without delay within the framework of existing Conventions and relevant fora toward a complete ban on dumping in the North Sea" (Oslo Commission 1986, 18). The Nordic countries also argued normatively by emphasizing that dumping represented an unacceptable disposal option in itself rather than due to its negative environmental effects.

The United Kingdom has preferred the environmental-quality approach based on the argument that money is not spent wisely if measures are taken before certainty of serious effects exists. The United Kingdom, supported by Spain, Portugal, and Ireland, has constantly opposed the Nordic position. On being recommended by the Commission's Standing Committee on Scientific and Technical Advice to phase out titanium dioxide dumping in 1985, the United Kingdom stated that "it would be a breach of fundamental principles to take such action when no sound scientific evidence had been presented to support such measures" (Oslo Commission 1986, 12). As for the Nordic countries, general values were also at stake for the United Kingdom since that country did not dump such waste. The Netherlands dumped more than the Nordic countries but less than the United Kingdom. Moreover, due to its geographical position, the Netherlands was liable both to import substances discharged by other countries and to export substances discharged by itself. In addition, the Dutch water-policy system was based on a combination of emission and quality standards. This placed the Netherlands in an intermediate position together with countries such as Germany and Belgium.

The polarized situation between the United Kingdom and the Nordic countries was reinforced by an asymmetrical distribution of basic game capabilities. The United Kingdom had full control over its own dumping activities, and the Nordic countries were not in a position to affect the United Kingdom by any activities due to the anticlockwise direction of the North Sea currents. Moreover, the United Kingdom also controlled the highest share of relevant scientific capabilities among the North Sea states (Skjærseth 1991). Thus, the stalemate could hardly be solved by means of information and persuasion based on a scientific line of argument directed against the United Kingdom. However, the dumping problem that was infected by both asymmetrical material interests and differences in values was in fact solved. In the next section we show how this was accomplished.

The Process of Regime Implementation: From Control to Cessation

The implementation of the Oslo Convention may be divided into two phases. In the first phase, running from the mid-1970s to 1987, the parties did not succeed in adopting joint commitments with "teeth." Though dumping activities were increasingly controlled, they did not change significantly. From 1987, various joint agreements were adopted that were aimed at stopping dumping and incineration operations.

The main objective of the Oslo Convention is to take all possible steps to prevent the pollution of the sea by substances that are liable to create hazards to human health, harm living resources and marine life, damage amenities, or interfere with other legitimate users of the sea.[4] Membership is restricted to European states, while others may be invited to participate. As of 1995, thirteen parties are members, and there are no signatories without ratification or accession. Geographically, the Convention covers the Northeast Atlantic and the North Sea, including the high seas, territorial seas, and internal coastal waters. The Convention is structured around a black-grey annex system. Dumping of substances listed in annex I (blacklist) is entirely prohibited, dumping of substances in annex II (greylist) requires a specific permit, while dumping of other substances (annex III) requires approval from national authorities. Annex IV deals with incineration at sea (Skjærseth 1992).

To implement the ambitions set out above, the Convention established an executive body—the Oslo Commission—where the parties meet annually. Parties to the Commission are required to submit to the Commission for consideration records of dumping permits and approvals they have issued. The Commission exercises overall supervision, reviews generally the condition of the sea and the effectiveness of control measures, and decides on other necessary measures. Thus, the level of collaboration was "medium." In 1983, the Oslo Convention was amended to include incineration of hazardous wastes at sea. The Commission may adopt decisions, recommendations, and other agreements to draw up programs and measures for the prevention of pollution. Decisions are regarded as binding within the framework of international law—that is, they oblige the parties to implement and act in accordance with the decision. Recommendations are prescriptions concerning how the parties should act in certain situations. Other agreements are mainly related to methods and procedures for classifying substances and selecting monitoring areas, as well as codes of practices concerning different activities. The Commission agreed at their Sixth Joint Meeting that, as a general rule, binding legal decisions were to be preferred to recommendations. The Commission was assisted by a Standing Advisory Committee for Scientific Advice (SACSA) and ad hoc working groups. Moreover, the Oslo and Paris Commissions shared a Joint Monitoring Group and a permanent secretariat in London.

The First Phase: Business as Usual
The Oslo Commission was formally based on unanimity. The requirement of unanimity had two important consequences within the Commission. First, the decision procedure provided the laggards with veto power over proposals that they opposed. Of the North Sea states, this procedure equipped the United Kingdom with the necessary power to block any initiatives. Of significant importance was also the fact that the Convention did not include any procedures that could differentiate among its members or allow others to proceed by fast-track options (Sand 1991). Spain and Portugal have no borders to the North Sea, but they could not be treated differently from the North Sea states even though parts of the Convention area had different needs from the North Sea owing to geographical, hydrographic, and ecological circumstances. Second, decision procedures led to protracted decision making due to the need to negotiate compromise agreements acceptable to all parties. Both the Netherlands and different systemic actors acted as mediators and contributed as bridge builders to overcome the polarized situation. The Netherlands acted as a mediator in the Oslo Commission by expressing sympathy with the Nordic proposals and then seeking to keep them at a politically realistic level. In response to a Nordic proposal to phase out dumping in 1987, the minutes of the meeting record that "The Netherlands delegation also expressed sympathy for the Nordic proposal, but intended to keep open the possibility of dumping relatively harmless wastes at sea in certain cases. . . . The Netherlands delegation expressed sympathy with the Nordic proposal . . . but the date proposed by the Nordic countries was too early" (Oslo Commission 1988, 7, 18–19).

The group of chairs and vice chairs had responsibility for keeping all matters under review intersessionally and for taking initiatives and putting forward proposals to the Commission that could promote their efficient operation. If we look at the officers of the Oslo Commission, we see that if the chair represented a laggard country, the vice chair would normally represent a pusher country and vice versa. Moreover, of the eight chairs of sessions up to 1993, three represented one of the Nordic countries, three represented one of the laggards, while the Netherlands chaired one session. This practice contributed to balancing proposals from the chair group. The tiny secretariat for the Oslo and Paris Commissions operated under more restrictive formal procedures in terms of political initiatives, and its main task was purely administrative. In practice, however, it frequently acted as mediator. In some cases, the secretariat came up with compromise proposals based on the parties' preferences. In other cases, the secretary would propose solutions (Sætevik 1988). Especially in those cases where the secretary and the chair were able to work closely together, their behavior could promote outcomes that would otherwise probably not have materialized.[5]

Permanent systemic actors also contributed to a cooperative climate that was based on mutual confidence and diffuse reciprocity (Keohane 1986). The contracting parties focused extensively on issues of common interest. Information collection and sharing—through the Joint Monitoring Program, the standing scientific groups, and ad hoc working groups on technical issues—represented two of the major tasks. The first decade of the Oslo Commission may be described as an exploration period where the parties concentrated on mapping who dumped what, where, and how much. Consequently, the level of serious conflict was low, and the confidence of the parties in the good faith of others was fairly great. According to the chair of the Paris Commission that worked closely with the Oslo Commission: "we disagreed during the day and sang together in the evening."[6] Critics have even described the early years of the Oslo and Paris Commissions as a "travel club" (Andresen 1996, 74).

The evolution of this cooperative climate influenced collective norms of dumping behavior. The Nordic states argued consistently and explicitly that dumping at sea represented an unacceptable disposal option. Although some negative environmental effects were detected, their arguments were more related to the notion that the sea should not be used as a "garbage can" simply because it was "wrong." This was brought to the forefront of the agenda at the OSCOM meeting in 1985. The Netherlands informed the Commission in 1986, prior to any joint decision having been taken, that they had decided to phase out dumping of industrial waste.

Following a Nordic proposal in 1987 to phase out dumping, others followed suit: "Some delegations considered that the Nordic proposal for a Recommendation to terminate dumping was perhaps the most important proposal ever placed before the Commission" (Oslo Commission 1988, 8). Belgium, Germany, France, and the Netherlands reduced their aggregate amounts of industrial waste dumping from 2.3 million tons in 1989 to zero by December 1989. Moreover, Germany decided to phase out sewage sludge dumping unilaterally, rendering Ireland and the United Kingdom as the only remaining dumpers of such waste. However, the United Kingdom continued to argue that dumping represented the best environmental option. A position paper reveals that most affected domestic actors—including governmental departments, parliamentary committees, industry, waste management and water industries, as well as marine scientists—argued that dumping at sea represented the "best environmental option" ("Lords Tell" 1986). This massive bloc of domestic consensus was challenged internally only by the green organizations, particularly Greenpeace.

A number of specific institutional factors helped to create a sense of stability, mutual understanding, and confidence. The institutional framework of the Commission was altered very little the first decade, and annual meetings were convened based on explicit conventions that forged permanent relationships. These relationships were further strengthened by a stable number of contracting parties, low level of participation, and high degree of consistency in national delegations. In essence, there was always a core of key persons that carried the history with them and ensured that new initiatives were based on earlier experience. The cooperation process was also entirely state driven, the only exception being that international governmental organizations (IGOs) were invited as observers. Nongovernmental organizations (NGOs) had no formal access to Commission meetings or working groups. Last but not least, the problem-solving procedures were incremental and bureaucratic. New initiatives were frequently met with the establishment of an ad hoc working group that reported to the standing scientific groups that proposed decisions or recommendations to the Commission.

Thus, even though the Commission did not succeed in changing the collective practice of dumping toxic waste at sea, the development of a highly cooperative climate and dumping norms was important.

The Oslo Commission held its first meeting in October 1974, and the joint output of the Commission up to 1987 does not seem very impressive (see table 3.1). We should first note that the parties were unable to reach any form of commitment at seven of the meetings. Most of the agreements adopted were directed toward

Table 3.1
Commitments adopted within OSCOM, 1974 to 1986

	Decisions	Recommendations	Agreements
1977	—	1	1
1980	—	1	3
1983	—	—	1
1984	—	—	1
1985	2	—	—
1986	—	1	1
Total	2	*3*	7

Source: *Oslo Commission Procedures and Decision Manual* (1986).

establishing cooperative procedures aimed at controlling current behavior rather than changing this behavior. One of the three recommendations adopted was directed at scrap metal covered by annex II and thus not believed to be particularly harmful. The only commitment made with direct relevance for inputs of annex I substances was the 1980 recommendation to take all possible steps to reduce at source the contamination of sewage sludge with heavy metals. This recommendation appears to have had some impact particularly on the United Kingdom, which dumped huge amounts of sewage sludge at sea.

Table 3.2 shows that the amount of heavy metals in sewage sludge decreased significantly between 1976 and 1986, while the total amount of sludge dumped actually increased. This development may to a large extent be traced back to more restrictive licenses issued by the U.K. Ministry of Agriculture, Food, and Fisheries (MAFF) (MAFF 1989). In spite of some positive results concerning sewage sludge, the total amount of industrial waste dumped by the contracting parties was actually higher in 1983 compared to 1976 but decreased somewhat up to 1987. In addition, more toxic waste was incinerated at sea in 1987 as compared to 1980 (Skjærseth 1992).

The Second Phase: Toward Cessation of Dumping and Incineration at Sea
The breakthrough in collective dumping policy did not originate within the Oslo Commission but at the International North Sea Conferences (INSCs). In 1984, Germany took the initiative to arrange the first INSC at ministerial level in Bremen. The aim of the conference was not to create a new set of international agreements

Table 3.2
Contaminants in sewage sludge dumped by the United Kingdom in British waters (tons)

	1976	1986	Change
Mercury	4.0	1.1	−72.5%
Cadmium	10.2	3.5	−66.0
Copper	225.0	150.0	−33.0
Zinc	730.0	490.0	−33.0
Total	*7,200,000*	*8,200,000*	*+13.8%*

Source: Ministry of Transport and Public Works (1990).

but to provide political impetus for the intensification of the work of the existing international bodies. References to OSCOM and other international bodies are sprinkled throughout the Bremen Declaration. However, the Declaration did not significantly strengthen international commitments in terms of behavioral change required. As Pallemaerts (1992, 6) correctly points out, the elasticity of the concepts such as "as far as possible," "practicable," and "economically feasible" imply that the Declaration hardly contains any substantive commitments. On the other hand, the Bremen Conference focused intensively on UES-EQS dispute and reached an agreement that both approaches have the common aim of improving the marine environment. Notwithstanding the meager substantial results, the ministers also drew particular attention to the need for reduction of pollution at sea through dumping. The Bremen Conference was initially envisaged as a unique event, but the ministers welcomed the invitation of the United Kingdom to host a second INSC to review implementation and adopt further measures.

The 1987 London Declaration represented a turning point, particularly concerning dumping at sea, compared to the Bremen Declaration, as well as the Oslo Commission. For the first time, decisions were adopted to impose significant targets on dumping and incineration at sea within fixed time limits. The key elements of the sixteen sections of the London Declaration related to dumping and incineration concerned the adoption of a general principle stating that dumping of polluting materials should end at the earliest practical date. More specifically, the ministers agreed to phase out the dumping of industrial wastes in the North Sea by December 31, 1989, and to reduce the use of marine incineration by not less than 65 percent by January 1, 1991, and to phase out the practice totally by December 31, 1994. The 1987 London Declaration was also the first international environmental text ever to explicitly incorporate the principle of precautionary action (O'Riordan and Cameron 1994, 267). In contrast to many other potentially harmful activities, this principle eventually became both concrete and important in the Oslo Commission. The next North Sea Conference—the 1990 Hague Conference—strengthened the London Declaration. The termination date for marine incineration was brought forward to December 31, 1991. Moreover, ministers agreed that dumping of sewage sludge should stop as soon as possible and that programs should be drawn up by the end of 1990 to phase out this practice completely, at the very latest by the end of 1998.

An important aim of the North Sea Conferences has been to speed up the work of existing international bodies. The direct consequences of *international implementation* of the North Sea Declarations would be twofold. First, the North Sea Declarations would be transformed into legally binding commitments within the framework of international law. Second, the geographical coverage of the Declarations would be extended to the Northeast Atlantic area. The specific pathways through which international implementation occurred can be understood in the light of the concept "nested regimes" implying a deliberate effort to link the activities of different institutions (Levy, Young, and Zürn 1995, 317). As we have seen, the INSC addressed issues and adopted commitments explicitly aimed at subsequent action within other "competent bodies." The key to understanding how this worked lies in (partially) overlapping participation. The Oslo Commission was not invited to participate in the preparatory work of the INSCs, although they participated as observers (Hayward 1990, 94). On behalf of OSCOM, the Secretariat responded to the Declarations by preparing documents on the implications of the Declarations for the Oslo Commission.

The London Declaration made a strong impact on the Oslo Commission. The first decision adopted by OSCOM in 1988 stated that the riparian states of the North Sea will apply the principles on the reduction and cessation of dumping of polluting materials as set out in the North Sea Conference Declaration. The aim of previous regulations adopted by the Commission was changed from control to termination of dumping and incineration operations. This represented a fundamental change in dumping policies. In addition to the general 1988 agreement, three decisions were taken on the reduction and cessation of industrial waste and sewage sludge dumping and incineration at sea within fixed time limits. Thus measured in terms of stringency and the number of substantial commitments adopted, the Oslo Commission achieved significantly more between 1987 and 1990 than from 1974 to 1987. Moreover, the Oslo Commission also adopted the Prior Justification

Procedure in 1989. This procedure represented an operationalization of the principle of precautionary action in the sense that the burden of proof was reversed. From now on, it was up to the dumpers to prove through complicated and expensive laboratory tests that the waste did not harm the environment. The initiatives taken within the North Sea Conferences on the UES-EQS approaches led to the establishment of an ad hoc working group under the Paris Commission on the evaluation of these approaches. Prior to the London Conference, the secretariat presented a paper emphasising the *complementary* nature of the emissions-standard and quality-standard approaches. At the second meeting of the ad hoc working group in 1988, the U.K. government informed the parties: "since 1985 . . . the UK has modified its position in relation to the most dangerous substances" (EQUSa 1988, 9).

EU responses to the 1987 Conference would also become important. The European Union had not succeeded in adopting a directive on dumping at sea even though the first proposal was submitted to the Council in 1976 (Suman 1991). The EU Commission participated in the INSC, and in 1988 the Council adopted a resolution specifically related to the protection of the North Sea, requesting the Commission to take specific action on urban waste-water treatment. This Directive was adopted in 1991, and it introduced a ban on the dumping of sewage sludge from the end of 1998.[7] In September 1992, the ministers representing the parties to the Oslo and Paris Commissions as well as Luxembourg and Switzerland met in Paris to sign a new Convention for the Protection of the Marine Environment of the Northeast Atlantic. The new Convention codifies the prohibition of dumping of industrial waste, sewage sludge, and incineration at sea. The Convention entered into force in March 1998 and replaced the Oslo and Paris Conventions. Over time, the process of international implementation has significantly strengthened the North Sea commitments on dumping. Similar activities were covered by different types of commitments adopted by different institutions. This significantly reduced discretion since implementation was monitored and controlled by different bodies simultaneously.

The development toward the cessation of dumping leaves us with three crucial questions. First, why did the breakthrough originate within the North Sea Conferences and not within OSCOM? Second, how could OSCOM and the European Union transform the North Sea Declarations into binding decisions, which had been impossible to prior to 1987? Third, why did the United Kingdom—contrary to its economic interests and principles—decide to phase out dumping and incineration at sea? The North Sea Conferences were composed of the North Sea states only.

Consequently, Spain and Portugal did not participate at the conferences and could not therefore block proposals for agreement. However, this change is insufficient to explain the decisions since the United Kingdom in particular remained as a powerful laggard within the North Sea regime. One important reason that the United Kingdom gave in can be found in the institutional set-up of the conferences as compared to OSCOM, although the basic decision rule remained the same. The North Sea Conference process significantly changed the negotiation climate concerning *specific reciprocity*, implying higher latitude for politics and political pressure. The 1984 Bremen Conference triggered a conference process characterized by tough bargaining over substantial problems. Hence, the attention of the North Sea states was diverted from the scientific and procedural cost-efficiency aspects of the dumping problems toward actual behavioral change characterized by asymmetrical interests. The organization of the conference process deviated substantially from OSCOM in the sense that there was more real bargaining, more action, and less stability due to their ad hoc nature. Of particular importance was the fact that mutual expectations were, compared to OSCOM, based less on personal relationships due to high-level ministerial representation, a less stable number of parties, and lower consistency in national delegations. In particular, the shift from lower-level participation to ministerial representation magnified pressure at high levels of government where the grand policy decisions are made. Moreover, the INSCs immediately began developing closer contact with NGOs, thus increasing transparency. The first official attempt to consult with NGOs took place in preparation for the 1984 Conference. In 1987, six representatives from NGOs were permitted to attend the opening session. At the 1995 North Sea Conference in Esbjerg, sixty-two representatives from NGOs were present. In addition, there were no permanent systemic actors although the political and scientific committees and the secretariats bridged the gap between the conferences. The problem-solving procedures were less incremental and less bureaucratic compared to OSCOM. Consequently, solutions could not be postponed by delegating unresolved problems to standing subordinate bodies. Moreover, since the North Sea Declarations were based on "soft law," they could take immediate effect.

However, the key to understanding the significant impact of the INSC and OSCOM lies in the symbiotic relationship between the two institutions. While the Oslo Commission provided a stable and legally binding regime with a relatively long history based on mutual confidence, the INSC provided swift action that could take immediate effect. A necessary condition for swift INSC action was based on the groundwork accomplished by the Oslo Convention and Commission. The INSC

Declaration could not have been implemented if national legislation and licensing procedures had not been in place prior to the agreement.

OSCOM could adopt the declarations due to changes in the rules of the game. Previously, agreement had to be reached among all thirteen contracting parties. At its fourteenth meeting in 1988, the Commission agreed that it should continue to work toward common goals for protecting the marine environment to be reached by consensus among all parties. However, agreements within the Oslo Commission, should, where necessary, reflect the fact that different routes and different timetables for compliance may be acceptable in other parts of the Convention area than those pertaining to the North Sea. Consequently, Spain and Portugal could be treated differently than the North Sea states. This fast-track option made the OSCOM Decisions possible. When the initiatives were first taken, the European Union could also adopt the declaration due to its supranational qualities involving both higher aggregation and integration capacity. The latter could be provided through funding mechanisms and a wide scope for issue linkages, while ability to adopt true majority decisions carried the potential to cut through lack of agreement.

However, there are clear indications that the United Kingdom did not seriously plan to change its behavior in accordance with the international commitments. Until 1990, there were few signs that the United Kingdom had significantly changed its practice except for incineration at sea, which was phased out in 1989. The reasons that it eventually did so lie first in external political pressure generated through both the implementation procedures linked to the INSC process as well as OSCOM. Second, internal political pressure increased significantly over time. In late 1989—at the time when industrial dumping was supposed to have ceased—the Ministry of Agriculture, Food, and Fisheries decided to support applications for licensing renewals for 50,000 tons of toxic waste through the OSCOM Prior Justification Procedure. Several North Sea states protested against the decision, and Greenpeace brought the case to the media's attention. An extraordinary OSCOM meeting of the ad hoc working group on dumping waste was convened. To MAFF's embarrassment, one of the firms involved, Fisons (a major chemical company in the United Kingdom), withdrew its application because it had suddenly discovered a land-based disposal option ("Ever-Widening Agenda" 1990). The decision to phase out dumping of industrial waste and sewage sludge was taken by the Agriculture Minister John Gummer in 1990.

The U.K. government has clearly been pushed externally to change its policy on dumping at sea. The major shifts in U.K. policy on dumping before the 1987 and

1990 North Sea Conferences have been accompanied by statements showing that the U.K. decisions have had nothing to do with change in the perception of the seriousness of the problem. The government has consistently argued that there were no scientific reasons to phase out dumping at sea. Moreover, there was definitively nothing to gain economically since phasing out dumping would imply net costs. In addition, the United Kingdom still had superior control over events important to it. Internally, however, significant changes occurred that weakened the government's position.

Companies and authorities came under increasing attack by the green organizations in the latter part of the 1980s when the British environmental organizations grew into a major domestic force. By 1990, 4 million Britons—nearly 8 percent of the population—had joined the green movement (Rawcliffe 1995, 20). Between 1980 and 1993, the strongest growth occurred in the politically active organizations such as Greenpeace and Friends of the Earth. In 1986, Greenpeace launched its first large-scale North Sea campaign. Even though the campaign was directed at the North Sea and the North Sea states in general, dumping and incineration conducted by the United Kingdom was picked out as one of the main targets (MacGarvin 1991). Greenpeace used these actions to expose the United Kingdom to the outside world as the "Dirty Man of Europe." Before the North Sea Conference in 1987, a joint Greenpeace and National Union of Seamen campaign was launched. The group threatened the U.K. government with direct actions against vessels operating from the United Kingdom, if the United Kingdom did not change its policy. Moreover, the group succeeded in persuading local authorities to oppose plans for new terminals to serve marine disposal operators ("New Campaign" 1987). Shortly after, it was reported that "Ocean Combustion Services (OCS), operators of the incineration vessel *Vulcanus II*, had been losing the public perception battle during a two-week fight in the North Sea with Greenpeace and Danish fishermen" ("Shakier Outlook" 1987) (see table 3.3).

Inert materials of natural origin represented an exception included in the 1989 OSCOM decision on dumping of industrial waste. In the run-up to the 1990 North Sea Conference, the British National Power vessel *MVA* was forced by Greenpeace to give up its attempt to discharge power-station fly ash in a government-approved zone, 5 miles off the northeast coast of England. Fishermen joined Greenpeace a few days later when twenty-two ships gathered to harass the *MVA* when it again tried to discharge waste ("Greenpeace Stops" 1990, 106). As part of the government's announcement just before the 1990 Conference, the government declared that National Power would lose its license to dump 550,000 tons of fly ash ("Ever-

Table 3.3
Liquid industrial wastes licensed and disposed of at sea in the United Kingdom, 1988 to 1993

Year	Licenses Issued	Licensed Quantity (tons)	Tonnage Deposited
1988	19	311,411	249,744
1989	16	292,968	248,454
1990	5	228,000	209,961
1991	2	205,000	191,945
1992	2	75,604	180,725
1993	—	0	0

Source: MAFF (1994).

Widening Agenda" 1990). National Power's remaining license to dispose of power station ash was terminated at the end of 1992 (MAFF 1994). An example of more exotic and sporadic intervention in the United Kingdom's dumping policy is that of the U.K. surfers, who have launched high-media campaigns highlighting the dumping of raw sewage into the areas in which they surf (Maloney and Richardson 1994).

The norms concerning dumping at sea have also gradually changed in the United Kingdom since the 1970s as they have in the other countries that relied on dumping at sea. There are clear indications that the norms concerning dumping at sea have changed within the MAFF itself since the signing of the Oslo Convention in 1972. Recognition has evolved around the idea that using the sea as a trash can is wrong in itself. The licensing disposal at sea section today underlines that it is found increasingly difficult to license "nasty substances." This is also emphasized by MAFF's scientific laboratory, which stresses that dumping should be the "last option" while simultaneously emphasizing that no scientific studies have ever shown any negative impact from dumping at sea (Skjærseth 2000). The control, monitoring, and inspection systems were also improved in respect of the Dumping at Sea Act of 1974 and as determined in the Food and Environmental Protection Act of 1985.[8] Only two cases of illegal dumping of solid industrial waste were reported to the MAFF in 1986 and 1987 (MAFF 1989).

Concluding Comments: Principal Determinants of Success

Given the moderately malign structure of the dumping problem, the determinants for success lie largely in the combined problem-solving capacity of various international institutions (see table 3.4). The Oslo Convention itself was particularly

Table 3.4a
Summary scores: Chapter 3 (OSCOM)

Regime	Component or Phase	Effectiveness		Level of Collaboration
		Improvement	Function Optimum	
OSCOM	1972–1987	Very low	Low	Intermediate
OSCOM	1987–1992	Major	High	Intermediate

Note: For information on variables and scores underlying all summary tables (chapters 3–16), see appendices B and C. In these summary tables all variables included in the index of problem-solving capacity have been trichotomized. Malignancy refers to variables 15 only.

important since it was implemented by national legislation covering previously uncontrolled activities. Due to this Convention, the parties established national licensing systems that led to national control over dumping and incineration operations. The work within the Oslo Commission was important in three ways. First, it established a highly cooperative climate based on mutual confidence. When severe conflict arose in 1989 concerning the United Kingdom's intention to rescind its decision on industrial waste, the long history of the cooperation and the cooperation climate made it unlikely that the United Kingdom would choose the exit option. Second, this was reinforced by the transformation of the soft-law Declarations into legally binding decisions within the framework of international law. The agreement to phase out sewage sludge by 1998 has been adopted by OSCOM, the North Sea Declarations, and even the European Union. The collective pressure from these bodies made it very difficult for the United Kingdom to withdraw the agreement. Third, the incremental work of the Commission and its standing subordinate bodies gradually stimulated norms that made it increasingly difficult for especially the "intermediate" parties to use the sea as a trash can. However, the North Sea Conferences were crucial in excluding laggards and thereby arriving at joint decisions on phasing out of dumping. The conferences provided the high-level political impetus due to its ministerial level of representation as well as soft law that could take immediate effect. In essence, each institution could provide crucial functions that the others could not provide. The key to understanding the impact of the international institutions lies first and foremost in the combined effect of OSCOM and INSC. However, the picture remains incomplete without internal factors. The growth of the U.K. green movement occurred simultaneously as the commitments were implemented and as campaigns launched

Table 3.4b

Problem Type		Problem-Solving Capacity			
Malignancy	Uncertainty	Institutional Capability	Power (basic)	Informal Leadership	Total
Moderate	Intermediate	Low	Laggard	Weak	Low
Moderate	Intermediate	Intermediate	Laggard	Weak	Low

by the green organizations weakened the bargaining power of the United Kingdom compared with other North Sea states. Eventually, the external and internal political costs of blocking and defecting from the agreement combined with changes in norms, outweighed the social and economic costs of phasing out dumping and incineration at sea.

Notes

1. For those interested in a comprehensive analysis of the Oslo and Paris Commissions, as well as the International North Sea Conferences, see Skjærseth (2000).

2. See the Oslo and Paris Commissions (1984) for an account of the origin of the Oslo Convention.

3. Interview with John A. Campbell, head, Fisheries Laboratory, Ministry of Agriculture Fisheries and Food, December 5, 1996.

4. Oslo Convention, art. 1.

5. Interview with D. Tromp, December 10, 1996. Tromp was director of the North Sea Directorate, Ministry of Transport, Public Works and Water Management; first secretary of OSPAR; and chair of the Paris Commission, 1984 to 1986 and 1991 to 1994.

6. Ibid.

7. E.C. Directive on urban waste-water treatment (91/271/EEC).

8. Data in this section are from MAFF (1989).

References

Andresen, S. 1996. "Effectiveness and Implementation Within the North Sea/North East Atlantic Cooperation: Development, Status and Future Perspectives". Nordic Council of Ministers, NORD 18, Copenhagen.

Clark, R. B. 1990. Review of W. Salomons et al., eds., *Pollution of the North Sea: An Assessment* (Berlin: Springer Verlag, 1988). In *International Environmental Affairs* 2(3) (Summer): 279–281.

Danish Environmental Protection Agency, Ministry of the Environment and Energy. 1995. Ministerial Declarations: Bremen 1984, London 1987 and the Hague 1990. Copenhagen.

EQUSa, 1988. Second Meeting of the Ad Hoc Working Group to evaluate the EQO and UES Approaches, London, 22–23 November 1988. Environmental Quality Objectives and Standards and Uniform Emission Standards: Complementary Approaches. EQUS 2/Info.I–E.

European Community. 1991. *EC Directive on Urban Waste Water Treatment*. 91/271/EEC.

"The Ever-Widening Agenda for the North Sea Environment." 1990. ENDS *Report* 181 (February), 13–16.

"Greenpeace Stops Ash Dumping." 1990. Marine Pollution Bulletin. 2(3): 106.

Hayward, P. 1990. "The Oslo and Paris Commissions." In D. Freestone and T. IjIstra, eds., *The North Sea: Perspectives on Regional Environmental Co-operation*. Special issue of the *International Journal of Estuarine and Coastal Law* (London).

Kay, D. A., and H. K. Jacobsen. 1983. *Environmental Protection: The International Dimension*. Osmund, NJ: Allanheld.

Keohane, R. O. 1986. "Reciprocity in International Relations." *International Organization* 40(1) (Winter): 1–27.

Levy, M., O. R. Young, and M. Zürn. 1995. "The Study of International Regimes." *European Journal of International Relations* 1(3): 267–330.

"Lords Tell EEC to Drop Legislation on Waste Disposal at Sea." 1986, *ENDS Report* 138 (July), 11–12.

MacGarvin, M. 1991. *Greenpeaceboka om Nordsjøen*. Oslo: Aschehoug.

Maloney, W. A., and J. Richardson. 1994. "Water Policy-Making in England and Wales: Policy Communities Under Pressure?" In H. T. A. Bressers, L. J. O'Toole, Jr. and J. Richardson, eds. "Networks for Water Policy: A Comparative Perspective." Special issue of *Environmental Politics*, vol. 3, no. 4, Winter.

Ministry of Agriculture, Fisheries, and Food (MAFF). 1989. *Report on the Disposal of Waste at Sea*. London: HMSO.

Ministry of Agriculture, Fisheries, and Food (MAFF). 1994. *Monitoring and Surveillance of Non-Radioactive Contaminants in the Aquatic Environment and Activities Regulating the Disposal of Wastes at Sea, 1992*. Directorate of Fisheries Research, Lowestoft.

Netherlands Ministry of Transport and Public Works. 1990. *The Implementation of the Ministerial Declaration of the Second International Conference on the Protection of the North Sea*. The Hague.

"New Campaign Puts Waste Imports, Disposal at Sea Under Pressure." 1987. *ENDS Report* 151 (August), 8–23.

Norway Ministry of the Environment. 1991–1992. "Concerning Norway's Implementation of the North Sea Declarations." *Report No. 64 to the Storting*. Oslo.

O'Riordan, T., and J. Cameron, eds. 1994. *Interpreting the Precautionary Principle*. London: Earthscan.

Oslo and Paris Commissions. Various years. *Oslo Commission Procedures and Decision Manual.* Published and continuously updated by the Secretariat of the Oslo and Paris Commissions, London.

Oslo and Paris Commissions. 1984. *International Co-operation in Protecting Our Marine Environment.* London: Oslo and Paris Commissions.

Oslo Commission. 1986. *Tenth Annual Report.* London: Secretariat of the Oslo Commission.

Oslo Commission. 1988. *Twelfth Annual Report.* London: Secretariat of the Oslo Commission.

Oslo Commission. 1989. *Review of Sewage Sludge Disposal at Sea.* London: Secretariat of the Oslo Commission.

Oslo Commission. 1990. *Dumping at Sea.* London: Secretariat of the Oslo Commission.

Pallemaerts, M. 1992. "The North Sea Ministerial Declarations from Bremen to The Hague: Does the Process Generate Any Substance?" *International Journal of Estuarine and Coastal Law* 7(1): 1–26.

Rawcliffe, P. 1995. "Making Inroads. Transport Policy and the British Environmental Movement." *Environment* 37(3): 16–20, 29–36.

Sætevik, S. 1988. *Environmental Cooperation Between the North Sea States.* London: Belhaven Press.

Sand. P. H. 1991. "Lessons Learned in Global Environmental Governance." *Boston College Environmental Affairs Law Review* (18): 213–277.

"Shakier Outlook for Ocean Incineration." 1987b. *ENDS Report* 153 (October), 7–8.

Skjærseth, J. B. 1991. *Effektivitet, problem-typer og løsningskapasitet: En studie av Oslo-samarbeidets takling av dumping i Nordsjøen og Nordøstatlanteren.* Oslo: Fridtjof Nansen Institute.

Skjærseth, J. B. 1992. "Towards the End of Dumping in the North Sea: An Example of Effective International Problem Solving?" *Marine Policy* 16(2): 130–141.

Skjærseth, J. B. 2000. *North Sea Cooperation: Linking International and Domestic Pollution Control.* Manchester: Manchester University Press.

Suman, D. 1991. "Regulation of Ocean Dumping by the European Economic Community." *Ecology Law Quarterly,* 18(3): 559–618.

4

Sea Dumping of Low-Level Radioactive Waste, 1964 to 1982

Edward L. Miles

Introduction and Reasons for Success

This chapter presents an assessment of the effectiveness of the regime governing sea dumping of low-level radioactive waste from its early beginnings in 1964 to its termination in 1982. This regime was highly effective in both the creation and implementation phases and has a unique evolutionary history.[1] The reasons for its success can to a significant extent be traced back to problem-solving capacity—particularly, changes in the institutional setting. Because the problem became more malign over time in terms of diverging interests and preferences among the states, institutional changes promoted an end to sea dumping of low-level radioactive waste by 1982.

In spite of the fact that sea dumping of low-level radioactive waste was eventually terminated, we score this case as a success in which performance approached functional optimality. The purpose of the regime was to determine whether sea dumping of low-level radioactive waste could be conducted in an environmentally safe, efficient, and cost-effective manner in a process that maximized international control of sea-dumping operations according to stipulated criteria governing both packaging of wastes and control of dumping operations. Oversight by Organization of Economic and Cooperative Development/Nuclear Energy Agency (OECD/NEA) Annual Committees combined with three quinquennial scientific reviews sponsored by the NEA confirmed that these objectives had in fact been met. After several years of operations the process was binding on member-state participants. NEA's scientific findings were confirmed by an independent global review conducted by the contracting parties to the London Dumping Convention (LDC). Notwithstanding this fact, the internal OECD/NEA conflict having been globalized by a shift in arenas, the contracting parties to the LDC voted eventually to prohibit such disposal of low-level radioactive waste.

The major progression of the analysis is divided into three phases. The *preregime phase* began in 1960 and lasted until 1964. The institutional setting is critical here since the Organization for European Economic Cooperation (OEEC) had created the European Nuclear Energy Agency (ENEA) for the purpose of expanding cooperation among member states in the nuclear field. Consequently, the initial question posed was not "What is the problem?" but "What shall we do?" This search for a cooperative focus predetermined the choice of a (then) benign problem.

The *regime-creation phase* began informally in 1964 with the choice of problem and produced the elaboration of detailed standards, rules, and procedures within the frame of a minimal formal regime as defined by article 25(1) of the 1958 U.N. High Seas Convention and the roles of the International Atomic Energy Agency (IAEA) (designated) and the International Committee for Radiation Protection (ICRP) (assumed). This elaboration was completed by 1967, but the operations were increasingly affected by the planning for the Stockholm Conference from 1969. The Conference itself was held in 1972, and this date marks the formal beginning of the *regime-implementation phase* because the Conference spawned the London Dumping Convention of 1973, which entered into force in 1975.

The Dumping Convention, in turn, had a significant impact on ENEA (later the Nuclear Energy Agency, with the addition of Japan as a member state in 1967) operations such that a formal mechanism was negotiated in 1977 under which rules dumping was conducted until 1982. After 1977, therefore, the fully fledged regime consisted of article 25(1) of the 1958 U.N. High Seas Convention, the combined work of IAEA/ICRP, the London Dumping Convention, and the NEA Sea-Dumping Mechanism.

In this constellation, the London Dumping Convention was not critical in a formal legal sense since it only ratified what had been done before. However, it was *politically* significant because it gave new voice to a large number of countries that were either not members of NEA or even if they were, like the Scandinavians, had limited voice on the nuclear waste-dumping issue. This Convention, therefore, globalized the issue, politicized it, and broke the control exercised by the dumping states in NEA. The antidumping coalition eventually gathered enough strength to terminate the activity altogether even though it was legal under the London Dumping Convention.

Problem Definition: The Preregime Phase, 1960 to 1964

The disposal at sea of low-level radioactive waste packaged in special containers began in 1946 but was not regulated internationally until the first U.N. Conference on the Law of the Sea in 1958. Article 25(1) of the Convention on the High Seas was interpreted to refer to the IAEA as the international standard-setting organization. Consequently, a scientific panel was created, and its first report was issued in 1961 (IAEA 1961).[2] In 1958 the ENEA was also created with the primary objective of fostering international cooperation among OEEC countries. That such cooperation was intended to encompass operational activities is reflected in the original organizational design.[3]

The early documentation of the ENEA shows clearly that the sea-dumping idea evolved as one of two major options considered by the Health and Safety Subcommittee (HSS) of the Steering Committee for Nuclear Energy, which provided oversight of *all* ENEA activities, from 1960 to 1964.[4] The driving force for HSS was the need to operationalize "international cooperation" among OEEC member states. One potential subject of cooperation concerned disposal of radioactive waste into the North Sea.

The initial discussion of this agenda item, which occurred in Paris on November 24, 1960 (OEEC/ENEA 1961b), was couched in terms of possible regional action by ENEA in the light of the report of the scientific panel of IAEA (the Brynielsson Report). It is interesting to note that Norway sought to persuade the members of HSS that ENEA jointly undertake a study "of the technical, biological and administrative aspects of the disposal of radioactive waste in the North Sea" (OEEC/ENEA 1960). Sweden, however, was concerned about the disposal of radioactive waste in the Baltic, and its opposition to a regional approach was intended to postpone any joint ENEA project until IAEA could do more work.[5] In deference to Sweden's concern, HSS was required to put forward "concrete proposals concerning the object of the study and the procedure to be adopted" (OEEC/ENEA 1960).

The meeting of HSS to respond to the mandate from the Steering Committee was held on October 19, 1961, and though seven countries and one international organization, the European Atomic Energy Agency (EURATOM), were in attendance, the discussion showed that most countries were not at all clear on the need for such a study (OEEC/ENEA 1961a). The clearest views were held by Norway and the United Kingdom. Norway's concern was that the North Sea could be contaminated by accidental or controlled disposal of isotopes and that the shallowness of the North Sea combined with its circulation patterns would mean that all North Sea states would be affected. This was therefore the problem that Norway wished studied.

The United Kingdom, on the other hand, was of the view that sufficient knowledge existed to allow a limited preliminary assessment to be made of the quantity of radioactivity that could safely be discharged into the North Sea as a whole. The United Kingdom also thought that the North Sea was unsuitable for discharges of higher-activity liquid wastes.

The set of questions surrounding the issue of how much radioactive waste of what types could safely be disposed into the North Sea constituted Option One for expanding collaboration among and between ENEA members. Such a study was approved by the Steering Committee for Nuclear Energy on December 5, 1961 (OEEC/ENEA 1962). From the beginning of these discussions, the ENEA Secretariat took an active part in facilitating a collaborative approach.

By 1964, the issue of the use of the North Sea as a disposal site for radioactive waste was coming into focus as the primary issue of the study. The United Kingdom insisted that any recommendations on restrictions on such use be based on the (IAEA) Brynielsson Report, and the Secretariat was instructed to take the lead in preparing a meeting of experts on this subject.[6] But nothing more came of this study, and after 1964 HSS shifted its focus to Option Two—sea dumping of low-level radioactive waste (LLW).

What actually happened was that, in the face of consistent U.K. skepticism that such a study was needed, the job was assigned to a Secretariat official, Jean-Pierre Olivier. Olivier eventually completed his report in 1968 (OEEC/ENEA 1968b), but the whole idea was allowed to die because his calculations showed that very large amounts of LLW could safely be discharged into the North Sea. The Steering Committee did not want at that time to raise political fears and refused to authorize publication of the report.

The ENEA in 1964 was deflected into Option Two as a result of a proposal by the Federal Republic of Germany (FRG). The Germans were anticipating experimental disposal of LLW in containers at a depth of 2,000 meters (OEEC/ENEA 1964). Responding to a specific query posed by Sweden, the German delegate to HSS made it clear that terrestrial disposal in salt domes was Germany's main option but that it was necessary to gather data on the safety and cost of sea disposal as an alternative to compare with land-based disposal. Since the Germans thought that many ENEA members states would face the same need, ENEA seemed to be the perfect mechanism for facilitating collaboration on an experimental dump.

At the moment, this German proposal was informal and delivered orally. It was supported by the United Kingdom, Norway, and the Netherlands, but substantial differences of view were expressed within the Subcommittee, and a consensus could not be achieved. A formal proposal to the director-general was made by the FRG on January 18, 1965.[7] Proposing a new agenda item for HSS entitled "Joint Undertaking of ENEA for the Discharge of Radioactive Waste into the Atlantic Ocean, as an Experiment in International Cooperation" and foreseeing the creation of a Restricted Group of Experts, the Germans argued as follows:

It would be useful, therefore, if ENEA would study the possibility to dispose of radioactive waste into the sea as a joint enterprise, in order to give a practical test to international cooperation in this field.

In many Member States of ENEA, there exist unfavorable demographical and geographical conditions for waste disposal into the ground, and safety requirements for this cost much. A joint undertaking for the discharge of low-level waste into the sea would enable these countries to unburden their collecting points. Furthermore, studies on the economy and safety of disposal into the sea could be initiated. Any risk can safely be excluded by confining the operation to low-level waste and by selecting a suitable and generally accepted area for the discharge.

The foregoing summarizes the evolution of the problem to be solved as perceived by the participants at that time. We note that this evolution was not driven by the *prior existence of a pressing external problem* on which all were immediately agreed. The evolution was instead driven by the *institutional setting*—that is, the ENEA, the primary objective of which was to foster collaboration in the nuclear field among member states of OEEC. That the problem turned out to be sea dumping of LLW instead of *how much* LLW could safely be discharged into the North Sea was the result of a change in German policy that brought Germany in line with a number of other states in facing a common need to compare the primary terrestrial disposal option with sea-based disposal as an alternative.

The benefits of international collaboration on the latter issue were immediately apparent. The purpose and promise of collaboration was to expand the knowledge base of member states about alternative disposal options and to share the costs of doing so—a variable-sum game. The problem, therefore, was characterized politically by synergy relationships that provided obvious opportunities for increasing cost efficiency by coordinating efforts. In other words, the problem was largely *benign* but, as we soon show, it contained incongruities that were at first muted and limited. The danger here is that *as problems evolve over time and in changing political contexts, their political characterization can change from benign to malign or vice-versa.*

The Regime-Creation Phase, 1964 to 1972

The regime-creation phase of the sea dumping of LLW case study constitutes a major anomaly. First, as we have already noted, the entire process was driven by the institutional setting and not by the type of problem, and second, the characteristics of the institutional setting combine those we expected for the regime-creation phase with those we expected for the regime-implementation phase. This development occurred because the ENEA work focused on the scientific, technical, legal, and policy requirements applicable to sea-dumping *operations* and this work antedated by eight years the formal, legal process of regime creation culminating in the London Dumping Convention of 1972. This process, in turn, was heavily influenced by the body of norms and procedures developed by ENEA. We should also note that various technical rationality requirements linked to dumping operations—such as review of oceanographic data, development of standards and procedures for operation, and control as well as monitoring—not surprisingly gave rise to the formation of an epistemic community that was embedded into the units created by ENEA to oversee dumping operations.

The Evolution of the Regime-Creation Phase

We recall that the initial, informal German proposal for an experimental dump was made on July 7, 1964, followed by a formal, written proposal on January 18, 1965. After discussion within HSS, the proposal was passed to an Anglo-German group of specialists known as "the Dunster Group."[8] The Dunster Group stated the objective of the operation more positively than in fact it had been decided in the Steering Committee: "the main aim of the ENEA operation is to establish a safe and economic means of disposing of such waste into the Atlantic."[9] Among the jobs to be done, the choice of site later raised intense political sensitivities despite the strict criteria initially proposed by the Group.[10]

German experts suggested a list of areas that, on the basis of current information, appeared to be of great natural isolation and clear of submarine cables, and they also proposed to conduct a program of investigation in the area with the R.V. *Meteor* during spring 1966. The proposal of the Dunster Group was adopted by the Steering Committee on January 12, 1966, and member states were canvassed on their desires with respect to participation in the oversight Committee of Experts.[11]

The first meeting of the Committee for the Experimental Disposal of Radioactive Waste in March 1966 was perhaps the critical event of the regime-creation phase since all the operational decision rules were established at this time. It should be noted also that these decision rules were based primarily on the U.K. experience and the Committee itself was chaired by Ian G. K. Williams of the Health and Safety Branch, U.K. Atomic Energy Authority.[12] Ian Williams was later to be appointed deputy director-general and later director-general of the ENEA/NEA and to serve in those capacities through most of the period during which dumping operations were carried out.

The most important policy questions considered by the Committee related to national and international control of operations. In the event that centralized international control managed by ENEA was chosen, the Working Paper recommended the appointment of an escorting officer for each dumping operation, empowered, among other things, to stop the dumping operation if, in his or her judgment, the safety and requirements for discharge could not be guaranteed. The Committee, in turn, recommended to the Steering Committee of ENEA that an experimental sea-dumping program be organized by ENEA under specific terms and conditions.[13] The role for ENEA that evolved out of the Committee deliberations, assisted by the output of the groups of specialists, is made clear from the scope of the decision rules that were recommended. These included national authorization in each case, risk assessment, criteria concerning choice of disposal site, and common standards for operation.

Clearly, the care that went into the formulation of these decision rules, and the consistent preference displayed by the Committee for the international rather than national option, is testimony to the fact that all representatives were aware that the experiment would occur on the high seas beyond national jurisdiction. The procedures that were developed were therefore intended to make the experiment a credible and legitimate exercise of high-seas freedoms by those states participating. Credibility as well as legitimacy also came from the oversight established at every stage of the process by ENEA.

The magnitude of the first operation was estimated at 33,000 to 35,000 drums of solid LLW offered by Belgium, France, Germany, the Netherlands, Norway, Sweden, and the United Kingdom. This activity represented approximately 6,000 to 8,000 curies.[14] The risk assessment assumed that there would be no release from containers for ten years, at which time all activity would be released at once. These were extremely conservative assumptions, but even so, the group of specialists concluded that disposal at a rate of 10,000 curies per year could be undertaken at a risk to humans many orders of magnitude below the levels recommended by the ICRP. The only marine biota that would be adversely affected would include "individual members of species in the immediate vicinity of the disposal area."[15] In the ensuing debate, the target date of spring 1967 was identified for the first operation (OEEC/ENEA, 1966a).

In its report to the Steering Committee, the Committee for the Experimental Disposal recognized that public relations concerning dumping operations would constitute a serious policy issue for the ENEA and participating states, and it had made some sensible suggestions to the Steering Committee in this respect. Events beginning with the Steering Committee meeting on June 9, 1966, were to prove them correct, and the problems raised as a result were a foretaste of the major conflagration that would eventually engulf all sea-dumping operations for LLW.

The site chosen for the dumping operation was about 450 kilometers west of Cape Finisterre in northwest Spain.[16] However, the Portuguese representative at the Steering Committee meeting took the opportunity to stress the concerns of his government relative to the single operation under discussion and to seek clarification on the question whether the Steering Committee was discussing a single experimental operation or the first step of a process of radioactive waste disposal at that particular site (OEEC/ENEA 1966, 9). Portugal's concern arose from the fact that the dump site was situated about 350 km off the Portuguese coast. The representative of Portugal was particularly frank. He said: "*Portugal is in a very delicate position.* We have an important fish industry and strongly developing touristic industry. Even if all the Portuguese authorities turn out to realize that the operation has no danger, public opinion in Portugal, or elsewhere, may feel otherwise."[17]

In response to the Portuguese statement, Williams "indicated that only one pilot sea-dumping program was envisaged, that there was no implied intention to follow up this operation with others, and that if similar operations were to take place in the future there was no commitment to carry them out in that particular area" (OEEC/ENEA 1966, 9). The Steering Committee then authorized the experimental program, subject to the condition expressed by the representative of Portugal that had received wide support.

By the end of September 1966, Belgium, France, Germany, the Netherlands, Norway, and the United Kingdom had declared their intention to participate in a joint waste-disposal operation now set for June 1967 (OECD/ENEA 1966). But by November 1, 1966, informal word had reached the Secretariat to the effect that the political winds were shifting and that a Board of Radiation Hygiene in Norway was not in favor of the operation due to "political considerations."[18] A formal decision to prohibit Norway's participation came in December 1966.[19]

The meeting to discuss the results of the German investigation was held in Lisbon from April 12 to 14, 1967, where matters became intense. One problem was a prior report in two French newspapers (*L'Aurore* and *Le Figaro*) about the dumping operation that stated that the site was near Portugal. ENEA faced as a result a determined effort by Portugal and Spain to force ENEA to move the dumpsite. This effort failed, and the operation was conducted as authorized from May 17 to August 21, 1967 (OECD/ENEA 1968a).

By the end of the 1967 experimental sea dumping, ENEA had achieved enhanced cooperation on policy and technical issues, expansion of the institutional design to provide the required expertise, as well as site-specific investigations that did not invalidate the very conservative risk assessment that the group of specialists had produced. On the other hand, two policy problems of significance emerged, although member states did not yet fully appreciate the considerable difficulties these portended—(1) third-party liability issues raised by the shipowner in which unlimited rather than limited liability was demanded and (2) sensationalized reports by the press, often incorrect, leading to scares in populations that perceived themselves to be at risk. The liability problem was temporarily solved until the 1971 to 1972 operation.

It is worth noting that the decision rules elaborated by the Committee for the Disposal of Radioactive Waste achieved a level of collaboration that is extremely rare in the experience of intergovernmental organizations—coordination through fully integrated planning and implementation, including centralized appraisal of effectiveness (level 5 in our evaluation scale in chapter 1, pages 9–10). However, we have to admit that this achievement was facilitated by factors that occur together relatively infrequently—an institutional setting that actually drives the search for cooperative action and finds something of singular importance to governments that they could not achieve by any other path; a relatively benign problem (at least in the beginning); physical characteristics that permit fully joint operations; a symmetrical distribution of capability across participating states, even if the United Kingdom had the greatest operational experience; similar and overlapping actor interests and positions; and high empirical uncertainty.

Not surprisingly, this combination of factors triggered primarily problem-solving decision modes that scored very high on technical rationality and yet avoided opposing coalitions and their negotiating strategies and tactics. Finally, empirical uncertainty triggered the creation of an embryonic epistemic community on a primarily transgovernmental basis involving experts from participating governments and the ENEA Secretariat.

1969 to 1972

At a meeting of the Steering Committee for Nuclear Energy held in Paris on April 25, 1968, the deputy director-general of the ENEA reported that informal discussions were proceeding with various countries on the possibility of organizing another dumping operation in 1969.[20] The primary criterion for deciding whether to dump would be the *economic advantage* to repeating this exercise, given what had been learned in 1967.[21] Three new developments occurred at this time in response to the statement of the deputy director-general:[22] Japan, which joined the OEEC in 1967, indicated its interest in participating in the 1969 operation; the IAEA reported that it wished to use the data generated by the ENEA dumping operations to reach international agreement on recommendations concerning sea-based disposal of LLW; and the Portuguese representative insisted that if the operation were to be repeated, a different site would have to be chosen.

On the basis of an informal meeting among interested states held on June 24 and 25, 1968, and a meeting of the Site Selection Group on July 3, 1968, various recommendations were made to the Steering Committee.[23] The most important in political terms were that seven states (Belgium, Italy, the Netherlands, Sweden, and Switzerland in addition to France and the United Kingdom) indicated their interest in participating in the 1969 operation, whereas Spain and Portugal did not wish to participate but indicated their intention to be active on questions concerning choice of site and control arrangements. In addition, a new site was recommended about 600 km from the Irish coast and 900 km from the northwest coast of Spain. There were no compelling scientific or technical reasons to recommend establishment of monitoring studies.

The Steering Committee accepted these recommendations but characterized the 1969 operations as *routine* rather than experimental (OECD/ENEA 1968a). At this meeting, Norway noted the apprehension of the fishing community about these operations, and Denmark expressed the hope that approval of the 1969 operation "would not be interpreted as permanent and general authorization for this type of operations." Germany, on the other hand, expressed full support for the operation, even though it did not intend to participate.

The 1969 operation faced serious policy problems of which the repeat of the Portuguese public relations problem, this time with Ireland, was the most serious.[24] The Irish problem had surfaced initially during the meeting of the Executive Group in Paris on October 17 to 18, 1968, but in a muted fashion (OECD/ENEA 1968b). The Irish government issued an aide-mémoire on the subject on March 19, 1969, declaring that the government of Ireland "is not satisfied that the program of radioactive waste disposal drawn up by ENEA devotes adequate attention to the political, psychological, and economic repercussions which these operations can have in coastal States." In an attempt to head off a confrontation, the Secretariat arranged an informal consultation between an Irish delegation headed by its ambassador to OECD and oceanographic, marine biological, and radiological specialists on April 22 1969.[25]

The pattern of discussions during this seven-hour meeting is instructive because it sets in bold relief the opposing forces that were eventually to swallow up the ENEA/NEA sea-dumping operation. While the Portuguese at the Lisbon seminar in April 1967 tried to cast their argument in scientific as well as political and psychological terms, the Irish ambassador brushed aside the former and went straight to the latter. He acknowledged that the Irish government had been convinced by the scientific and technical explanations that had been given to it but explained that its problem was essentially political. In turn, the combined response of the Secretariat and scientific specialists emphasized issues of technical rationality stressing that the quantities of waste involved were such that the disposal operation under ENEA supervision could safely be carried out within a few kilometers of the coast.

While the meeting came to no formal conclusion, the Irish ambassador later informed the deputy director-general of the ENEA that he was prepared to accept the assurances given him and recommend to his government that they accept the site chosen for the 1969 operation, with "the understanding that there was no commitment, even implied, to use the same site for any future operations under ENEA auspices."[26] Those assurances, including the possibility of carrying out an annual review of amounts of waste and hazards involved, were indeed given to Ireland by the Steering Committee for Nuclear Energy on April 24, 1969.[27] Ironically, while this last concession led eventually to the establishment of a scientific program for the periodic review of site suitability for LLW, the latent effect it produced was to expand significantly the size of the epistemic community attached to ENEA/NEA on this question and a continued reaffirmation of the safety and effectiveness of sea disposal of LLW.

On the basis of the evidence provided, we think the developmental impact of the 1969 operation was important in two respects. First, the detailed operational planning represented a deliberate attempt to learn from the 1967 operation, particularly about making the operation cheaper. In this objective the 1969 operation was a complete success. Therefore, the 1969 operation provided a definite boost for organizing further operations with the next one targeted for 1971 (OECD/ENEA 1970a, 1970b).

The second implication was that site selection was of major political and psychological importance and that technical rationality by itself was an inadequate response to the problem. The system had grudgingly agreed to move the site once again on the basis of the Irish objections, but the evidence is abundantly clear that neither the Secretariat nor the groups of specialists and experts fully appreciated how important this issue really was.

For the 1971 operation, a Group of Experts recommended a site approximately 750 km from both Ireland and the Iberian Peninsula.[28] In August 1970 it was not clear how many countries were willing to participate in a disposal operation for either 1971 or 1972. On one extreme, the ENEA Secretariat did not wish to provide collective legitimacy for only a few countries.[29] On the other extreme, the United Kingdom wanted the ENEA cloak if only *one* member state was participating. Four countries participated: Belgium, the Netherlands, Switzerland, and the United Kingdom, with the Steering Committee accepting the Secretariat's recommendation that the operation be spread over 1971 and 1972 (OECD/ENEA 1970b).

Sweden did not intend to participate and raised no objection to the operation, but at the Steering Committee meeting the Swedish representative did say that sea disposal of nuclear wastes "was not necessarily a valid solution for the future (OECD/ENEA 1970b). In response to the circular issued by the Secretariat, Denmark declined to participate because it had not accumulated enough waste to make it worthwhile. No objection on principle was raised. Moreover, the Spanish Junta de Energia Nuclear stated a willingness to participate in the operation in 1972. While Ireland, Portugal, and Sweden had no waste to contribute, they sought to be "associated" with the Operations Executive Group, and for the first time a representative of IAEA was formally involved in the Group's deliberations.

Following the first of two dumps planned for the 1971 to 1972 operation, the issue of sea dumping of radioactive waste became increasingly linked to the preparations for the Stockholm Conference on the Global Environment to be held in Stockholm in 1972. The linkage was made jointly by the delegates of Norway and Sweden, both of whom began to raise serious reservations on the acceptability of sea disposal of radioactive waste at the Steering Committee.[30] This position intensified the closer the time came to the Stockholm Conference. Denmark joined Norway and Sweden in 1972 and prohibited all Danish flag vessels and Danish vessel owners from participating in these operations.[31] Such was the transition that turned the three Scandinavian supporters of, and even participants in, sea disposal of LLW into principled opponents. At this point, the opposing coalition within the Nuclear Energy Agency (*European* was dropped from the title after Japan became a member in 1972) grew from a potential three-member group (Spain, Portugal, and Ireland) to a potential six-member group (Spain, Portugal, Ireland, Norway, Sweden, and Denmark). The source of opposition was domestic in each case: Spain,

Portugal, and Ireland were concerned about the political effects of *perceptions* by their fisheries and tourism industries that they would be adversely affected, whether or not these perceptions had any basis in fact; Norway's opposition came primarily from the fisheries constituency and the personal views of the Minister of Fisheries; Sweden's primary concern was to protect the Baltic; and Denmark's position was changed by the growing domestic preeminence of the green movement as a political force in relation to the Stockholm Conference.

The second half of the 1971 to 1972 operation was politically more difficult than any that had preceded it. First, the unlimited-liability issue precluded the participation of Italy and Germany in the operation. The options developed for solving this problem, pending the entry into force of the new Intergovernmental Consultative Maritime Organizations (IMCO) Convention on the subject, included a requirement for participants to provide the shipowner with a guaranty of unlimited liability for their portion of the load and to give Switzerland the same guaranty for the mixed load.[32] Both Italy and Germany were unable to participate because domestic law forbade giving a guaranty of unlimited liability to either the shipowner or to Switzerland in the event of a nuclear accident. Consequently, only Belgium, the Netherlands, Switzerland, and the United Kingdom took part.

Second, Sweden informed the Steering Committee that a new law had been passed by the Swedish Parliament banning sea disposal of radioactive waste; Denmark reported on a similar law by the Danish Parliament; and Norway argued strongly for the formalization of a sea-disposal system under new international legislation, setting out permissible levels of waste disposal and creating a system for international registration of sea-disposal operations (OECD/NEA 1972b).

Third, public controversy over the continuation of sea-disposal operations increased in France, the Netherlands, and elsewhere. As a result, the Steering Committee decided against divulging any information about the operation, including its location, prior to completion (OECD/NEA 1992b). This decision stimulated a strongly negative response on the part of member states of NEA that were nonparticipants in the 1972 operation.

In the face of this increasing conflict, the Secretariat became more and more uneasy about running informal operations, and they floated the idea of suspending operations given the imminence of the London Dumping Convention (LDC).[33] Suspension would provide a period of reflection during which the Secretariat could convene a group of specialists to elaborate the formal arrangements under which operations could resume. The United Kingdom emphatically disagreed with this idea and gave notice to the Secretariat of its intention to propose a 1973 operation that would be entirely compatible with the LDC as it was being developed (OECD/NEA 1972a).

It should be noted that the Secretariat had fully grasped the need for treating the public relations aspects of dumping operations as a major policy question and for adopting a far more nuanced response than the narrowly technically rational version the United Kingdom preferred.[34] In spite of the increasing political controversy, the 1972 operation was again executed without major safety or radiological incidents.

Evaluating Effectiveness

The Regime-Creation Phase, 1964 to 1972

Our test of effectiveness for the regime-creation phase required us to focus on *outputs* with a particular view to how much international collaboration has been achieved. We judge that the first phase of the ENEA sea disposal of LLW process surpasses our most demanding test for a variety of reasons as specified previously (see pages 9–10). The policy procedures developed by ENEA called for

· Clear national authorization for each dump;
· A risk assessment prior to each operation;
· Stipulated criteria for choosing dump sites;
· Development of common criteria relating to design of waste-package containers, waste form, amounts of waste, operating procedures aboard ship;
· Design characteristics of the disposal ship;
· Approval of all national plans and procedures that had to conform to explicit criteria; and
· The presence of independent ENEA observers aboard ship during disposal operations with the authority to stop the operations if necessary.

Following these procedures, various expert and political groups and committees were created, and successful dumping operations were organized in 1967, 1969, 1971, and 1972. Each performance was centrally appraised by the Steering Committee.

The solution to the low-level radioactive waste-disposal problem devised by ENEA was the technically rational and efficient solution that was politically feasible at the time. Additionally, it contained the seeds of its own destruction because it activated mass public opposition. These opponents later went on to form a global transnational coalition, allied with certain governments, to produce a change in the regime prohibiting sea dumping of low-level radioactive waste altogether.

As a result of preparations for the Stockholm Conference, a convention regulating pollution of the oceans by dumping, popularly known as the London Dumping Convention (LDC) was signed in 1972 and entered into force in 1975.[35] This Convention not only defines dumping (article III) but regulates substances, like high-level radioactive waste, that may *not* be dumped (annex I, known as the blacklist) and substances, like low-level radioactive waste, that *may* be dumped only on the basis of permits issued by national governments under stipulated conditions (annex II, known as the graylist). The LDC also identifies IAEA as the competent international organization to formulate standards and recommendations and NEA as an observer member of both IAEA and the annual consultative meetings of parties to the LDC.

As perceptions of negative externalities emerged particularly in Norway, Spain, Portugal, and Ireland, the problem for which the ENEA had devised a solution began to take on malign characteristics, but the full implications of this development were not yet apparent. Conflict increased still further in 1972 to 1973, prior to and as a result of the Stockholm Conference. Given increasing conflict, the NEA Secretariat proposed to suspend sea-dumping operations until after the Stockholm Conference had concluded, but the United Kingdom refused to agree. We now have to treat the phase of sea dumping following 1969 as a separate issue because the problem changed its nature over time and became malign, combining characteristics of externalities and competition. But it has been demonstrated that not only did the ENEA sea-dumping system and process achieve the highest level ever achieved by an international organization—that is, coordination through fully integrated planning and implementation, including centralized evaluation of effectiveness—but the NEA experience substantially influenced the creation of the global regime that came to be applied.

The evidence from the ENEA experience between 1964 and 1972 also demonstrates a substantial amount of learning on the part of both member states and the Secretariat gained in the sea-dumping operations organized during that time (Miles 1989).

The Impacts of Regime Change on Operations, 1973 to 1982

In the course of events, the epistemic community expanded, the opposing coalition grew stronger than the supporters of sea dumping, and a variety of negotiating strategies came into play. But perhaps the most significant change was the addition of a new institutional setting—the annual consultative meetings of parties to the LDC—thereby introducing an arena that was fully transnational in nature.

The 1973 to 1976 sea-disposal operations were routine; consequently, there is no need to consider these operations in detail. We focus only on those events that had developmental significance. Each year, the United Kingdom, Belgium, and the Netherlands continued to propose annual operations. Sometimes they were joined by Switzerland and sometimes not. Both Germany and Italy remained interested in participating in the operations, but they continued to be inhibited by their inability to provide anyone a guaranty of unlimited liability.

The eagerness of the United Kingdom, Belgium, and the Netherlands to continue annual operations was matched by the reluctance of the Secretariat to do so without formalizing the arrangement under which these operations were conducted. It is interesting to contrast the Dutch position here with the change in Scandinavian positions. The Dutch kept their nuclear interests separate from their general preferences about the North Sea, while the Scandinavians did not. This difference is explained by which governmental agency determines a state's position on the issue in question. Matters eventually came to a head in April 1975 at the Steering Committee to begin review of the future role of NEA/OECD in radioactive waste disposal in view of the imminent entry into force of the LDC on August 30, 1975, and the publication of the IAEA recommendations concerning application of the LDC (NEA/OECD 1975).

This decision came three years after the opposing coalition (Norway, supported by Denmark, Portugal, and Sweden) had made a similar proposal at the Steering Committee meeting on February 1, 1973 (NEA/OECD 1973). At that time Norway urged suspension of operations in 1973 until IAEA could produce the first set of recommendations under the LDC. This proposal was not accepted by the Steering Committee.

While the 1973 to 1976 operations were routine, it is instructive to note that the political intensity of the dumping issue, which had already increased as a result of the imminence and aftermath of the Stockholm Conference in 1972, continued to increase after 1973. There had already been several incidents in which newspaper reporters had referred to a dump site 350 km from Portugal as "near to Portugal" with very serious public relations problems for the operation.[36] What these incidents revealed was that, quite apart from the technical rationality of the operations and their consistent avoidance of any serious safety or radiological accidents, the degree of public opposition was intense, and there was deep distrust of what NEA was doing. External public opposition, however, split and polarized the internal NEA community, and those increasingly opposed to continuing the operations began to be increasingly resentful that their views received only short shift within NEA. Public opposition combined with an alternative and legally superior global regime (the LDC) and eventually gave these opponents the leverage they sought to stop the operations altogether.

In such a politically charged atmosphere, two incidents that occurred in the course of the 1976 operations assumed a significance out of all proportion to their actual impact. Both incidents occurred on June 26, 1976, during the first phase of the operation and involved two containers from the Eurochemic Company in Mol, Belgium. One container floated, and the other burst on discharge into the ocean.[37] Irish authorities requested a suspension of operations pending a full investigation of the incident and its implications. Such an investigation was carried out three days later on July 9 by a group of specialists from eight member states, Eurochemic Company, and the NEA Secretariat. The investigation concluded that the incidents were caused by human error and lack of compliance with prescribed specifications.[38]

On the basis of these conclusions, Ireland withdrew its objection to continuing the operation, and the work resumed. However, the period between July 9 and August 2 was virtually consumed with managing the public relations aspects of these incidents, a matter that vitally interested the governments of Ireland, the Netherlands, and Switzerland. The Netherlands also protested that it was not informed immediately after the incident had occurred. From a purely political point of view, of course, the incidents were fraught with danger for NEA and all its European member states, not only those that were participating in the 1976 disposal operation.

Establishment of the Formal Sea-Dumping Mechanism, 1976 to 1977

Given the conflict within NEA between the proponents and opponents of sea disposal of LLW, the negotiations over the formalization of a Multilateral Consultation and Surveillance Mechanism for Sea Dumping of Radioactive Waste were a very delicate affair. Clearly, the whole issue was driven by the Secretariat,[39] but it was necessary to design an arrangement that would simultaneously meet the needs and interests of those countries participating in sea disposal as well as those countries that objected in principle to the sea disposal of radioactive waste yet still wished to ensure that if sea disposal were conducted, it would be done solely under international control. This position was consonant with the LDC, which permitted sea disposal of LLW by individual countries acting alone, provided that such operations conformed to IAEA regulations and recommendations and provided that they were conducted pursuant to an explicit national permitting procedure.

The approach initially recommended by the Secretariat drew a distinction between the operational and supervisory roles of NEA.[40] The former was developed between 1964 and 1976, providing a framework for OECD member countries to cooperate on radioactive waste disposal and to share in the accumulation of practical experience. Since the entry into force of the LDC, however, parties to the LDC were able to dispose of radioactive waste themselves, under conditions stipulated in the Treaty and according to recommendations issued by IAEA. These recommendations had been derived largely from NEA experience, but all OECD member states could organize sea-dumping operations by themselves.

On the other hand, the supervisory role of NEA continued to be of particular importance. The involvement of NEA as an international organization ensured, and was seen to ensure, that disposal operations were carried out safely, through appropriate international controls. However, the role of the Steering Committee would be to approve the overall agreement and no longer to approve every operation. If the arrangement were based on voluntary submission, the Secretariat thought that it could be approved by the Steering Committee only. If, however, the compulsory option were chosen, then it would have to be approved at least by the OECD Council if not by a formal diplomatic conference.

Negotiation of the Sea-Dumping Mechanism within NEA consumed much of the time between June 8, 1976, when the first meeting of the Ad Hoc Working Party on Sea Disposal of Radioactive Wastes convened, and July 22, 1977, when the OECD Council formally approved the agreement. We do not attempt to describe in detail the course of these negotiations because the details are not necessary to the argument. The stickiest issue turned out to be the preamble, where the opposing coalition insisted that the impression not be given that NEA was encouraging resort to sea disposal of LLW and demanded that the coalition's principled objections to this method be referred to. Apart from that, most of the negotiation concerned the details and the need to seek appraisal and endorsement from other *OECD units*, in particular the Radioactive Waste Management Committee, the Committee on Radiation Protection and Public Health, and the Environment Committee.

The provisions desired by the Secretariat at the beginning of the process matched well with the final outcome approved by the OECD Council, although certain modifications were required. Member states were obliged to notify NEA concerning implementation of LDC, IAEA, and NEA requirements, their intention to dump LLW, as well as any new dump sites. Citations of the full documentation on the internal negotiations are provided below.[41]

Compared with the decision rules implemented during the regime-creation phase, the new rules implied that (1) the Secretariat would remove itself from the execution of operational details, restricting itself to a review function to ensure that LDC, IAEA, and NEA requirements were being met; (2) Secretariat control over the identification of new dump sites was somewhat diminished; and (3) Secretariat control over the actual conduct of operations while they were underway was somewhat diminished. This diminution of control is not surprising given the fact that the LDC legitimized national sea-disposal operations. What is noteworthy is that so many of the stringent original rules were retained since member states were acutely aware that joint operations under international control would be the surest route to credibility and legitimacy.[42]

The 1977, 1978, and 1979 sea-disposal operations were also routine, but certain events in 1977 led both to changing the dump site once again and to the initiation of the first site suitability review. But far more ominous, 1978 saw the first appearance at the dump site during disposal operations of the Greenpeace vessel *Rainbow Warrior*, although no attempt at direct interference was made. Subsequent to the operation, a letter of concern was sent by Greenpeace to the U.K. Secretary of State for the Environment. Thereafter, events began to overtake the routine execution of sea-disposal operations.

In 1979, intense antidumping demonstrations occurred in the Netherlands, and Greenpeace attempted to interfere with U.K. dumping operations on July 10, 1979 (NEA/OECD 1979). In 1980, major interruptions by Greenpeace were anticipated in the United Kingdom, Holland, and Belgium. In 1981, there was concerted opposition to Japanese plans to dump LLW in the North Pacific by South Pacific island states, Japanese fishermen, and Greenpeace. Greenpeace also appealed to an Administrative Court in The Hague to stop the 1981 operation in Holland, and a temporary injunction was granted but later lifted. This event did delay the joint Dutch, Belgian, and Swiss operation. In the United Kingdom, Greenpeace attempted seriously to interfere with U.K. dumping operations with one near fatality of a Greenpeace individual.[43]

The year 1982 proved to be the denouement. A major Greenpeace disruption of the U.K. and Dutch operations occurred (NEA/OECD 1982). At the same time, Greenpeace succeeded in mobilizing the population of the Spanish coastal provinces, and because these provinces were primarily Basque, the sea-dumping issue became linked to the question of a *modus vivendi* between the new socialist government of Spain and the Basque population. Up until 1982, the Spanish representative to the Steering Committee routinely commented annually that Spain had no legal objections to sea-disposal operations of LLW. Given the transformation of the political issue, however, the position of the Spanish government changed drastically in 1982, and Spain now called for complete termination of ocean disposal of radioactive waste within NEA.

Also in 1982, the seventh Consultative Meeting of the Contracting Parties to the LDC (LDC7) voted to impose a moratorium for two years on all sea disposal of radioactive wastes in response to an initiative jointly sponsored by Nauru, Kiribati, Spain, Norway, Denmark, and others.

In 1983, Spain in the NEA and the OECD Council raised major objections to U.K. plans to dump given public opinion in some countries and the vote at LDC7.[44] The Spanish delegation asked for advice from an ad hoc group of specialists on the ecological impacts of accumulating radioactive waste disposal in the Northeast Atlantic. At the same time, the Nordic countries and Ireland raised objections. This pressure led to termination of the Dutch, Swiss, and Belgian operations, but the United Kingdom insisted on going ahead alone.[45] However, Greenpeace mobilized the Seamen and Transport Workers Unions and the unions refused to load the U.K. ship.[46] Already involved in a bitter and prolonged confrontation with the coal miners unions, the Thatcher government decided against simultaneously confronting the Seamen and Transport Workers. No dumping operation therefore occurred either in 1983 or since.

Greenpeace was the catalyst that mobilized mass opinion in Western Europe and the Western Pacific, thereby fusing the South Pacific island states, the Nordic countries, Ireland, and others into a superopposing coalition in the annual consultative meetings of the LDC, and this development allowed the internal NEA opposing coalition far more leverage.

Regime Change and Implementation

Our approach to evaluating effectiveness regarding outcomes and impacts requires us to ask two separate questions: did the regime make a difference, and did the regime solve the problem? We proceed seriatim, without attempting to combine both measures into a single, composite measure.

The answer to the first question is clearly yes: the regime made a difference but in different ways in different phases. In the creation phase, the regime fostered international collaboration on LLW dumping as a feasible disposal option. The regime achieved a high level of cooperation through the creation of sophisticated policy procedures, including centralized appraisal. In the implementation phase, however, the regime prompted an end to dumping, particularly by the establishment of the LDC. The growing opposing coalition pushed for the establishment of a new global arena that contributed to braking the control exercised by the dumping states in NEA.

Turning to the second question, the evaluation becomes difficult especially in the second phase due to fundamental conflict concerning values and goals. In the first phase, however, we would argue that the regime contributed to solving the problem. Formalizing the sea-dumping experiment under NEA auspices after entry into force of the LDC completed the global regime and aligned NEA squarely under LDC and IAEA. Sea dumping was therefore lawful, and not even the opposing coalition within NEA could deny that fact. The agreement continued to demonstrate that an effective solution to the LLW disposal problem was available, that it was cheaper to operate than the land-based option, and that it was safe and reliable based on technical and scientific criteria. It must be acknowledged that the design of the operational system was without major flaws and that no significant hazards made themselves felt during the eighteen years the system was in operation.

There were, however, scientific limitations to the early choices of sites, to the risk assessments, and to the inability of the models used to predict short- and medium-term transfer effects.[47] While site and risk assessments were required in the regime-creation phase, they were crude, and the oceanographic models underlying this work were far too simple. As a result of the formalization of the arrangement, however, NEA was required to do site assessments every five years, and participation in these reviews was opened up to other member states of OECD not participating in the sea-dumping mechanism. As a consequence, empirical uncertainty declined substantially, findings and risk assessments became increasingly robust, and the NEA Coordinated Research and Environmental Surveillance Program Related to Sea Disposal of Radioactive Waste (CRESP) grew in scientific credibility just as it was being killed politically.[48] On the basis of all relevant reviews, we must conclude that sea disposal of low-level radioactive waste, under regulations imposed by the LDC, IAEA, ICRP, and NEA, was a safe and effective disposal option.

Since the late 1960s, the case has developed from consensus to conflict concerning basic values and goals. This change eventually led to the phasing out of LLW dumping. Thus, if we stick to the original goal of fostering collaboration on dumping, we would have to score the implementation phase low in terms of problem solving. On the other hand, if we include in our evaluation the change witnessed in values and goals, we would have to attribute a higher score. Here, we see a clear parallel to the whaling case (see chapter 15). The original goal of the International Whaling Commission—focusing on resource utilization—changed over time toward preservation.

Instead of pushing the evaluation any further, let us take a closer look at the political dynamics underlying the significant change concerning LLW dumping. In the beginning those participating in the planning for the experimental dumps included Norway, Sweden, Denmark, Germany, the United Kingdom, the Netherlands, France, Belgium, Switzerland, and Italy. However, this group disintegrated over time with proponents becoming eventual opponents or at least nonparticipants.

Defections began with Norway in 1967 followed by Sweden and Denmark. It was only after the scheduling of the Stockholm Conference that Scandinavian opposition was generalized and became firm and united. Germany and Italy remained nonparticipants only because domestically neither could resolve the unlimited liability problem. Following this, there was a late shift in the German position to opposition as a result of internal political impacts of the green movement. Spain, Portugal, and Ireland were opposed from the late 1960s, not on issues of principle but only on location of the site and the psychological impacts of the location on their fisheries and tourism communities.

Those member states opposing dumping were reflecting the emotional and psychologically important preferences of significant domestic constituencies. Within the NEA, opponents found themselves constrained by the official decision rules. They could state their opposition to dumping, but they could not prevent it. To do so would have required making a factually based argument that sea dumping of LLW according to NEA rules was dangerous and harmful to humans and marine ecosystems. Such an argument would not have survived scrutiny. The opponents therefore shifted their attack to another arena—the LDC—not only to globalize the issue and construct a winning coalition but also to avoid the constraints of argument based only on scientific and technical considerations. Their tactic in the LDC Consultative Meetings was to seek a moratorium on all sea-dumping operations and to precipitate a major assessment of terrestrial and marine disposal options, including economic, social, and political considerations in addition to scientific and technical ones. This strategy was ultimately successful.

We should note, though, that while the drama unfolded in two international arenas—the NEA and LDC Consultative Meetings—the really consequential actions were taking place at the *domestic* level even among the proponents. Among the participating states, it was clear that the NEA framework significantly impacted their behavior on all issues covered by the decision rules and that the conflict between proponents and opponents actually strengthened the hand of the Secretariat toward proponents. Only through compliance with the multilateral decision rules independently verified by the Secretariat, the Radioactive Waste Management Committee, and the Steering Committee for Nuclear Energy could the benefits of legitimacy, economies of scale, and cost-effectiveness in relation to terrestrial alternatives be achieved.

For opponents, on the other hand, an alternative global arena receptive to different decision rules could be successfully deployed only by raising the political salience of the issue domestically to the point where member states of NEA could actively and successfully seek to defeat an NEA program with which they disagreed by destroying the legitimacy of continued dumping operations even when such operations were permissible under the regime then in place.

We can say, therefore, that from the point of view of technical rationality, sea disposal of LLW is a benign problem requiring primarily coordination and problem solving. Politically, however, it became an increasingly malign problem, particularly after 1979, filled with presumed externalities and real incongruities. The latter absolutely cut off the political impact of the expanding epistemic community and the rapid decrease in empirical uncertainty after 1977. Consequently, political feasibility as a whole declined drastically. In this context, the configuration and weighting of the primary variables changes significantly. The institutional setting still predominates, but now the institutional setting of the LDC subsumes a different configuration of actor interests than the one that exists in the NEA. Moreover, type of problem has changed from benign to malign.

In essence, we are facing a complex interplay among factors that are changing over time. Type of problem changes from benign to malign, the institutional setting changes by the establishment of the LDC, and a different distribution of interests and capabilities develops at a different arena. For the intervening variables, the political significance of the epistemic community declines even as empirical uncertainty in the knowledge base decreases and coalitions and negotiating strategies, particularly in the LDC, increase tremendously in importance.

Conclusions

Sea dumping of low-level radioactive waste is an interesting case for several reasons (see table 4.1). First, it emerged as a benign problem, but after several years became a malign problem as conflict intensified over presumed externalities. Second, the activity preceded the elaboration of a regime, achieved the highest level of

Table 4.1a
Summary scores: Chapter 4 (RADWASTE)

| Regime | Component or Phase | Effectiveness | | Level of Collaboration |
		Improvement	Function Optimum	
RADWASTE	0: 1964–1972	NA	NA	High
RADWASTE	1: 1972–1977	Major	High	High
RADWASTE	2: 1972–1982	Major	High	High

cooperation, largely shaped the regime that followed, but then was brought to a halt by a coalition of governments and nongovernmental organizations whose membership and representatives opposed the decision rules on the basis of fears about negative externalities. The problem attracted high media attention, public concern, and attention from the green movement, particularly Greenpeace. Dumping of LLW was framed in terms of values and symbols, and its psychological importance was important for changes in the policy of governments. In this sense, the LLW case shares important similarities with other cases like dumping of hazardous substances in the North Sea and whaling. Third, the mechanism that was created initially to oversee the activity also served as a powerful vehicle for the realization of joint gain. Consequently, it created both an epistemic community and consensual knowledge. Governments did not meddle in the production of the latter since divisibility of the collective good among the participants was low. This was not a resource to be carved up.

The regime-implementation phase, when sea dumping of low-level radioactive wastes became routine, extended from 1973 to 1982 with a shift in 1977 when the sea-dumping mechanism in NEA was formalized. (With the addition of Japan in 1972, OEEC became OECD—Organization for Economic Cooperation and Development—and ENEA became NEA.) It was considerably more turbulent than the regime-creation phase because the problem had by then become malign.

Since the combination of the global regime and the decision rules within NEA did not provide the opposing coalition much leeway to stop sea-disposal operations, the coalition sought to change the global regime. Increasing politicization of the issue and mobilization of public opinion were the main avenues for achieving this end. Coincidentally, aided by Japanese plans to dump in the Western Pacific and by mobilization of the Spanish coastal provinces, the global coalition as well as the

Table 4.1b

Problem Type		Problem-Solving Capacity			
Malignancy	Uncertainty	Institutional Capability	Power (basic)	Informal Leadership	Total
Benign	Intermediate	High	Pushers	Strong	High
Moderate	Low	High	Pushers	Strong	High
Strong	Low	High	Pushers	Strong	High

issue exploded simultaneously. Accordingly, the opposing coalition achieved outside of NEA what it was unable to achieve inside. Thus, change in the institutional setting—the establishment of the LDC—is perhaps the single most important problem-solving capacity factor for explaining why dumping of LLW ended.

Notes

1. Both the regime-creation and regime-implementation phases contain two subphases; the first divides into 1964 to 1969 and 1969 to 1972; the second divides into 1972 to 1977 and 1977 to 1982. However, both regime-creation subphases contain strong implementation characteristics because operational decision rules and standards to govern dumping were worked out by 1967.

2. This report set out principles governing all forms of radioactive waste disposal and discharges into the sea.

3. A Radioactive Waste Management Committee was created along with a Committee on Radiation Protection and Public Health to provide oversight on matters related to reactor safety and radiation protection. The concept of operations was not specifically targeted at sea dumping of low-level radioactive waste; it was simply one of the domains in which international cooperation could be achieved.

4. During the academic year 1986 to 1987, the author served as a nonremunerated consultant to the NEA Secretariat, and during that period his access to these files was facilitated.

5. Letter from E. Saelund, deputy director of ENEA, to T. Hvinden of Norway, December 12, 1960, Ref. EM/S/9888.

6. See Doc. #NE/SAN (64)1, August 21, 1964. This was a revised version of Doc. #SEN/SAN (63)14 and (64)6 reporting on discussions of a restricted group of HSS on July 7, 1964.

7. Letter signed by ?Schulte-Meermann, chief, International Relations Division, Federal Ministry for Scientific Research, to E. Saelund, then director-general of ENEA.

8. See Doc. #SEN/SAN (65)17, September 16, 1965. This document was later submitted to the Steering Committee as SEN/SAN (66)1, January 12, 1966.

9. Ibid., 3.

10. Ibid.

11. Of the states canvassed, Austria, Belgium, France, Germany (FRG), Italy, Netherlands, Norway, Portugal, Spain, Sweden, Switzerland, and the United Kingdom, accepted the invitation; Turkey, Denmark, and Luxembourg declined; and no response was received from Greece and Ireland. Doc. #SEN/SAN (66)1.

12. See the Working Paper submitted by the United Kingdom to the Committee on the Experimental Disposal: *Joint Study on the Possibilities of Radioactive Waste Disposal into the Atlantic Ocean*, March 22, 1966.

13. Doc. #SEN/SAN (66)12, April 20, 1966, p. 5.

14. Doc. #SEN/SAN (66)12, p. 3.

15. Doc. #SEN/SAN (66)12, app. 1, p. 17.

16. Doc. #SEN/SAN (66)12, p. 7.

17. Statement by the Representative of Portugal at the ENEA Steering Committee, June 9, 1966 on the Experimental Disposal of Radioactive Waste. Copy of the text in the NEA files. DEEC/ENEA (1966a, 9).

18. Memo of December 6, 1966, from J.-P. Olivier of the ENEA Secretariat to E. Wallauschek (ENEA) and ?I. G. K. Williams (U.K.) reporting on discussions concerning Escorting Officers held at Harwell from November 28–30, 1966.

19. Interviews by the author with U.K. and Norwegian officials involved at the time suggest that the decision was made by the then Norwegian Minister of Fisheries. Oddmund Myklembust on the basis of real or anticipated pressure from constituents.

20. Doc. #NE/M (68)I, Scale 6, May 22, 1968.

21. Ibid.

22. Ibid.

23. Doc. #NE (68)10, July 19, 1968.

24. The second was the unlimited liability problem faced by the shipowner.

25. Memo from I. G. K. Williams to T. Kristensen, secretary-general of OECD, April 22, 1969.

26. Memo from I. G. K. Williams to Thorkil Kristensen, secretary-general of OECD, April 24, 1969.

27. Minutes of the 33rd Session, Doc. #NE/M (69)1, item 5.

28. Doc. #SEN/AT (70)1, June 30, 1970.

29. Memo from E. Wallauschek to I. G. K. Williams, "Sea Disposal Operation in 1971 or 1972," October 13, 1970.

30. Doc. #NE/M (71)2, p. 6 (NEBAM).

31. Letter from the Dutch agents (NEBAM) for the ship M.V. *Alice Bewa*, withdrawing the vessel from the charter for the 1972 operation; sent to W. J. Kolstee, inspector of public health, the Netherlands, February 1972.

32. Summary Records of the Operations Executive Group for meetings of November 25–26, 1971 and January 26, 1972, Docs. #SEN/AT (71)8, December 30, 1971 and (72)1, February 8, 1972; plus letter from DG/NEA to all participants of December 7, 1971.

33. See the exchange of internal memos between E. Wallauschek (July 6, 1972) and I. G. K. Williams (July 13, 1972).

34. Memo from I. G. K. Williams to J.-P. Olivier on the "Public Relations Aspects of the 1972 Atlantic Operation," August 1, 1972.

35. The formal name of the LDC was the Convention on the Prevention of Marine Pollution by Dumping of Wastes and Other Matter, 1972.

36. See Doc. #SEN/AT (73)3, November 20, 1973.

37. *1976 Operation for Disposal of Radioactive Waste into the Atlantic* (Note to Members of the Operations Executive Group), July 16, 1976.

38. Ibid.

39. See the exchange of memos between Pierre Strohl, then legal advisor of NEA, and the deputy director-general of NEA, I. G. K. Williams, November 4–5, 1975.

40. Memo by Pierre Strohl to I. G. K. Williams, November 4, 1975, and Note by the Secretariat to the Steering Committee of NEA on the *Role of NEA in Relation to Sea Disposal of Radioactive Wastes*, Doc. #NE (76)3, Scale 4, March 9, 1976.

41. See Working Document No. 1, *Ad Hoc Working Party on Sea Disposal of Radioactive Waste*, June 30, 1976; Corrigendum to the *Draft Agreement on the Establishment of an International Consultation and Surveillance Mechanism for Sea Dumping of Radioactive Waste*, August 5, 1976; Steering Committee for Nuclear Energy, *Establishment of an International Consultation and Surveillance Mechanism for Sea Dumping of Radioactive Waste*, Doc. #NE (76)19, scale 4, September 15, 1976; *Opinion of the Committee on Radiation Protection and Public Health and of the Radioactive Waste Management Committee*; Addendum to Doc. #NE (76)19, September 15, 1976; *Consultation with the Environment Directorate*, Internal memo from P. Strohl to I. G. K. Williams, October 4, 1976; Committee on Radiation Protection and Public Health, *Summary Record of the Special Meeting*, held in Paris on September 9–10, 1976, Doc. SEN/SAN (76)16, October 21, 1976; Steering Committee for Nuclear Energy, Doc. #NE (77)8, scale 4, March 31, 1977; and documents of the OECD Council:

• Report of the Secretary-General, Doc. C (77)115, scale 1, June 22, 1977; and Addendum, July 7, 1977;
• Executive Committee, *Summary Record of the 420th Meeting*, Paris, July 5, 1977, Doc. #CE/M (77)9 (Prov.), scale 3, July 13, 1977;
• *Summary Record of the 421st Meeting*, Paris, July 18, 1977, Doc. #CE/M (77)10 (Prov.), scale 3, July 28, 1977;
• *Minutes of the 449th Meeting*, Paris, July 22, 1977, Doc. #C/M (77)17 (Prov.) part I, scale 3, August 8, 1977.

42. Two reports were issued detailing the implementation of the new arrangements by NEA. See Steering Committee for Nuclear Energy, *Implementation of the Decision of the Council Establishing a Multilateral Consultation and Surveillance Mechanism for Sea-dumping of Radioactive Waste*, Doc. #NE (78)4, scale 4, April 5, 1978; and *Report by the Secretariat*, Doc. #NE (79)1, scale 4, April 1979.

43. NEA, Steering Committee for Nuclear Energy, *Report on the 1981 Radioactive Waste Sea Disposal Operations*, Doc. #(81)16, drafted September 29, 1981, p. 6. See also the Report of the NEA Escorting Officers: Memo from Z. Chéghikian to P. Strohl,

August 3, 1981 on "Completion of the U.K. Sea Dumping Operation (13th–23rd July 1981)."

44. See a series of letters from H. E. Tomás Chávarri, head of the Spanish delegation to OECD, to Howard K. Shapar, director-general of NEA, March 11, March 29, and April 20, 1983. See also NEA, Steering Committee for Nuclear Energy, Minutes of the 66th Session held in Lyon on April 20–21, 1983. Doc. #NE/M (83)1, drafted July 22, 1983, pp. 16–17; and OECD Council, *Multilateral Consultation and Surveillance Mechanism for Sea Dumping of Radioactive Waste: Information Note on Recent Developments in Relation to the Mechanism*, Doc. #C(83)118, July 13, 1983.

45. See Remarks of the U.K. representative, G. M. Wedd, to the Steering Committee on Nuclear Energy in response to the points made by the Spanish delegation, Summary sent by G. M. Wedd to J. de la Ferté, NEA Secretariat, April 25, 1983; and *Second Notification Relating to the 1983 Sea Dumping Operation by the U.K. to the NEA*, n.d.

46. Letter from F. R. Terry, Ministry of the Environment, U.K., to J.-P. Olivier, NEA/OECD, January 6, 1984.

47. Detailed interviews with the U.K. scientists involved in the NEA sea-disposal operations since 1973 and with some of the U.S. scientists involved in the CRESP program since 1980.

48. See the following reports:

• OECD/NEA, *Review of the Continued Suitability of the Dumping Site for Radioactive Waste in the North-East Atlantic* (April 1980).
• OECD/NEA, *Interim Oceanographic Description of the North-East Atlantic Site for the Disposal of Low-Level Radioactive Waste* (Paris 1983).
• OECD/NEA, *Co-ordinated Research and Environmental Surveillance Programme Related to Sea Disposal of Radioactive Waste* (Paris 1984), Progress Report at the end of 1983.
• OECD/NEA, *Review of the Continued Suitability of the Dumping Site for Radioactive Waste in the North-East Atlantic* (Paris 1985).
• OECD/NEA, *Interim Oceanographic Description of the North-East Atlantic Site for the Disposal of Low-Level Radioactive Waste* (Paris 1986), vol. 2.

See also J. M. Bewers and C. J. R. Garrett, "Analysis of the Issues Related to Sea Dumping of Radioactive Wastes," *Marine Policy* (April 1987), 105–124.

References

International Atomic Energy Agency (IAEA). 1961. "Radioactive Waste Disposal into the Sea." *Safety Series* No. 5.

Miles, Edward L. 1989. "Scientific and Technical Knowledge and International Cooperation in Resource Management." In Steinar Andresen and Willy Østreng, eds. *International Resource Management: The Role of Science and Politics*. London: Belhaven Press.

Nuclear Energy Agency/Organization for Economic Cooperation and Development (NEA/OECD). 1973. Steering Committee for Nuclear Energy. "Minutes of the 43rd Session held at the Chateau de la Muette in Paris on 1 February 1973. Doc. #NE/M (73)1, Paris, February 27.

Nuclear Energy Agency/Organization for Economic Cooperation and Development (NEA/OECD). 1975. Steering Committee for Nuclear Energy. Meeting of April 25. Doc. #NE/M (75)4.

Nuclear Energy Agency/Organization for Economic Cooperation and Development (NEA/OECD). 1979. *Report on 1979 Radioactive Waste Sea Dumping Operations.* Doc. NE (79)19, scale 4, September 19.

Nuclear Energy Agency/Organization for Economic Cooperation and Development (NEA/OECD). 1982. Steering Committee for Nuclear Energy. *Report on the 1982. Radioactive Waste Sea Disposal Operations.* Doc. #NE (82)15, drafted September 28, pp. 6–8.

Organization for Economic Cooperation and Development/European Nuclear Energy Agency (OECD/ENEA). 1966. Steering Committee for Nuclear Energy. Experimental Disposal of Radioactive Waste into the Atlantic Ocean. *Summary Record of the First Meeting of the Executive Group, held in Paris on September 26–27,* p. 1.

Organization for Economic Cooperation and Development/European Nuclear Energy Agency (OECD/ENEA). 1968a. Steering Committee for Nuclear Energy. *Summary of Decisions Taken at the Thirty-fourth Session of the Committee on Thursday, 19th September 1968.* Doc. #SUM/DEC. 34, September 23, p. 2. *Minutes of the Thirty-fourth Session Held at the Château de la Muette in Paris on 19 September 1968.* Doc. NE/M (68), October 29, item 5.

Organization for Economic Cooperation and Development/European Nuclear Energy Agency (OECD/ENEA). 1968b. Steering Committee for Nuclear Energy. Disposal of Radioactive Waste into the Atlantic Ocean. *Summary Record of the Meeting of the Executive Group held in Paris on 17 and 18 October 1969.* Doc. #SEN/AT (68)4, November 8, p. 1.

Organization for European Economic Cooperation/European Nuclear Energy Agency (OEEC/ENEA). 1966. Steering Committee for Nuclear Energy. *Minutes of the Twenty-ninth Session Held at the Château de la Muette in Paris on 9 June 1966.* Doc. #NE/M (66)1, July 6.

Organization for European Economic Cooperation/European Nuclear Energy Agency (OEEC/ENEA). 1968a. Report Submitted to the Steering Committee for Nuclear Energy. *Disposal of Radioactive Waste into the Atlantic Ocean.* Doc. #NE (68)3, scale 4, March 21, p. 21.

Organization for European Economic Cooperation/European Nuclear Energy Agency (OEEC/ENEA). 1968b. *Theoretical Study of the Total Capacity of the North Sea to Receive Radioactive Material.* Doc. #SEN/SAN (68)5, June 12.

Organization for European Economic Cooperation/European Nuclear Energy Agency (OEEC/ENEA). 1960. Steering Committee for Nuclear Energy, Health and Safety Subcommittee. Doc. #NE (60)8, scale 4, November 3.

Organization for European Economic Cooperation/European Nuclear Energy Agency (OEEC/ENEA). 1961a. *Draft Summary of Restricted Meeting of Health and Safety Subcommittee on 19th October 1961: Problems Relating to Radioactive Waste Disposal in the North Sea.* Informal document.

Organization for European Economic Cooperation/European Nuclear Energy Agency (OEEC/ENEA). 1961b. Steering Committee for Nuclear Energy. *Minutes of the Eighteenth Session.* Doc. NE/M(61)1, scale 6, January 24, pp. 9–10.

Organization for European Economic Cooperation/European Nuclear Energy Agency (OEEC/ENEA). 1962. Health and Safety Subcommittee. *Study of Problems Relating to Radioactive Waste Disposal into the North Sea.* Doc. #SEN/SAN (62)4, February 14.

Organization for European Economic Cooperation/European Nuclear Energy Agency (OEEC/ENEA). 1964. Health and Safety Subcommittee. Radioactive Waste Disposal into the North Sea. *Minutes of the Restricted Meeting Held in Paris, 7 July 1964.* Doc. #SEN/SAN (64)9, August 10.

Organization for Economic Cooperation and Development/European Nuclear Energy Agency (OECD/ENEA). 1970a. *The HSS Final Report on the 1969 Operation.* Doc. #SEN/SAN (70)2, February 13. Steering Committee for Nuclear Energy, *Consultations on a New Waste Disposal Operation into the Atlantic Ocean.* Doc. #NE (70)3, April 7.

Organization for Economic Cooperation and Development/European Nuclear Energy Agency (OECD/ENEA). 1970b. Steering Committee for Nuclear Energy. *Minutes of the Thirty-eighth Session, Paris, 15 October 1970.* Doc. #NE/M (70)2, November 27, pp. 3–4.

Organization for Economic Cooperation and Development/Nuclear Energy Agency (OECD/NEA). 1972b. Steering Committee for Nuclear Energy. *Minutes of the Forty-first Session, 20 April.* Doc. #NE/M (72)1, May 23, p. 10.

Organization for Economic Cooperation and Development/Nuclear Energy Agency (OECD/NEA). 1972a. Health and Safety Subcommittee, Summary Record of the Twenty-second Session, November 16–17. Doc. #SEN/SAN (72)13, December 28.

Organization for Economic Cooperation and Development/Nuclear Energy Agency (OECD/NEA). *Review of the Continued Suitability of the Dumping Site for Radioactive Waste in the North-East Atlantic* (Paris, April 1980).

Organization for Economic Cooperation and Development/Nuclear Energy Agency (OECD/NEA). *Interim Oceanographic Description of the North-East Atlantic Site for the Disposal of Low-Level Radioactive Waste* (Paris 1983).

Organization for Economic Cooperation and Development/Nuclear Energy Agency (OECD/NEA). *Co-ordinated Research and Environmental Surveillance Programme Related to Sea Disposal of Radioactive Wate* (Paris 1984), Progress Report at the end of 1983.

Organization for Economic Cooperation and Development/Nuclear Energy Agency (OECD/NEA). *Review of the Continued Suitability of the Dumping Site for Radioactive Waste in the North-East Atlantic* (Paris 1985).

Organization for Economic Cooperation and Development/Nuclear Energy Agency (OECD/NEA). *Interim Oceanographic Description of the North-East Atlantic Site for the Disposal of Low-Level Radioactive Waste* (Paris 1986), vol. 2. See also J. M. Bewers and C. J. R. Garrett, "Analysis of the Issues Related to Sea Dumping of Radioactive Wastes," *Marine Policy* (April 1987), 105–124.

5

The Management of Tuna Fisheries in the West Central and Southwest Pacific

Edward L. Miles

Introduction and Reasons for Success

This chapter evaluates the effectiveness of the regime for managing the tuna fisheries of the West Central and Southwest Pacific Ocean. We begin by describing the important role of four new variables that are critical for explaining the eventual success of this regime. These new variables are (1) a nested regime—that is, the international law relating to highly migratory species in the context of a new regime for the oceans, which recognized the sovereign rights of the coastal state over all living resources within the exclusive economic zone (EEZ); (2) the life histories and migration patterns of the tunas, skipjack, billfish, and shark fisheries of the West Central and Southwest Pacific; (3) the politico-geographic context; and (4) the alignment of interests as determined by the divisibility of payoffs between coastal states and distant-water fishing nations (DWFNs).

These variables are new, but they are powerfully connected to the ones specified in our model and modify their weighting in significant ways. For instance, we show later how the transformation of the world ocean regime, and particularly the law relating to the management of anadromous species, changed the nature of the Pacific salmon problem by definitively tipping the scales in favor of states of origin. Similarly, the new ocean regime definitively tipped the scales in favor of coastal states in relation to the management of highly migratory species. This fact, when combined with the life history and migratory patterns of the species in question and the politico-geographic context, turned the region into a potentially gigantic EEZ encompassing the entire range of the skipjack resource and a very large portion of the ranges of the other species. The alignment of interests of Pacific island states created the possibility of a coalition of like-minded states strongly united against the DWFNs and gaining strength by pooling their resources and harmonizing their respective policies.

This combination, from the point of view of the Pacific island states, changed an actually malign problem into a potentially benign one. Distribution of capabilities was strongly asymmetric, lending an "us-versus-them" cast to the situation and emphasizing the necessity of collaboration among Pacific island states, all of which benefited from a resilient preexisting infrastructure of institutionalized regional cooperation. The life history and migratory patterns of the species in question, combined with the specifics of the politico-geographic context, determined the configuration of actor interests and positions of the island states and strongly reinforced the coalition. Uncertainty in the knowledge base was initially high, which facilitated the emergence of an epistemic community. The strong coalition, combined with the institutionalized regional infrastructure, eventually facilitated the development of robust collective negotiating strategies.

How all this came to be is assessed in two phases—the process of regime creation from 1976 to 1979 and the process of regime implementation from 1979 to 1995. Until 1983 the performance of the regime was weak, but a major leadership change in 1983 led to the implementation of a series of policies that resulted in a very high order of problem-solving capacity. These changes focused on the development of negotiating capacity, the production of vital information, the emergence of the Nauru Group as the spearhead of a cartel within the island state coalition, and resolution of interorganizational rivalry between the South Pacific Forum Fisheries Agency and the South Pacific Commission. This surge in problem-solving capacity turned the regime into a highly effective one.

A Nested Regime

This regional coalition could have existed without the larger transformation of the world ocean regime facilitated by the Third United Nations Conference on the Law of the Sea (UNCLOS III), (see the existence of the South Pacific Economic Commission [SPEC]), but it would not have been effective. *No DWFN would have accepted the jurisdiction of the Pacific Island states* if the operative regime had been the Grotian one. Not only did all Pacific island states and territories participate in UNCLOS III, but they were organized into a coherent coalition, the Oceania Group, with a special interest in the issues of the EEZ and coastal-state authority to manage highly migratory species within the zone. By 1976 these issues were largely settled, even though the United States chose to see in the rules related to highly migratory species far more ambiguity than other delegations did.

Article 56 of the 1982 U.N. Convention on the Law of Sea, reflecting similar provisions set down in the Revised Single Negotiating Text (RSNT) of 1976, recognizes coastal states' sovereign rights over all living resources within the EEZ. Article 61(1) further specifies that coastal states shall determine the allowable catch of the living resources in their EEZs, and article 62(2) makes it clear that coastal states shall determine their own capacity to harvest the living resources of the zone. Article 297(3) closes this circle by declaring that these decisions, which fall within the discretionary authority of the coastal state, are nonreviewable by any third party. Most coastal states, except the United States and Japan, took the view that their sovereign rights encompassed even highly migratory species when they were within the EEZ.

The United States, on the other hand, seeking to protect the interests of the U.S. tuna fleet, put most emphasis on article 64, which related specifically to highly migratory species:

Article 64. Highly migratory species

1. The coastal States and other States whose nationals fish in the region for the highly migratory species listed in Annex I[1] shall cooperate directly or through appropriate international organizations with a view to ensuring conservation and promoting the objective of optimum utilization of such species throughout the region, both within and beyond the exclusive economic zone. In regions for which no appropriate international organization exists, the coastal State and other States whose nationals harvest these species in the region shall cooperate to establish such an organization and participate in its work.

2. The provisions of paragraph 1 apply in addition to the other provisions of this part.

The U.S. interpretation was that article 64 implies that highly migratory species could not be managed unless this was accomplished through an international organization that gave states whose nationals fished for the stocks in question equal voice with the coastal states through whose zones the stocks migrated. It was always difficult to see how this position could be legally sustained, given the provisions of articles 56, 61(1), 62(2), and 297(3) and given the formulation in article 64(2) (Burke 1994). Eventually, the U.S. case collapsed in the face of overwhelming state practice but not before it triggered intense conflict in the Western and Eastern Tropical Pacific. Conflict was made particularly intense as a result of the passage of the Magnuson Fisheries Conservation and Management Act (FCMA) of 1976, which contained prohibitions on U.S. recognition of coastal-state jurisdiction over tuna and provisions to embargo tuna and tuna products originating in the EEZs of any coastal states that enforced their national laws against any U.S. tuna vessels fishing within their zones. Since the United States at that time controlled 70 percent of the canned tuna market, this was a potent threat indeed.

In any case, the UNCLOS III negotiations triggered the move among the South Pacific island states and territories to create a regional regime for the South Pacific and to attach it to the South Pacific Forum and its programs. The initial connections were made by Fiji and Papua New Guinea (PNG) (Gubon 1987; Kent 1980).

The Life History and Migration Patterns of the Species in Question and the Alignment of Interests

The resource complex of the West Central and Southwest Pacific exhibits two main patterns. North of 30 degrees South, skipjack (*Katsuwonus pelamis*) is by far the predominant species, followed by yellowfin tuna (*Thunnus albacares*) and bigeye tuna (*Thunnus obesus*). In addition, in much smaller abundance at depth, throughout the entire region there are striped marlin (*Tetrapturus audax*), black marlin (*Makaira indica*), blue marlin (*Makaira nigricans*), and swordfish (*Xiphias gladdius*). South of 30°S the primary species mix changes to a combination of skipjack tuna, albacore tuna (*Thunnus alalunga*), and southern bluefin tuna (*Thunnus maccoyii*). A large variety of sharks permeates the entire region at depth just like the skipjack on the surface but in lesser abundance.

This species composition, in turn, determines the type of fisheries that pursue not only different combinations of species but also at different points in their life cycles. The surface fisheries in tropical waters, primarily pole-and-line and purse-seine, targeted skipjack and juvenile yellowfin. They also caught incidentally a moderate amount of juvenile bigeye tuna.

The longline fisheries, on the other hand, targeted adult yellowfin, bigeye, southern bluefin, and albacore tuna at depth. They also caught incidentally a variety of billfishes and sharks. While roughly two-thirds of the catch is derived from the surface fisheries (primarily purse-seine), the reverse is the case with respect to market value.

Two characteristics of the dominant species in the West Central and Southwest Pacific are particularly important for the management of fisheries in this area. These species are all highly fecund, and most migrate tremendous distances. In all species of tuna, a female at sexual maturity produces about 100,000 eggs per kilogram of body weight (Joseph, Klawe, and Murphy 1979; Sund, Blackburn, and Williams 1981). In skipjack in the tropics, a female at sexual maturity produces about 300,000 eggs per 1.18 nights year round (Kearney 1991). The combined result of these two biological characteristics is that large quantities of fish migrate annually through the jurisdictions of twenty-two states and territories, from the Eastern Carolines (Federated States of Micronesia) and the Marshall Islands in the north to Australia and New Zealand in the south. These fish, however, are hunted primarily by the foreigners (the DWFNs), thereby creating an immediate us-versus-them cast to the situation.

Without major indigenous capability in any of the primary fisheries, the coastal states of the region had an identical interest in banding together to control and derive revenues and other benefits from the fisheries that occurred within their new EEZs. The only questions were how the arrangement would be designed and how it would be implemented. That it could be implemented was never in question since the coastal states, on the basis of developments in UNCLOS III, would be holding all the trump cards.

The Politico-Geographic Context
The area of the region extends from 122°W to 132°E and from 22°N to 42°S. It excludes the Philippines and Hawaii but includes twenty-two separate island states and territories of which fifteen are either atolls or small islands. Fish and coconuts are the primary consumable or saleable resources of many island states, but other island states also possess valuable nonfish resources, like Fiji (gold), Nauru (phosphates), and Papua New Guinea (a variety of mineral resources). The total area of the region amounts to 29 million square kilometers, but the total land area, excluding Australia and New Zealand, consists of 551,000 km² of which 84 percent is Papua New Guinea. Figure 5.1 shows the geographic effects of a 200-mile EEZ regime. Virtually the entire region is enclosed, except for a few high-seas enclaves or areas uncovered by coastal-state jurisdiction.

Within the region, there are three broad cultural groupings. These are (1) Melanesia in the southwest, consisting of New Caledonia, PNG, the Solomon Islands, Vanuatu, and Fiji. This grouping encompasses 98 percent of the land area, with 84 percent in PNG alone and 28 percent of the sea area; (2) Micronesia in the north, consisting of Guam, Kiribati, Nauru, the Northern Mariannas, Palau, the Federated States of Micronesia, and the Marshall Islands. This grouping encompasses 0.6 percent of the land area but 34 percent of the sea area. However, Guam as a U.S. territory and the Northern Mariannas, which chose commonwealth status with the United States, cannot be included as belonging to the South Pacific island-state political coalition, and (3) Polynesia in the east, which consists of French Polynesia, the Cook Islands, Tonga, Tuvalu, Western and American Samoa, Tokelau Island, and Pitcairn Island. This grouping holds 1.4 percent of the land area but 38 percent of the sea area.

The total population of the region is in excess of 5 million, with Melanesia accounting for 84 percent (PNG alone has more than 3 million), Micronesia 6

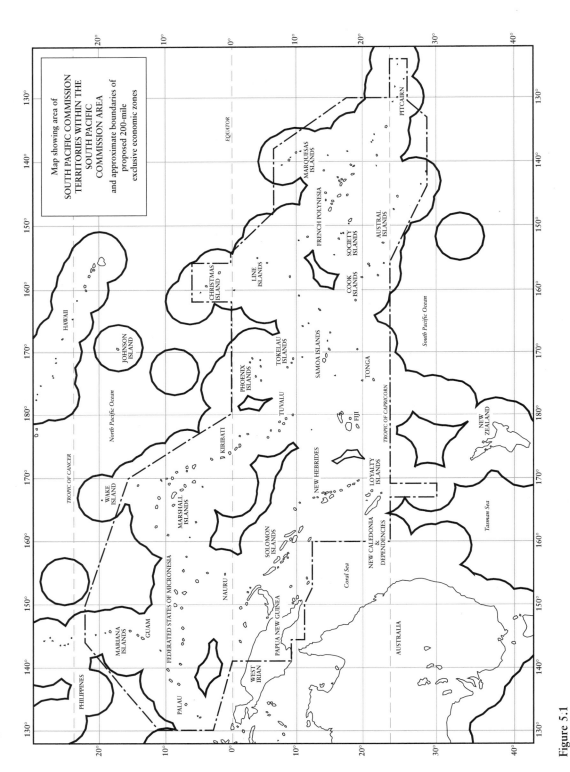

Figure 5.1

Geographic effect of a 200-mile EEZ regime in the southwest Pacific Ocean

Based on a map of estimated boundaries after 1976 by courtesy of ORSTOM. Reprinted by permission.

percent, and Polynesia 10 percent. In terms of regional collaboration one would have to note a considerable regional infrastructure consisting of the South Pacific Forum, based on representation of all heads of government; its Secretariat, formerly the South Pacific Economic Commission (SPEC) now called the Forum Secretariat, situated in Suva, Fiji; and the South Pacific Commission (SPC), with a focus on economic, cultural, and scientific and technical collaboration and containing as members metropolitan powers with territories in the region like the United States, the United Kingdom, and France. During the late 1970s and the 1980s the region spawned two more collaborative entities—the South Pacific Forum Fisheries Agency (FFA) and the South Pacific Regional Environmental Program (SPREP).

The point here is that cultural forces underlie a pattern of interisland cooperation and the development of a self-conscious ideology of cooperation based on consensus called "the Pacific Way" and also that a powerful regional political infrastructure is anchored in the South Pacific Forum. As the Forum expanded its reach into fisheries and other areas of activity, the center developed the capability to make package deals, thereby avoiding much zero-sum conflict as a result of being restricted to trading on a single dimension. This, then, is the context that was linked to UNCLOS III developments and spawned the FFA. The problem-solving capacity of the region was therefore *potentially* great.

The Process of Regime Creation, 1976 to 1979

Let us return to the UNCLOS III negotiations as the crucible of the regional coalition. The Oceania Group—consisting of Australia, the Cook Islands, Fiji, Micronesia, Nauru, New Zealand, PNG, Tonga, and Western Samoa—was deeply involved in negotiating the EEZ framework of the Convention as well as article 64. The group was struck both by the size of geographic area that would be jointly controlled by Pacific Island states and the potential power of maintaining a united front with common fisheries policies regarding the DWFNs (Ramp 1977).[2] They were also struck by the example of a *regional* EEZ as imposed by the European Economic Community (Ramp 1977).

Three political entrepreneurs made the link between the UNCLOS III negotiations (the Revised Single Negotiating Text [RSNT] 1976 and the Informal Composite Negotiating Text [ICNT] 1978) and the South Pacific context—at the political level, the Fijian and PNG delegations; at the scientific level, an Australian fisheries scientist, Robert E. Kearney, then serving as director of the South Pacific Commission's Skipjack Survey and Assessment Program (Gubon 1987; Kearney 1976,

1977). Kearney was then the most authoritative regional fisheries scientist at the interface of both an incipient epistemic community and governmental representatives at the highest levels. Indeed, during and after the 1976 Forum meeting, his writings were highly influential because they provided the basis for consensual knowledge within the island-state political coalition. Not only was it the most comprehensive knowledge available to the politicians, diplomats, and their fishery officers; it was also written in an easily assimilable style for nonbiologists.

What then were the problems to be solved at the regional level? They involved applying the new law of the sea to the South Pacific context; coordinating extensions of coastal-state sovereign rights over all living and nonliving resources to the maximum extent provided by international law; establishing a united front regarding DWFNs and the payment of access fees; establishing the means for cooperation in the exchange of data and other information relevant to conservation and management of the tuna and billfish resources of the region; devising a means for achieving regional surveillance and enforcement mechanisms; and assisting coastal states in their own attempts at fisheries development. However, before any of this could be done, decision makers within island states needed to increase their understanding of what resources were involved, their distribution and abundance, and their migratory patterns. This knowledge was being created at the time by the immensely important Skipjack Survey and Assessment Program of SPC. From this knowledge base political leaders could then realistically conceive of coordinated extension of sovereign rights over living resources as the principal solution for maximizing returns to themselves.

The institutional setting was a temporary bridge between UNCLOS III and the South Pacific Forum with the distribution of capabilities to fish almost exclusively in the hands of the DWFNs. This division created a clear conflict between coastal states and DWFNs, with the former seeking to place controls on the latter, based on the new law, of a scope and detail never before experienced. While the coastal-state coalition in the region was a concrete entity, outside UNCLOS III this was not true for the DWFNs. Each played a separate game, although at times one could detect a tacit alliance between the United States and Japan. This was not to last long. Uncertainty of all kinds was high, and there was at the time only a nascent regional epistemic community developing around the Skipjack Survey and Assessment Program in SPC. Negotiating strategies were as yet unformed.

The spring 1976 meeting of the Forum agreed to convene a special meeting in October to consider the whole range of issues identified above (Gubon 1987). That meeting was held in Suva, Fiji, in October 1976, at which time delegates issued on

October 15 a Suva Declaration. This declaration stated, *inter alia*, their intention to establish 200-mile EEZs at appropriate times and after consultation with one another; to harmonize regional fisheries policies and adopt a coordinated approach in negotiations with DWFNs; to establish a South Pacific Fisheries Agency; and to request the director of SPEC to prepare specific proposals for consideration at the next session of the Forum and to examine ways of cooperating with respect to surveillance and enforcement of fishing vessels in the region (Suva Declaration 1976).

From this point on, negotiations became highly controversial and very murky on the subject of membership in the agency and on the subject of the organization's prerogatives toward member states. On the latter issue, it was not surprising that just having gained extended jurisdiction over large sea areas, coastal states were eager to protect that authority from all potential challengers, even their own organization. Coastal states at the Port Moresby meeting in 1977, which followed the Suva meeting, stressed the inviolability of their sovereign rights and insisted that the new agency could assist member states only in the exercise of those rights. This sentiment would later lead to restricted powers for the FFA.

On the former issue of membership, trouble occurred because the Port Moresby Declaration was somewhat loosely worded. Instead of explicitly limiting membership to coastal states of the region, the declaration stated an intention "to establish a South Pacific regional fisheries agency whose membership would be open to all forum countries and all countries in the South Pacific with coastal state interests who supported coastal state sovereignty to conserve and manage living resources—including highly migratory species—in the 200-mile zones" (Gubon 1987; Kent 1980).

This broader formulation allowed France, the United Kingdom, the United States, and Chile to participate, the three metropolitan powers presumably representing the interests of their Pacific territories. In fact, however, each had its own hidden agenda. The United States, for instance, argued against a strictly coastal-state organization and for an article 64 consultative body including both coastal states and DWFNs. The U.S. claim was buttressed by the requirements of the Magnuson Act and by the need to maintain consistency in its negotiating position toward the Eastern Tropical Pacific negotiations with Mexico.

The U.S. position derailed the momentum of the coastal-state coalition by splitting the coalition. Reportedly, Kiribati (then the Gilberts) and Western Samoa wanted the United States in the agency as a means of getting U.S. funding.[3] Papua New Guinea, on the other hand, threatened to go its own way. The French took a strong coastal-state position ostensibly on behalf of Tahiti. Tahiti itself wished colonies to have voting rights in the agency. France seemed happy to support this position since it gave colonies official status and added to the votes France would control. The United States also split the Micronesians and isolated the Congress of Micronesia, which had taken a hard-line coastal-state position.

The outcome of these negotiations, not surprisingly, was a document that seemed to argue for an article 64 type body and not a purely coastal-state coalition, and the document was carried over to the next meeting of the Forum to be held in Suva in June 1978 (Kent 1980). The U.S. view dominated that meeting as well, and a draft agreement was presented to the Forum in Niue in September 1978 (Kent 1980). At that point support for the U.S. position collapsed, and the coastal-state view prevailed: "this Forum decides to establish forthwith a South Pacific Forum Fisheries Agency with the Port Moresby Declaration as a guide but restricted to Forum members at this stage."[4]

While the conflict over membership of the Agency temporarily overshadowed the insistence of Pacific Island states that their sovereignty remain inviolate, that latter desire once again came to the fore in the actual design of the Agency. Forum members made a distinction between the Forum Fisheries Committee (FFC) and the Forum Fisheries Agency (FFA). The former was a mechanism for consultation between the parties with respect to harmonizing fisheries policy, cooperating on surveillance and enforcement, coordinating relations with DWFNs, onshore fish processing and marketing, and allowing other parties access to the EEZs. In addition, the FFC also gives direction to FFA on an annual basis, sets policy, approves the operational budget, and sometimes meets to act on specific issues during the year.

The latter was designed purely as a service agency to assist member states in collecting, analyzing, evaluating, and disseminating data on the status of stocks, management procedures, legislation and agreements, prices, shipping, processing and marketing of fish and fish products, and seeking to establish working arrangements with "relevant regional and international organizations, particularly the South Pacific Commission."

One would therefore have to acknowledge that while the view of a strong coastal-state coalition prevailed, the reality as defined by the Convention establishing the FFA in 1979 was far from that. At best, the outcome was a level 2 organization. However, initial pledges by Australia, New Zealand, the Solomon Islands, the Commonwealth Secretariat, the United Kingdom, the Cook Islands, and United Nations Development Program/Food and Agriculture Organization (UNDP/FAO) amounted to Fijian $300,000 of extrabudgetary funding to get the new agency started (SPEC 1979). On a routine basis, appropriations would be required from member states.

It is worth noting here that the composition and roles of the FFC had important unintended side-effects that later facilitated the emergence of a far stronger organization than was reflected in the official design. FFC membership comprised official representatives from member states from both fisheries and foreign affairs departments. This was the source of considerable continuity, and it therefore became a powerful training ground for a new cohort of specialists.

The Process of Regime Implementation, 1979 to 1995

The new organization began without any information base of its own, so the Forum appointed Robert E. Kearney to a three-month consultancy to develop a regional view of tuna resources, catches, and so on and to report on possible future developments in the tuna industry in the Western Pacific. Kearney produced the first set of estimates of local and DWFN catches and a comprehensive view of the present status and possible future of this fishery (Kearney 1979).

Given the major changes that have occurred in the region, it is easy in hindsight to lose sight of how useful Kearney's publication was at the time it was published.[5] But we should remember that the first foreign-fisheries access agreements in the West Central and Southwest Pacific were signed only in 1978 between Japan and Australia, then New Zealand, and then PNG. Most of the coastal states at that time knew almost nothing of the status of stocks and fisheries. Limited information was held by the DWFNs—Japan, Taiwan, and Korea. But Japan was by far the most knowledgeable since its fleets had fished the area for thirty years by that time. In 1979, someone charged with managing fisheries for most Pacific island states could turn only to Kearney (see table 5.1) and to the FAO (see table 5.2) for a gross idea of what was going on. The Federated States of Micronesia was better off than most island states with respect to information since it could turn both to Kearney and to Richard Shomura, then director of the Honolulu Laboratory of the U.S. National Marine Fisheries Service, for assistance. Shomura and Jerry Wetherall provided not only analyses of historical data but also assessments of current events. In addition, MMA hired Wilbur van Campen, a former State Department fisheries official, to provide summaries of the major Japanese fisheries publications.

One would also have to admit that in 1979 and 1980 the outlook for the coastal-state coalition was not good. FFA was weak. Apart from the weakness in organizational design, from the perspective of a participant observer some staff did not appear to be especially competent; there were interpersonal difficulties within the staff; rivalry with the Skipjack Survey and Assessment Program increased as a result of the insecurities operative within FFA; separate island-state traditions seemed to

Table 5.1
Local catches and catches by distant-water fleets in the waters of the countries and territories of the South Pacific Commission

in Country or Territory	Local Total Fish Catch (tons)	Local Tuna Catch (tons)	Long-Line Catch[h] in 200-Mile Zone by Foreign Fleets in 1976 (tons)	Pole-and-line[h] Catch by Japanese Fleet 200-Mile Zone in 1976 (tons)
American Samoa	220 (1978)[a,b]	20 (1978)[b]	387	29
Cook Islands	—	—	2,866	10
Fiji	11,594 (1977)[a,c]	7,262 (1977)[a,c]	1,553	233
French Polynesia	2,386 (1974)[d]	1,293 (1974)[e]	7,264	0
Guam	—	—	—[f]	—[f]
Kiribati	1,344 (1977)[a]	786 (1977)[a]	11,349	16,570
Nauru	0	0	1,845	8,224
New Caledonia	499 (1977)[a,c]	186 (1977)[a,c]	1,800	58
New Hebrides	10,500 (1976)[g]	10,000 (1976)[g]	1,012	93
Niue	20 (1978)[a]	10 (1978)[e]	289	4
Norfolk Island	—	—	700	2
Pitcairn	—	—	1,090	0
Solomon Islands	17,444 (1976)[e]	15,787 (1976)[h]	2,709	17,248
Tokelau	—	—	450	1,645
Tonga	1,117 (1977)[a]	300 (1977)[e]	816	18
Trust Territory of the Pacific Islands	10,000 (1976)[e]	5,284 (1976)[h]	20,601	38,360
Tuvalu	80 (1978)[a]	40 (1978)[a]	1,886	7,611
Wallis and Futuna	—	—	386	155
Western Samoa	1,700 (1976)[a]	850 (1976)[a]	160	24
Total	*56,904*	*41,818*	*57,163*	*90,284*

Source: Kearney (1979).
a. Figures from Crossland and Grandperrin (1979).
b. Excluding unloadings to the Pago Pago canneries.
c. This includes only the catches that passed through markets.
d. From Kearney (1977).
e. Estimated by the author.
f. Catches included under Trust Territory of the Pacific Islands.
g. Mainly long-line catches transshipped at Santo.
h. From Kearney (1979).

Table 5.2
Estimated catches of tunas and tunalike species of major significance from FAO statistical areas 71 and 81 (all figures in tons)

Species	1970	1971	1972	1973	1974	1975	1976	1977
Skipjack (*Katsuwonus pelamis*)	70,000	115,000	120,400	198,600	304,615	211,771	258,845	278,060
Albacore (*Thunnus Alalunga*)	20,300	27,400	29,500	34,800	50,433	33,273	14,863	11,225
Southern bluefin tuna (*Thunnus maccoyii*)	20,800	19,600	19,500	14,100	9,196	9,646	10,126	3,521
Yellowfin tuna (*Thunnus albacares*)	19,200	37,000	42,400	45,400	105,590	101,363	103,963	122,909
Bigeye tuna (*Thunnus obesus*)	15,500	19,700	30,700	26,200	26,200	39,119	36,401	35,030
Total major tuna species	145,800	218,700	242,500	319,100	506,873	395,172	424,198	450,745
Swordfishes, marlins, and billfishes	10,400	10,400	12,400	11,700	15,241	13,829	11,555	10,341
Total tunas and billfishes	*156,200*	*229,100*	*254,900*	*330,800*	*522,114*	*409,001*	*435,753*	*461,086*

Note: Figures are from FAO (1975, 1978).

be resistant to Secretariat attempts at policy harmonization; and some Secretariat staff pushed too hard for harmonization, thereby angering member states.

In addition to weaknesses in the regional mechanism, the coastal states themselves were weak. They had minimal management infrastructure in place, little or no scientific capabilities, and precious little surveillance and enforcement capabilities. On the other hand, not only did the Japanese hold most of the information; they had a monopsony.

And yet things changed, and this regional experiment turned out to be highly successful. How could this come about? *The type of problem, institutional setting, and configuration of actor interests and positions remained constant until 1982 to 1983, when there was a major expansion in institutional capacity as a result of leadership changes in the FFA.* But even prior to 1983, major changes were in the making regarding the other variables. Between 1979 to 1981 the Pacific Island states individually substantially developed their negotiation capabilities against the DWFNs at the same time that substantial reductions of empirical uncertainty in the knowledge base occurred as a result of the work done by Kearney's group in the SPC. Simultaneously with these developments, major improvements in the infrastructure of the South Pacific coalition developed in 1980 and 1981 with the formation of the Nauru Group. From the conclusion of the Skipjack Survey and Assessment Program in 1981 a *modus vivendi* became possible between SPC and FFA, thereby resulting in the creation of a genuine epistemic community. And finally, out of the major leadership changes that occurred in 1983, significant improvements in island-state negotiating strategies ensued. Consequently, the regime as a whole experienced quantum improvements in problem-solving capacity. Let us follow these developments in detail.

Expanding Island-State Capabilities and Reducing Uncertainty in the Knowledge Base, 1979 to 1981

After the first three bilateral access negotiations occurred in 1978 (Japan serially with Australia, New Zealand, and PNG), most Pacific Island states began to negotiate with DWFNs from 1979 for the first time. Only the Solomon Islands, PNG, and Fiji had had prior experience negotiating with the Japanese on fisheries joint ventures, exclusive of right of access, so that most Pacific Island states had to learn how to negotiate fisheries access agreements from scratch. All of them learned primarily from Japan, which maintained bilateral relations with sixteen island states. Some island states were involved in relationships with either Taiwan or Korea or both, but these were, in most cases, of minor significance. U.S. vessels (purse-seiners) did not constitute a significant presence up to 1980 because there were only eight as compared to fourteen large (500 gross tons) Japanese purse-seiners and two Korean vessels of the same kind.

The dominant fleets consisted at that time of about 1,000 Japanese small and large long-liners and pole-and-liners, combined with about 680 Taiwanese and 568 Korean long-liners. Many of the Taiwanese and Korean vessels were fishing out of American Samoa on contract to U.S. canners, while Taiwanese vessels also fished on contract to Japanese suppliers out of Fiji and Vanuatu.

As part of the negotiating experience, the Pacific Island states learned that they needed information on the deployment, operational procedures, and catches of the different types of fishing vessels themselves and, equally important, up-to-date information on price by species in the dominant tuna markets of the world. It was therefore necessary to learn that the Japanese market was dominant in fresh-fish sashimi products, which were the most valuable, covering primarily large yellowfin, bigeye, southern bluefin, and all billfish. However, the price differed depending on quality and on where the fish were landed in Japan. The U.S. market was dominant in canned tuna (at that time the U.S. market accounted for 70 percent of canned production), and the U.S. market also dominated Japanese skipjack production.

American domination of the Japanese skipjack production had its origin in the eating habits of the Japanese population, which preferred skipjack fresh in the spring and summer when the pole-and-line fleet operated around Japan. The Japanese market for canned tuna, while growing, is and was small and is really based primarily on yellowfin; therefore, to survive, the Japanese skipjack fleet needed to export large amounts of skipjack to the United States. The Japanese *Katsuobushi* (smoked skipjack) market is also small and suffers from a low price elasticity of demand.

The Japanese delegation, under the leadership of the late Norio Fujinami, maintained a grueling schedule of bilateral negotiations on an annual basis between 1978 and 1984 because it was Japanese strategy to teach the South Pacific island states in detail about the structure and operations of Japanese tuna fisheries in the Pacific as a whole and the nature of the tuna market in Japan. The island-state negotiators proved to be apt pupils because they had so much at stake. But they could not help notice that they were steadily losing ground every year from 1978 in terms of access fees received.

The problem was that each island state faced Japan alone with varying amounts of information and negotiating skill. Conflicts arose over the quality and quantity of information reported by the Japanese fleets, and the region did not then have a single logbook format for reporting catch-per-effort data. In fact such a format was actually agreed to in 1978 and 1979 at SPC, but the Japanese modified the format in different ways in subsequent negotiations with individual island states so that the data were no longer comparable. There was also considerable ignorance about the monthly price fluctuations for tuna by landing port in Japan and exactly what factors determined price. All of these factors were sources of dissatisfaction in the coastal-state coalition, which led to discussions beginning in 1980 to create the Nauru Group. Before we deal with these discussions, however, we must describe the SPC efforts to reduce both theoretical and empirical uncertainty in the Skipjack Survey and Assessment Program and its annual technical meetings between 1978 and 1981.

While the first cohort of island-state negotiators were being trained on the job at FFC and SPC meetings, they and the other members of the island-state fisheries administrations were also learning about resource distribution on a regionwide basis. Was there one stock of skipjack operating throughout the West Central and South Pacific, two stocks, or five? All of these possibilities had been proposed at one time or another (Fujino 1967, 1970a, 1970b; Sharp and Kane n.d.; Sharp n.d.). What was the extent of skipjack migration? What factors determined distribution and abundance? Where and when did spawning occur? What was the magnitude of skipjack biomass, and what annual yields could be taken without endangering the stocks?

Until the late 1970s, no one knew the answers to these questions. It was Kearney's magnificent contribution to the region to conceive of and successfully establish the Skipjack Survey and Assessment Program and to locate it in SPC to secure contributions not only from Australia and New Zealand but from the Metropolitan (French, U.K., and U.S.) powers and Japan as well. Moreover, over the course of five years, Kearney and his group organized annual meetings at which findings to date were reported and issues discussed in detail. These sessions were occasions for intense learning as well as regional bonding of this first cohort of technical experts. This group emerged from the mid-1980s as a fully fledged epistemic community penetrating the centers of national and regional decision making.

The results of the Skipjack Survey and Assessment Program were startling (Kearney 1983). It was found that skipjack migrated extensively, as shown in Figure 5.2. Analyses of blood genetic data and of parasites countered those arguing for discrete subpopulations of skipjack. The data showed support for a single population on a regionwide basis. Spawning occurred across the whole of the tropical region, but its incidence was highest at the longitudinal extremes (Kearney 1983). The standing stock estimate was 3 million metric tons (mmt) at the 95 percent confidence level with a range of 2.5 to 3.7 mmt (Kearney 1983). The rate of turnover in the population at any particular location in the region was on the order of 17 percent per month (Kearney 1983). And the estimated annual throughput and therefore yield was in excess of 6 mmt per year (Kearney 1983). At that time (1981) the estimated total regional catch was 260,000 mt per year. Therefore, it was clear that

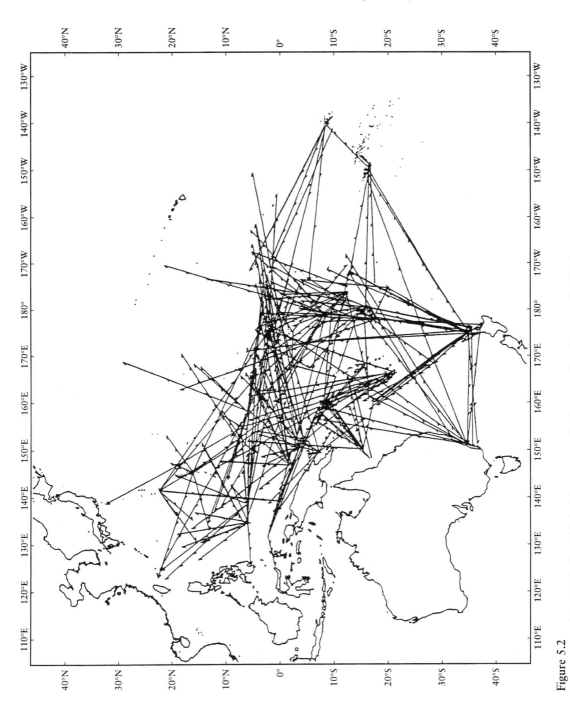

Figure 5.2
Assessment of the skipjack and bait fish resources in the central and western Pacific Ocean

the ability of the world market to absorb skipjack without stimulating a total collapse of prices was far less than could theoretically be produced by the combined fleets operating in the region. *The management problem for the island states, therefore, was maximizing benefits from the resources and not catch per se.*

Changes in Coalition Structure and Capabilities

Four events are germane to this analysis of the evolution of the coalition: the emergence of the Nauru Group from 1980 to 1981, the breaking of the Japanese monopsony as a result of agreements concluded with the American Tunaboat Association (ATA) between 1980 and 1984, the creation of the Regional Register as an enforcement tool of considerable power in 1982, and the emergence of a modern Taiwanese purse-seine fishery for skipjack, which further eroded Japanese control.

Doulman (Doulman 1987) has provided a concise analysis of the emergence of the Nauru Group. The Group consists of the Federated States of Micronesia, Kiribati, the Marshall Islands, Nauru, Palau, Papua New Guinea, and the Solomon Islands. These countries have contiguous EEZs, and they shared a powerful stimulus to create a cartel, i.e., the operations of the Japan-based distant-water tuna fleet then consisting of small and large long-liners, large pole-and-liners, and large purse-seiners. The combined area of the EEZs of the Nauru Group amounted to 72 percent of the total zone area of all FFA members (see Figure 5.3) and accounted for more than 75 percent of the catch of the DWFNs then operating in the region, *including the entire purse-seine catch* (Doulman 1987). The Nauru Group also accounted for more than 70 percent of all access fees paid by DWFNs to Pacific Island states (Doulman 1987).

The evidence summarized above illustrates the theoretical potential of such a cartel. All members were dissatisfied to varying degrees with the way in which the bilateral access arrangements with Japan were developing, and led by PNG and FSM and to a lesser extent Kiribati, they drafted objectives to provide for greater sharing of information among them to counter a divide-and-conquer strategy by the DWFNs and to agree on minimum terms and conditions governing access to increase the negotiating leverage of island states. For reasons internal to FFA, this move became controversial and even contentious.

To understand why such an obviously necessary move would become contentious within the coalition at large, it is necessary to remember that at the time FFA was new and weak. Indeed, out of frustration with its weakness, PNG wished to reach agreement with FSM, the Marshall Islands, and Palau at least to protect their own interests against Japan and later the United States. Other members of FFA, however,

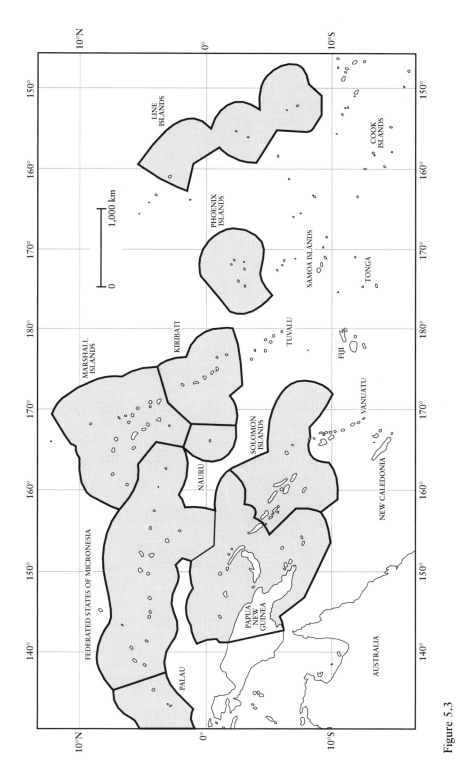

Figure 5.3
EEZs of the Nauru Group countries
Based on a map by Doulman (1987). Reprinted by permission.

saw this as a potentially disruptive move threatening the viability of the entire coalition. Initially, this was the view of Australia, New Zealand, and Fiji, in particular, all extremely influential members of the Forum.

Consequently, progress in the formation of the cartel was slow, and the internal negotiating process took two full years. FSM insisted, in spite of general overall disagreement, that the arrangement had to be in treaty form, binding in international law. Part of the reasoning behind this position related to the Micronesian struggle toward formal statehood against a recalcitrant U.S. government seeking to protect the interests of the American Tunaboat Association as much as possible. But part of FSM's position related to a consideration of negotiating tactics. Two such reasons were adduced in favor of a treaty: (1) a treaty binding under international law would remove any ambiguity surrounding the motives of participating countries, especially when each might separately face the possibility of a breaking off of negotiations with Japan, thereby disrupting the flow of critical income; and (2) to be credible to Japan with the minimum terms and conditions, these commitments had to pass into national legislation via a process of formal ratification. Eventually, these views prevailed, and the Nauru Agreement concerning cooperation in the Management of Fisheries of Common Interest was indeed signed as a treaty on February 11, 1982.

The Nauru Agreement is a framework agreement that envisages a number of subsidiary implementing agreements. The parties undertake to "coordinate and harmonize the management of fisheries with regard to common stocks within the Fisheries Zones" (article I), "to establish a coordinated approach [regarding] foreign fishing vessels," and to establish minimum terms and conditions (article II) relating to

(1) the requirement that each foreign fishing vessel apply for and possess a license or permit; (2) the placement of observers on foreign fishing vessels; (3) the requirement that a standardized form of log book be maintained on a day-to-day basis which shall be produced at the direction of the competent authorities; (4) the timely reporting to the competent authorities of required information concerning the entry, exit, and other movement and activities of foreign fishing vessels within the Fisheries Zones; (5) standardized identification of foreign fishing vessels; (6) the payment of an access fee, which shall be calculated in accordance with principles established by the Parties; (7) the requirement to supply to the competent authorities complete catch and effort data for each voyage; (8) the requirement to supply to the competent authorities such additional information as the Parties may determine to be necessary; (9) the requirement that the flag State or organizations having authority over a foreign fishing vessel take such measures as are necessary to ensure compliance by such vessel with the relevant fisheries laws of the Parties; and (10) such other terms and conditions as the Parties may from time to time consider necessary.

Even the possibility of creating a centralized licensing system for foreign vessels fishing within the region was envisaged, FFA's assistance was sought in the design of the regional logbook system and in administering the agreement, and an annual meeting of Parties to the Nauru Agreement (PNA) was stipulated in conjunction with annual meetings of the Forum Fisheries committee.

At the 1982 PNA meeting in Honiara, an implementing subsidiary agreement was signed that contained a critical institutional innovation. This was the concept of the Regional Register, which worked in the following way. To be licensed by any of the parties, a foreign fishing vessel had to be inscribed on the regional register and be in good standing. This register was to be maintained by FFA, and later its scope was expanded to cover the entire region. Maintenance of good standing required compliance with the laws of the parties in whose zones the vessel fished. Removal of any vessel for cause from the regional register required the concurring votes of all parties that the vessel was banned from fishing in all the zones of all the Parties until withdrawal of all outstanding charges by the party or parties entering the complaint.

Initially, this innovation was strongly resisted, especially by Japan and the United States, on the grounds that it amounted to blacklisting and, as such, was discriminatory and illegal. The Pacific Island states, however, did not back down, and later both Japan and the United States came to support the concept of the regional register since they perceived that it gave them more effective control over their own fleets.

The Nauru Agreement established the legitimacy of subregional groupings based on common interest fisheries. Another potential grouping was seen to be the Cook Islands, Tuvalu, and French Polynesia regarding the Korean and Taiwanese fleets that in the early 1980s fished in those zones. Such an agreement did not materialize, but two others did: one in the form of a Polynesian Agreement between the ATA, the Cook Islands, Tuvalu, and Niue; and the other in the form of the three Micronesian entities (FSM, to the Marshall Islands, and Palau) and the ATA. The Polynesians later became very dissatisfied with their arrangement since very little fishing activity occurred in their zones while most ATA-member activity occurred in the Micronesian zones.

From the point of view of the ATA, an agreement with the Pacific Island states was moderately desirable in 1980 because the U.S. fishery was just on the cusp of a momentous shift in the Eastern Tropical Pacific fishery as a result of a combination of political conflict with Mexico, Ecuador, Peru, and other Latin American states and the changing economics of the fishery as seen by U.S. processors. Consequently, a few pioneers from the U.S. fleet were interested in moving their base of operations to the Western Pacific. The U.S. government was also subsidizing this shift by sponsoring trial fishing efforts in the Western Pacific for several years using Saltonstall-Kennedy funds under the former Pacific Tuna Development Foundation. These funds amounted to about $1.2 to $1.5 million per year.

On the other hand, seen through the eyes of the Micronesians two reasons were strong incentives for making agreement with the ATA their highest priority at that point. One reason was that since the Micronesians were then involved in a difficult series of long-running negotiations with the United States concerning the termination of the Trusteeship, an agreement with the ATA would remove one potentially serious source of conflict with the U.S. Congress. In addition, of the available candidates (the United States, Korea, and Taiwan), only the Americans would be a credible threat to the Japanese monopsony, which had to be broken before the value that Pacific Island states would receive from their EEZs would increase. Since the fishing fleet was only reluctantly turning toward the Western Pacific, it would have to be enticed. The initial incentive could only be an access agreement comparable to the ones with Japan but *ipso facto* on terms that were less than the island states hoped to achieve.

No subsidiary agreement relating to the level of access fee under the Nauru Agreement was yet in force, but the region had been aiming at 5 percent of the landed value of fish caught in the region as a whole but landed in Japan. Japan was stoutly resisting this figure and was holding actual value to between 2.5 and 3.0 percent. The reasoning behind the Micronesian position was that a two-year agreement with ATA would draw in many more U.S. vessels than the eight operating in the region by 1980. The expectation was that these vessels would do well and constitute the beacon for the others to follow. Even though settlement would have to be on terms below what the parties then sought, agreement with the ATA would allow them to drive up the 1981 price to Japan, and then they would be able to make up lost ground on the renewal of the arrangement with the ATA in 1982.

Events occurred largely in conformity with these expectations, but it proved not to be possible to make up as much ground with ATA in 1982 as had been hoped. This was a source of irritation on the part of political authorities, which was intensified by the behavior of particular vessels in the U.S. fleet between 1982 and 1985. Several U.S. vessels were seized by island-state authorities for illegal fishing, particularly in PNG, the Solomon Islands, and FSM. The conflicts with PNG and the Solomon Islands were especially acrimonious, with the United States issuing a threat of embargo against PNG and actually imposing an embargo against the Solomon Islands.

For the second agreement with the ATA, the Marshall Islands dropped out and was replaced by Kiribati. Discussions were also held with PNG to constitute a foursome, but all difficulties could not be resolved before the start of negotiations with ATA. Two results of bringing the ATA into the region are noteworthy. The first is that Japan was no longer the only buyer. They remained so only for the sashimi trade, but soon internal splits on the Japanese side between the large fishing associations and the trading companies opened up substantial competition. The second is that continuing conflicts between island states and the United States over the behavior and subsequent seizures of U.S. vessels became acute. In the face of this conflict, the Soviet Union made its appearance into the tuna fisheries of the Western Pacific by making an agreement first with Kiribati and then with Vanuatu.

In response to the Soviet appearance, Australia and New Zealand orchestrated substantial pressure on the United States, including visits to Secretary of State George Schulz by all island-state foreign ministers. The United States then agreed to negotiate a regional agreement with FFA members that was concluded in 1987 at a price tag of $50 million over five years with only $10 million coming from industry. This was in effect a package deal combining commercial access fees with government foreign aid. The region then attempted to impose the same approach on Japan but so far without success. The U.S. treaty was renewed for a further five years in 1992.

It is appropriate at this juncture to say a word as well about the principal U.S. negotiator, who played a large role in the events related above up to but not including the regional agreement between the United States and the Pacific Island states. That person is August J. Felando, then general manager of the ATA, who led his organization through a difficult economic transition and whose association also made substantial contributions to the region.

However, as in a classical Greek tragedy, several member-owners of the ATA behaved in a manner that outraged political authorities in the region. When these vessels were caught fishing illegally, the owners rejected the jurisdiction of the island states and called on the U.S. government to invoke the import ban provisions of the Magnuson Act. The U.S. government on each occasion responded as the ATA wished. At that time the ATA was highly influential both with the U.S. Congress and with the Reagan administration, to whose election they had made major contributions. In addition, Felando was personally exceedingly well connected to the upper echelons of the Reagan administration and consequently had access to the president on issues affecting the ATA.

Over the period 1980 to 1984, as ugly incidents with U.S. vessels in the Western Pacific became more frequent, Felando adopted an increasingly hard-line confrontational stance with island authorities, eventually to the point where he became *persona non grata* to all island-state governments at the highest levels. When, therefore, the U.S. government determined on a basic shift in policy and the negotiation of the first regional agreement, the island-state authorities made it clear to Secretary of State George Shulz that they would not deal with Felando.

The fourth and last event portending major shifts in the political and operational landscape in the 1980s was the emergence of a significant Taiwanese purse-seine fleet in the early 1980s. While this fleet most probably was built with Taiwanese capital, it did receive Japanese financial and technical aid but not from the Japanese government. Instead, the impetus for the fleet construction and the financial and technical assistance came from large Japanese trading companies that wanted cheaper sources of supply to offset rising costs in Japan and thereby to boost fish sales in supermarkets as opposed to the traditional small-family fish shops controlled by large fishing companies and fishing associations. A similar pattern with the Korean long-line fleet had in fact occurred in the 1960s.

The same economic shift that generated large-scale U.S. disinvestment from tuna fleets and canneries based in the United States was here making itself felt in the Japanese context, but the political effect was to erode the position of both the large fishing associations and the Japanese government, whose fisheries agency maintained strict control over the licensing of all fishing vessels landing their catches in Japan and allocated areas of the world ocean in which different types of Japanese fishing vessels could fish.

By one stroke, the trading companies removed themselves from under the control of large fishing associations by stimulating the creation of fleets that were not under the control of the Japan Fisheries Agency, did not land their catch in Japan, but either landed their catch in Thailand for cheaper processing or air-freighted it to Japan. This lesson was not to be lost on the small long-liners in Japan, who through joint ventures in the region could realize similar arrangements with the trading companies, this time for the sashimi market. All of these cascading events would, in time, significantly weaken the hand of the traditional coalition of large fishing associations and the Japanese government relative to fishing operations in the Western Pacific.

Between the years 1980 and 1986 enormous changes occurred in the fishery and the economics of the fishery in the Western Pacific. In 1980, as we have seen, there were twenty-four purse-seiners operating in the region, of which only eight were

U.S. vessels. By 1982 there were 102 seiners, of which sixty-one were U.S. vessels, along with thirty-two Japanese, four Korean, two Taiwanese, two Mexican and Spanish joint-venture vessels, and two Philippine vessels. After 1984, U.S., Mexico and Spanish, and Philippine fishing efforts in the region declined, but the Japanese held steady, and Taiwan and Korea increased rapidly, the latter with private U.S. financial and technical assistance.

As the skipjack catch increased rapidly between 1980 and 1984, American Samoa and Guam increased in importance as transshipment points and soon became significant bottlenecks. American Samoa also became a processing bottleneck, while at-sea transshipment emerged to relieve the transshipment bottlenecks. Far more important, from 1985 to 1986 U.S. industry began a full-scale disinvestment from both fleet and canneries, keeping only American Samoa and Puerto Rico for the Atlantic fleet. The processing industry shifted to Thailand and the Philippines based initially on U.S. capital, so that transshipment problems no longer were significant for the purse-seine fleet. U.S. capital was later replaced by local investors in both countries involved.

In the 1990s particularly, the split between the trading companies and the large fishing associations in Japan allowed PNA members to make new kinds of deals with small Japanese long-liners fishing essentially for the trading companies and transshipping their catch by air into the Japanese market. Some island states could thereby increase the value derived from access fees by adding ship chandelling and bunkering as well. The first 5 percent access fee agreements with were Japan achieved by 1991.

Increasing Institutional Capacity in FFA: Leadership Changes

Leadership changes, which occurred between 1982 and 1983, constitute the final pieces of the puzzle and explain the shift in performance from failure to success. Both the director and the deputy director of FFA were replaced, and the new team, consisting of Philip Muller of Western Samoa and Les Clark of New Zealand, combined resources in a highly effective manner to make FFA and the system of cooperation it desired greater than the sum of its parts. Together they breathed life into the early operational objectives of FFA and secured strong extrabudgetary support from Australia, New Zealand, Canada, and the FFA's EEZ Program, which was financed by Norway.

The new leadership mobilized the Secretariat into a cohesive unit focused on four objectives: (1) continuing to harmonize coastal-state policies concerning access conditions; (2) assisting member states engaged in access negotiations; (3) assisting member states to develop local capabilities in harvesting, processing, and marketing of fish; and (4) developing the required database of catch and effort on a region-wide scale and collecting and disseminating information on annual negotiating experiences of member states.

By 1984 the importance of FFA to member-state access negotiations was critical. The Secretariat became indispensable by collecting and disseminating information on price indices in the Japanese market. In addition, it created a pooled database on catch-per-effort statistics in which the information contributed by each country could be shared only by the express permission of that country. But since each member state had an identical interest in the most comprehensive database on the broadest scale, individual contributions and permissions granted were never a problem.

FFA also provided real-time support for each negotiation to FFA member states both by seconding staff to all negotiating teams and being responsible for preparing the typewritten reports and performing calculations on the consequences of different options in real time with phone or fax links to Honiara. These methods of assistance were highly valued in the complicated and technical series of negotiations accompanying the introduction of the per-vessel and per-trip system to replace the lump-sum system of access fee payment after 1982. By being present in all negotiations and maintaining the record, the Secretariat thereby became the collective institutional memory that reinforced the institutional learning gained on negotiating access agreements. Finally, aware of the loss of negotiating capacity over time in member states as the original cohort was promoted, retired, or died, the Secretariat systematically arranged for the training of a second generation of negotiators in 1991.

Conclusions

In summary we argue that improving island-state capabilities to negotiate access conditions, reducing uncertainty in the knowledge base, developing an epistemic community with tight links penetrating both national and regional political levels, improving on coalition structure through the introduction of the Nauru Group, breaking the Japanese monopsony and forcing the United States into a sustained regional arrangement, creating the regional register, and improving the leadership capacity of FFA were all additive steps of enormous strength and resilience (see table 5.3). In fact, achieving all of this simultaneously moved the FFA's performance from the weak, ineffective category into the highly effective category, *even though it was stuck with a Convention that created only a level 2 type of organization.*

We argue, therefore, that the corrective steps taken amounted to significantly increasing the problem-solving capacity of the organization by expanding institutional capacity, expanding the power of the organization via the Nauru Group and the regional register, and recruiting a leadership structure of exceptional skill in political engineering.

The FFA, in fact, exploited the substantial room that existed for internal collaboration by focusing on approaches to the setting of fee levels, administrative provisions for the operations of foreign fleets, facilitating cooperation on surveillance and enforcement, facilitating annual exchanges of catch-per-effort data, and zealously facilitating exchanges of information on foreign violators.

At the beginning, while growing urbanization and cash incomes increased commercial demand for fish, total indigenous demand was still very limited. Island states focused for their own use essentially on inner- and outer-reef resources, plus the near-shore pelagics. Even in the 1980s collective local effort was small and rapidly growing to commercialization only in a few places, like Kiribati, the Solomon Islands, and parts of FSM. The major tuna resources were vast so that one did not find any significant conflict between coastal states and DWFNs. The region also did not suffer from proliferating incompatible development objectives in the tuna fisheries.

In this context, FFA assisted in the rapid development of small, competent national fisheries administrations; sensible in-shore fisheries development programs; a well-developed regional framework for managing foreign fleets; standardized regional logsheets and an improved database; and assistance to national fishing industries (Clark 1989). In addition, mindful of the need to maximize value and not catch in the purse-seine fishery, the Secretariat led the way in organizing a high-level consultation as early as 1985 to begin discussing the need for imposing limits on the growth of purse-seine fishing effort (Forum Fisheries Agency 1985).

Let us therefore explicitly answer the questions for determining effectiveness in the regime-implementation phase.

Expert Judgment: Did the Agreement Solve the Problem?
The answer to this question, as we have shown, is yes but not at first. Between 1979 and 1982 the FFA was largely weak and not very effective, but major developments were in the offing with the simultaneous development of the Nauru Group, the ATA agreements that broke the Japanese monopsony, and the creation of the Regional Register. Increases in institutional capacity by changing the leadership structure after 1982 complete the shift. The problem never changed: it was solved.

Table 5.3a
Summary scores: Chapter 5 (FFA)

| Regime | Component or Phase | Effectiveness | | Level of Collaboration |
		Improvement	Functional Optimum	
FFA	1976–1979	Low	Low	Intermediate
FFA	1980–1995	High	High	High

Political Judgment: Was the Agreement the Maximum Feasible Within the Institutional Setting?

The answer to this question must be no since the delegates creating the FFA in 1979 produced only a level 2 organization, which by itself would not have been able to solve the problem they faced. One can say, however, that the structure created did not stand in the way of evolutionary growth, which did produce the maximum feasible institutional design and performance.

Substantive Measurement: Are Conditions Improved as Much as Were Expected, at Costs That Were Projected, Without Producing Involuntary Losers?

The answer to this question is yes. The Pacific Island states jointly implemented changes in the Law of the Sea covering tuna in their respective zones. In doing so, they built the necessary management infrastructure, made major innovations in cooperative regional surveillance and enforcement, substantially boosted their negotiating capability and effectiveness, reduced both empirical and theoretical uncertainty affecting the stocks at issue, began to put in place ahead of time the policies required for effective conservation, and achieved a 5 percent of landed value return on the resource from Japan and substantially more than that from the United States. This is an enviable record that cannot be matched by any other regional fisheries management commission, including the Commission of the European Union.

This favorable assessment covers the period largely up to the early 1990s. At the same time, beginning in the late 1980s a desire for the island states themselves to get deeply into tuna harvesting, processing, and export trade raises some very troubling questions because this process proliferates incompatible development objectives that could erode the high level of cooperation that has been achieved. The us-versus-them character of the situation could thereby be transformed into a

Table 5.3b

Problem Type		Problem-Solving Capacity			
Malignancy	Uncertainty	Institutional Capacity	Power (basic)	Informal Leadership	Total
Mixed	High	Intermediate	Laggard	Intermediate	Intermediate
Mixed	High	High	Laggard	Strong	Intermediate

predominantly competitive relationship among Pacific Island states experiencing differential market access. Moreover, getting in only on the raw material end of the process has distinct economic disadvantages. All of these questions have been explored most instructively in considerable detail by Schurman (1998).

Would the Problem Have Been Worse Without the Regime?

The answer to this question is unquestionably yes. Without the regime, each island state would have been left to face the DWFNs on its own. The outcome is not hard to imagine under those conditions.

However, we return to the point made in the introduction to this chapter. While this was a malign problem as between island states and DWFNs, among island states it was a benign problem since each stood to gain from cooperation far more than would have been possible without it. Consequently, we have regime building and development in a cocoon created by changes in the Law of the Sea, a series of very large contiguous EEZs covering at least 70 percent of the stocks being targeted, an advanced regional infrastructure including a relatively dense network of associations and frequent fact-to-face meetings and rather large gains to be made from cooperation. This constellation of factors constitutes a veritable shower of selective incentives. Consequently, the Pacific Island state coalition existed in a highly favored situation, both naturally and politically. To the coalition's credit, it had the vision to see this situation and the leadership to capitalize on it.

Notes

1. Annex I, below, was actually compiled by Gordon Broadhead, Dayton L. Alverson, and August Felando (then general manager of the American Tunaboat Association), all of whom were U.S. nationals. The list therefore shows, to some extent, how influential the U.S. delegation was on parts of the draft text.

Annex I. Highly migratory species

1. Albacore tuna: *Thunnus alalunga*
2. Bluefin tuna: *Thunnus thynnus*
3. Bigeye tuna: *Thunnus obesus*
4. Skijpjack tuna: *Katsuwonus pelamis*
5. Yellowfin tuna: *Thunnus albacares*
6. Blackfin tuna: *Thunnus atlanticus*
7. Little tuna: *Euthynnus alletteratus; Euthynnus affinis*
8. Southern bluefin tuna: *Thunnus maccoyii*
9. Frigate mackerel: *Auxis thazard; Auxis rochei*
10. Pomfrets: Family *Bramidae*
11. Marlins: *Tetrapturus angustirostris; Tetrapturus belone; Tetrapturus pfluegeri; Tetrapturus albidus; Tetrapturus audax; Tetrapturus georgei; Makaira mazara; Makaira indica; Makaira nigricans*
12. Sailfishes: *Istiophorus platypterus; Istiophorus albicans*
13. Swordfish: *Xiphias gladius*
14. Sauries: *Scomberesox saurus; Cololabis saira; Cololabis adocetus; Scomberesox saurus scombroides*
15. Dolphin: *Coryphaena hippurus: Coryphaena equiselis*
16. Oceanic sharks: *Hexanchus griseus; Cetorhinus maximus*; Family *Balaenidae*; Family *Alopiidae; Rhincodon typus*; Family *Carcharhinidae*; Family S*phyrnidae*; Family *Isurida*
17. Cetaceans: Family *Physeteridae*; Family *Balaenopteridae*; Family *Balaenidae*; Family *Eschrichtiidae*; Family *Monodontidae*; Family *Ziphiidae*; Family *Delphinidae*

2. Ramp was then serving as legislative counsel to the Congress of Micronesia, a member of the Micronesian Delegation at UNCLOS III, and a participant in the Oceania Group.

3. The Kiribati position may also have been influenced by the fact the United States and the United Kingdom (on behalf of the Gilbert Islands) were then engaged in a negotiation on the disposition of the U.S.-claimed Line Islands.

4. SPEC (1979, Info. 4, Suva, 28.12.79).

5. The author must acknowledge that from January 1979 to September 1983 he served as the Joint Appointee of the Micronesian Maritime Authority (MMA) Federated States of Micronesia and that from 1981 to 1993 he served as the chief negotiator of foreign fisheries agreements of the MMA.

References

Burke, William. 1994. *The New International Law of Fisheries* (pp. 199–254). Oxford: Clarendon Press.

Clark, Les. 1989. "Trends and Implications of Extended Coastal State Sovereign Rights for the Management and Development of Fisheries: The West Central and Southwest Pacific." In Edward L. Miles, ed., *Management of World Fisheries: Implications of Extended Coastal State Jurisdiction* (pp. 201–209). Seattle: University of Washington Press.

Crossland, J., and R. Grandperrin. 1979. *Fisheries Directory of the South Pacific Commission Region* (Noumea, New Caledonia: South Pacific Commission).

Doulman, David J. 1987. "Fisheries Cooperation: The Case of the Nauru Group." In David J. Doulman, ed., *Tuna Issues and Perspectives in the Pacific Islands Region* (pp. 257–278). Honolulu: Doulman.

FAO. 1975. *Yearbook of Fishery Statistics, 1974: Catches and Landings*, vol. 42, (Rome: FAO Dept of Fisheries).

FAO. 1978. *Yearbook of Fishery Statistics, 1977: Catches and Landings*, vol. 44, (Rome: FAO Dept of Fisheries).

Forum Fisheries Agency. 1985. *Report of Proceedings: High-Level Meeting on Regional Co-operation in Fisheries Management and Development, Honiara, Solomon Islands, March 26–29*. Honiara: Forum Fish.

Fujino, K. 1967. "Review of Subpopulation Studies on Skipjack Tuna." In *Proceedings of the Forty-seventh Annual Conference of the Western Association of State Game Fish* (pp. 349–371).

Fujino, K. 1970a. "Range of the Skipjack Tuna Subpopulation in the Western Pacific." In K. Sugawara, ed., *Proceedings of the Second Symposium on Results of Cooperative Studies of the Kuroshio and Adjacent Regions, Tokyo, September 28–October 2* (pp. 373–384).

Fujino, K. 1970b. "Skipjack Tuna Subpopulation in the Western Pacific." In J. C. Marr, ed., *The Kuroshio: A Symposium on the Japanese Current* (pp. 385–393). Honolulu: East-West Center Press.

Gubon, Florian. 1987. "History and Role of the Forum Fisheries Agency." In David J. Doulman, ed., *Tuna Issues and Perspectives in the Pacific Islands Region* (pp. 245–256). Honolulu: East-West Center.

Joseph, James, Witwold Klawe, and Pat Murphy. 1979. *Tuna and Billfish: Fish Without a Country*. La Jolla, CA: Inter-American Tropical Tuna Commission.

Kearney, Robert E. 1976. "A Regional Approach to Fisheries Management in the South Pacific Commission Area." Paper presented at the South Pacific Forum Meeting on the Law of the Sea, Suva, October 13–14.

Kearney, Robert E. 1977. *The Law of the Sea and Regional Fisheries Policy*. Decasional Paper No. 2, April. Noumea, New Caledonia: South Pacific Commission.

Kearney, Robert E. 1979. *An Overview of Recent Changes in the Fisheries for Highly Migratory Species in the Western Pacific Ocean and Projections for Future Development*. South Pacific Bureau for Economic Cooperation, June. Doc. #SPEC(79)17.

Kearney, Robert E. 1983. *Assessment of the Skipjack and Baitfish Resources in the Central and Western Tropical Pacific Ocean: A Summary of the Skipjack Survey and Assessment Programme*. Noumea: South Pacific Commission.

Kearney, Robert E. 1991. "Extremes in Fish Biology, Population Dynamics, and Fisheries Management: Pacific Skipjack and Southern Bluefin Tuna." *Review in Aquatic Sciences* 4(2–3): 289–298.

Kent, George. 1980. "South Pacific Fisheries Diplomacy." *New Pacific* (January–February): 22–27.

Ramp, Frederick L. 1977. "Regional Law of the Sea: A Proposal for the Pacific." *Virginia Journal of International Law* 18(1) (Fall): 121–132.

Schurman, Rachel. 1998. "The South Pacific Island Countries in the 1990s." *Ocean Development and International Law* 29(4) (October–December): 323–338.

Sharp, Gary D. n.d. "Studies of Pacific Ocean Skipjack Tuna (*Katsuwonus pelamis*) Genetics Through 1978." Paper prepared for FAO, Fishery Resources and Environment Division.

Sharp, Gary D., and William P. Kane. n.d. "Biochemical Genetic Comparison and Differentiation Among Some Eastern Atlantic and Pacific Ocean Tropical Tunas." Paper prepared for FAO, Fishery Resources and Environment Division.

South Pacific Economic Commission (SPEC). 1979. *Meeting of Forum Government Officials on Regional Fisheries Policy.* Doc. SPEC (79) FA-3, January 30–February 1.

Sund, Paul N., Maurice Blackburn, and Francis Williams. 1981. "Tunas and Their Environment in the Pacific Ocean: A Review." *Oceanography and Marine Biology Annual Review* 19: 443–512, esp. 464–473.

Suva Declaration of October 15, 1976.

6

The Vienna Convention and Montreal Protocol on Ozone-Layer Depletion[1]

Jørgen Wettestad

Overall Performance: Considerable Achievements But Challenges Ahead

The stratospheric ozone layer shields life on earth from potentially disastrous levels of ultraviolet radiation. In 1974, the theory was launched that a group of man-made chemicals called chlorofluorocarbons (CFCs) could destroy the ozone layer.[2] By the late 1970s, the ozone-depletion issue was placed on the agenda of international environmental and scientific organizations like the World Meteorological Organization (WMO) and the United Nations Environmental Program (UNEP). Negotiations on a binding international political instrument were started in 1981, and in 1985 the Vienna Convention on the protection of the ozone layer was signed by twenty countries plus the European Community.[3] Progress in the following decade was quite impressive, not least so when we look at regulatory development. The 1985 Vienna Convention was a loose framework convention. The 1987 Montreal Protocol was a big step forward with regard to binding requirements and specificity. The protocol included specific measures and timetables for reducing production and consumption of ozone-depleting substances. The London and Copenhagen meetings of the early 1990s and the 1995 Vienna meeting speeded up and specified these obligations in several ways, both in the form of accelerated timetables and the addition of new substances. In terms of national behavioral impact, available national compliance data indicate that the required behavioral changes are being implemented, although more faithfully in the North than in the East and South. However, reporting problems and insufficient national implementation knowledge add a note of soberness and caution to this otherwise convincing success picture. With regard to the adequacy of these measures in relation to perceived problem-solving requirements at the time, an overall picture of high correspondence emerges. The evolution of science and reduction of uncertainty were steadily followed by increasingly stringent political decisions within a very short time span. Moreover, the first signs of effective regulations can already be detected, as growth rates in the atmospheric concentrations of controlled substances like CFCs and halons are slowing down.

The main challenge ahead lies in the latent fundamental asymmetry in every global regulatory effort—between the rich countries of the North that can afford greener, more expensive alternative substances and the poorer countries of the South that would like to keep present and future development costs to a minimum. Tellingly, the production and consumption of ozone-depleting substances (ODSs) in developing countries has increased steadily in the 1990s. This issue indicates that success so far does not ensure smooth sailing with regard to complete problem solving.

Brief Institutional Background

Before we substantiate the effectiveness assessment, let us briefly introduce the institutional background. This has, of course, developed over time in line with evolving scientific and political needs. However, a summary looks like this: on the scientific side, the Conference of Parties (COP) to the Vienna Convention established a Meeting of Ozone Research Managers, which meets every three years (every two years prior to 1993) and is composed of government experts on atmospheric research. This group reviews ongoing national and international research and monitoring programs and produces a report to the Conference of the Parties. Moreover, the Montreal Protocol has established three panels of experts, to be convened at least one year before each assessment—a Scientific Assessment Panel, a Technology and Economics Assessment Panel, and an Environmental Effects Panel. In addition, ad hoc groups on data reporting and destruction technologies have been in operation. On the political side, there is also a formal dual structure. The Conference of the Parties related to the Vienna Convention is held every three years (every two years up until 1993). The Montreal Protocol has an annual Meeting of the Parties (MOP). There is also a separate Implementation Committee, with representatives of ten parties, which was established in 1990. In addition, the Open-Ended Working Group of the Parties and the Bureau of the Montreal Protocol meet intersessionally to develop and negotiate recommendations for the MOP on protocol revisions and implementation issues. Secretarial functions are carried out by a Secretariat at the UNEP headquarters in Nairobi, which serves both the Vienna Convention and the Montreal Protocol. In 1990, a Multilateral Fund was established, financed by developed countries, to support phaseouts in developing countries. The Fund is governed by a fourteen-party Executive Committee and has its own Secretariat in Montreal.

Substantiating the "High Effectiveness So Far" Assessment

The impressive regulatory development can be specified and summed up in table 6.1.[4] In relation to these requirements, the developing countries generally have a ten-year grace period in terms of implementation.

Moving on to the question of *relative improvement*, given the very loose 1985 Vienna Convention baseline (as shown in table 6.1), the international regulative situation changed quite drastically over a seven-year time period—which, for instance, is roughly the same period of time the acid-rain and long-rang transport of air pollutants (LRTAP) participants used to come up with the first sulfur protocol. But what has this swift international process led to in terms of national behavioral impact—the "proof of the effectiveness pudding"?

Turning to the compliance picture, progress is on the whole satisfactory.[5] By 1995, the global production and consumption of CFCs, halons, other fully halogenated CFCs, methyl chloroform, and carbon tetrachloride had been reduced by about 75 percent from 1986 baseline levels—with the best data available on the two first groups of substances. Moreover, although data are shaky, the global production and consumption of HCFCs seems to have increased threefold from the 1989 baseline level. With regard to methyl bromide, data are even more uncertain, and reporting obligations are of a quite recent nature. With this caveat, reported data show a slowly declining trend, while estimates suggest a slight increase.

In relation to this overview picture, there are marked regional differences. Compliance in the industrialized countries is very good overall, despite slow progress with regard to HCFCs and methyl bromide and some scattered instances of noncompliance. The situation in the countries with economies in transition (CEITs) is far more problematic. Several countries, most important Russia, have been unable to meet the CFC and halon phaseout targets of the industrialized countries. In fact, they have themselves invoked the regime's noncompliance procedure, and a package including revised targets and global environment facility (GEF) assistance has been established (see Greene 1996). Russia has also been involved in illegal CFC production and smuggling to the West (Greene 1996). Production and consumption in the developing countries have increased substantially. In 1995, for instance, developing-country CFC consumption was nearly 40 percent higher than in 1986, and the general consumption of ozone-depleting substances was larger than in

Table 6.1
Overview of regulatory development

Substances (baseline)	Montreal 1987	London 1990	Copenhagen 1992	Vienna 1995	Montreal 1997
CFCs 11, 12, 113, 114, 115 (1986)	Mid-1989: freeze Mid-1993: −20% Mid-1998: −50%	Mid-1989: freeze 1995: −50% 1997: −85% 2000: −100%	Mid-1989: freeze 1994: −75% 1996: −100%	No change	No change
Halons 1211, 1301, 2402 (1986)	1992: freeze	1992: freeze 1995: −50% 2000: −100%	1992: freeze 1994: −100%	No change	No change
10 other CFCs (1989)		1993: −20% 1997: −85% 2000: −100%	1993: −20% 1994: −75% 1996: −100%	No change	No change
Carbon tetrachloride (1989)		1995: −85% 2000: −100%	1995: −85% 1996: −100%	No change	No change
Methyl chloroform (1989)		1993: freeze 1995: −30% 2000: −70% 2005: −100%	1993: freeze 1994: −50% 1996: −100%	No change	No change
HCFCs (1989 plus 3.1% of CFC consumption in 1989)			1996: freeze 2004: −35% 2010: −65% 2015: −90% 2020: −99.5% 2030: −100%	Baseline: 1989 plus 2.8% of CFC consumption in 1989	No change
H-BFCs			1996: −100%		No change
Methyl bromide (1991)			1995: freeze	1995: freeze 2001: −25% 2005: −50% 2010: −100%	1995: freeze 1999: −25% 2001: −50% 2003: −70% 2005: −100%

Note: This table is based on Gehring (1994) and Oberthur (1997b).

industrialized countries. Moreover, production and consumption have been particularly dynamic in the newly industrialized Asian countries. China alone accounts for half of the developing countries' consumption and even more of the production. However, the developing countries are in a special situation as they were given a ten-year grace period in the Montreal Protocol.[6] Still, as the activity of the Multilateral Fund has apparently had moderate influence on this increasing trend so far, the uncertainty related to the further development of implementation efforts in developing countries is the main black cloud on the ozone-regime horizon.[7]

How, then, can we interpret the many positive developments so far, especially in the industrialized countries? Do we have convincing evidence that the reported behavioral changes have been caused by the developing ozone regime—or would they have happened anyway? Although there is a general lack of detailed national-implementation and regime-influence studies so far, the following factors point toward the tentative conclusion that the ozone regime can be credited for a substantial portion of the reported behavioral changes, at least with regard to CFCs.[8] First, no "natural" phaseout due to general economic development could be expected from the late 1980s on, as the CFC production trend from 1960 to 1985 was steadily increasing, although with some natural market variations (see Benedick 1991, 27). Second, successful unilateralism either by the United States or the European Community seems unlikely; unilateral measures adopted by the United States in 1978 and the European Community had limited effects (in fact, CFC production increased from 1980 to 1985). Moreover, it is not plausible that the United States unilaterally provide (billion-dollar) incentives to persuade the rest of the world to adopt its regulations. In addition, U.S. commitment to international CFC control was highly uneven.[9] Third, and equally important, without binding international measures, there would have been fewer incentives to look for substitutes for the ozone-depleting substances; in other words, the Montreal measures and later amendments radically accelerated the industry's search for substitutes. After having received the crucial international regulatory signals, market forces pretty much drove the implementation processes (Parson 1996, 24–25). A fourth factor is the regime's contribution to general knowledge improvement and related national reassessments of the economic importance of the targeted ozone-depleting substances.

Let us then turn to the question of *distance to collective optimum*. As actual problem-solving is hard to assess due to several reasons, what can be said about the evolving match between scientific knowledge and advice and the international policy measures taken in response? Even if the measures agreed have failed to solve the problem, were they the best possible, given the knowledge situation at the time? Although, of course, these are complex assessments to make, an overall picture of high correspondence over time emerges.[10] Scientific progress and reduced uncertainty were steadily followed by increasingly stringent political decisions within a short time span. For instance, the 1987 Montreal Protocol can clearly be interpreted as a response to the discovery of the 1985 Antarctic ozone hole and alarming results from a comprehensive WMO/NASA (U.S. National Air and Space Administration) research program. Moreover, the 1990 London revisions were decided on against the background of confirmations by the Ozone Trends Panel of the "ozone hole" and "strong indications" that man-made chlorine compounds were primarily responsible for the observed reduction in ozone. Similarly, the 1992 Copenhagen amendments were based on conclusions from both the Scientific and not least the Economic and Technological Assessment Panels (Gehring 1994, 307–313). As can be noted in this brief overview, the regime played an increasingly important role over time as the main framework for knowledge improvement.

Is there an imminent solution to the ozone-depletion problems? Like most other environmental problems, there are substantial time lags between reductions in ozone-depleting substance emissions and ozone-layer effects. According to scientific assessment reports released in 1994, if countries stuck to their commitments under the Montreal Protocol, ozone depletion would continue to worsen until the end of this century. The peak total chlorine and bromine load in the troposphere was expected in 1994, with a stratospheric peak lag of about three to five years (NOAA, NASA, UNEP, and WMO 1994, 7–8). However, more recent information about higher than expected consumption of ozone-depleting substances in developing countries may mean a delay of this development (Oberthür 1997, 66–67). Meanwhile, the first indications of effective regulations can already be detected: there are declining growth rates in the atmospheric concentrations of controlled substances. Data from recent years clearly show declining growth rates of CFCs and halons. But as already indicated, it must be kept in mind that continued ozone-layer restoration is heavily dependent on the implementation of regulations that have been agreed on, not least by the developing countries.

Summing up, despite reporting problems and lacking national knowledge, the overall impression is that the international regulatory development has been impressive, and available data indicate that the required behavioral changes are largely being implemented by the industrialized countries, apparently greatly influenced by the international process. However, less impressive follow-up in the East and South must be noted. In terms of environmental problem solving, the mutual match between evolving knowledge and international policy responses has been significant, and the first atmospheric signs of effective regulations can be detected. Addressing implementation failures in the CEIT and increasing production and consumption of ODS in developing countries are the main challenges for upholding regime effectiveness in the years ahead.

How, then, has this development come about?

The Background for a Loose Framework Convention: High Uncertainty and Hesitant States[11]

The stratospheric ozone layer shields life on earth from potentially disastrous levels of ultraviolet radiation. Depletion of the ozone layer as a potential environmental problem appeared for the first time on the international agenda at the United Nations Conference on the Human Environment in Stockholm 1972. The Conference called for research on stratospheric transport (SST) and distribution of ozone. Then, in 1974, based on the work of two American groups of scientists—Stolarski and Cicerone, and Molina and Rowland—the theory was launched that a group of man-made chemicals called chlorofluorocarbons (CFCs) could destroy the ozone layer.[12] CFCs are a family of chemical compounds containing chlorine, fluorine, and carbon. Being nontoxic, noncarcinogenic, and nonflammable, CFCs were widely used in refrigeration and air conditioning (as blowing agents and aerosol propellants) and in solvents (for cleaning electronic parts). The disturbing ozone-layer-depletion theory led to the establishment of research and monitoring programs, conducted mainly by the United States in the mid-1970s. By the late 1970s, the ozone-depletion issue was placed on the agenda of international environmental and scientific organizations like the World Meteorological Organization (WMO) and the United Nations Environmental Program (UNEP).

How, then, can the ozone-depletion problems be characterized in the mid-1980s, in terms of knowledge and principal interests? And how can these initial problem characteristics shed light over the regime-creation phase and the establishment of the Vienna framework convention? On the knowledge side, the overall picture was one of high complexity. Moreover, the situation in the mid-1980s was still one of lacking empirical proof of ozone depletion. Measurements of ozone levels had been carried out for thirty years but had not yet demonstrated any statistically significant loss of total ozone. Evidence concerning harmful effects of ozone modification was also sparse. Hence, despite widespread U.S. acceptance, some scientists and the chemical industries questioned the validity of the basic theory. Efforts from the late

1970s and later to establish an international consensus by organizations like the WMO, UNEP, and the International Council of Scientific Unions (ICSU) were multiple. A comprehensive international research program had been launched in 1984, including 150 scientists from eleven countries. However, by 1985, an international consensus had failed to emerge. Technologically, experience from the 1978 spray-can ban in the United States and the industry's own research indicated that development of alternatives to CFCs was largely a question of the right market incentives, as several chief alternative gases had been identified. But no technological breakthrough was announced before 1985.

In terms of interests, as all states would be negatively affected by increased ultraviolet radiation (although the degree of vulnerability might vary somewhat), no nation could perceive any benefits from ozone depletion. Hence, with regard to environmental vulnerability, concepts like "winners" and "losers" and "pollution exporters" and "importers" were less relevant in this case. Moreover, on the regulatory side, CFCs were not critically important for industrial and economic development, either for countries or industries. Furthermore, promising substitutes were in the pipeline. Production and consumption were at this stage concentrated mainly in the industrialized countries, with about 35 percent of the total amount of CFCs produced and 30 percent consumed worldwide in both the United States and the European Community, respectively. Hence, at this initial stage, the ozone-depletion problems must overall be characterized as *moderately malign*. This is largely due to the situation being one of substantial intellectual uncertainty and complexity. Although CFCs were not critically important for industrial and economic development, there were potential competitive effects related to regulatory development, primarily between American and European producers of ozone-depleting substances.

In terms of major participants and their preferences, they were in this initial phase the European Community, the Toronto Group, and the United States. In addition, the USSR and Japan should also be mentioned. The Toronto Group was formed in 1983 by Canada, Finland, Sweden, Norway, and Switzerland and joined later that year by the United States. This group sought CFC reductions and a ban on nonessential use of CFCs in spray cans. However, the European Community was somewhat of a laggard in much of this regime-creation phase, together with the USSR and Japan. This role was played by the European Community for several reasons (see Jachtenfuchs 1990; Benedick 1991). However, it should be noted that the EC countries were divided on this issue. Belgium, Denmark, the FRG, and the Netherlands were increasingly inclined toward strong CFC controls (with only the FRG being a major producer). The United Kingdom, France, Italy, and Spain, all rather large producers, resisted stringent measures (Benedick 1991, 35). In the United States, the ozone-depletion problem had been a hot issue since the beginning of the 1970s. Still, the United States also experienced domestic political problems on this issue, as the antiregulatory Reagan government took office in 1981, and the new Environmental Protection Agency (EPA) administrator downplayed the ozone-depletion problems—complicating the U.S. negotiating positions. Hence, during the early 1980s, although being a definite scientific leader, the United States must still be characterized as something of a political laggard. This changed in 1983 and 1984, due among other things to a change of leadership in the EPA (Parson 1993, 38). Overall, against a background of scientific uncertainty and lack of empirical evidence, the factors briefly outlined above give us an idea of why the preferences among the main actors, the United States and the European Community, perhaps best can be described as somewhat hesitant at the time of the establishment of the Vienna Convention.

The Background for Rapid Regime Development: Reduced Uncertainty, Technological Progress, and Converging Preferences in the North[13]

The Regime Heyday: The Late 1980s and Early 1990s

Overall, uncertainty was gradually reduced beginning in 1985. Two months after the adoption of the Vienna Convention, Joe Farman and his research team from the British Antarctic Survey published their findings on the Antarctic ozone hole—the first empirical indication that the theories launched might be correct. Under Robert Watson's leadership, the 1986 WMO and NASA project was clearly important in terms of consensus building. However, the main breakthrough in terms of empirical evidence came shortly *after* the adoption of the Montreal Protocol. For instance, in March 1988, the Ozone Trends Panel released the results of a sixteen-month comprehensive scientific exercise. The panel concluded that the evidence "strongly indicates that man-made chlorine species are primarily responsible for the observed decrease in ozone" (Benedick 1991, 110). On the technological side, things developed quite quickly after the adoption of the Montreal Protocol. Four months after Montreal, several hundred industry representatives met in Washington to exchange information and stimulate research on alternatives to CFCs (Benedick 1991, 104). Moreover, on the basis of article 6 in the Montreal Protocol, a technology review panel was appointed consisting of over 100 international experts. The panel's 1989 synthesis report concluded that, based on the current state of technology, the five controlled CFCs could be reduced by over 95 percent by the year 2000. Moreover, the Economic Panel's part of this report concluded that the monetary value of the benefits would undoubtedly be much greater than the costs of CFC and halon reductions. But it was also acknowledged that developing countries would be less able to invest in this effort due to more immediate concerns such as food supply and economic development (Gehring 1994, 276–277; Skjærseth 1992, 34). Hence, a drastic cut in CFCs and halons seemed both technologically feasible and economically reasonable within a short period after the Montreal Conference.

Related to this development, perceptions of the magnitude of the problems changed. Mounting evidence suggested that increased ultraviolet radiation related to ozone-layer depletion induced skin cancer, cataracts, suppression of the human immune response system, and the development of some cutaneous infections such as herpes. Plants reacted adversely, and aquatic organisms would be negatively affected by increased ultraviolet-B radiation. This meant that potential regulatory benefits became increasingly specific, while costs were reduced in line with rapid technological progress. Hence, it has been argued that intellectual progress toward the end of the 1980s indicated that at least the industrialized countries in many ways faced a relatively benign cost-efficiency problem, where the main challenge was coordination and information perfection (Skjærseth 1992, 35). Still, this does not mean that the process leading up to the important Montreal Protocol break-through was uncomplicated. Following a domestic U.S. administrative battle between, on the one hand, the Office of Management and Budget and the Depart-ments of Commerce and Energy and, on the other hand, the EPA and the State Department, the latter coalition got support from the U.S. Senate, which passed a resolution calling for a 50 percent reduction and eventual phaseout. This was approved by President Reagan in the summer of 1987 and gained strong support from Canada and the Nordic countries, among others. Overall, the United States played an important part in the process leading up to Montreal, founded on threats of unilateral action and persuasion based on U.S. scientific, diplomatic, and politi-cal strength (see Parson 1993, 60; Benedick 1991, ch. 6). Meanwhile, discussions continued within the European Community, with Germany gradually moving toward the U.S. position but with the two other main CFC producers in the European Community—France and the United Kingdom—holding back. Only threats of a complete collapse of the Montreal negotiations and skillful entrepre-neurial diplomatic efforts, especially by the UNEP leader Mostafa Tolba, made the EC Commission cede its standpoint (see Benedick 1991, ch. 7).

After the Montreal Protocol and alarming signals from the Ozone Trends Panel, and following Du Pont's announcement of a CFC and halon manufacturing stop by the end of the century, the United States in March 1989 called for a complete phase-out of CFCs by the year 2000. Within the European Community, the U.K. stance was softening, due to pressure from environmental groups and Parliament, and the ICI (the United Kingdom's biggest CFC producer) was taking a more positive position toward strengthened regulations. At a meeting hosted by a greened Prime Minister Thatcher in March 1989, France gave in, and the European Community was able to support the United States by declaring it would eliminate CFCs completely by the end of the century. Hence, fifteen months after Montreal, the main discrepancies between the United States and the European Community had vanished. Moreover, although the developing countries at this stage were dissatisfied with distribution of finances and technology, they supported the ultimate objective of a total elimination of CFCs and halons.

The More Recent, Less Dynamic Period
The process of knowledge improvement has continued. Two subsequent rounds of assessment panels have been carried out—in 1991 to advise the 1992 Copenhagen meeting and in 1994 to advise the 1995 Vienna meeting. Overall, models have become more powerful, more substances have been identified, and there is growing empirical evidence (for instance, in the form of the ozone loss seasonally observed over Antarctica). However, despite the development of such a general-level consensus, ozone politics in the 1990s have not developed without strife and conflict. In connection with the 1992 Copenhagen meeting, there was little disagreement concerning an accelerated phaseout of already controlled substances (Gehring 1994). However, there was more conflict over the regulation of additional substances— hydrochlorofluorocarbons (HCFCs) and especially methyl bromide. Due to the role of this latter substance in agriculture (soil fumigation), the European Community and especially its Mediterranean members could not accept a U.S. proposal for a complete phaseout by the year 2000. Moreover, fierce protests came from Israel and several developing countries. With regard to the Vienna 1995 tenth anniversary meeting and the 1987 Montreal meeting, the issues of HCFCs and methyl bromide regulations have continued to be controversial.[14]

In general, the issue of developing countries and their special needs led to a more complex negotiating game again in the 1990s. In the mid-1980s, these countries produced and consumed only a small fraction (less than 10 percent) of the ozone-depleting substances compared to the OECD countries. Although there are variations also within the developing-country group, these countries envisioned substantial future increases in the production and use of the relatively cheap CFCs, and production and consumption limits could be perceived as potential barriers to their own economic growth, at least in a short-term perspective. Hence, production of CFCs in developing countries increased by 87 percent between 1986 and 1993, and exports increased seventeenfold. While the consumption of CFCs fell by 74 percent in industrialized countries during the same period, it increased by more than 40 percent in the developing countries (Brown et al. 1996, 68).

Problem-Solving Capacity: Strong Initial Leadership and Some Institutional Assets[15]

In terms of *level of collaboration*, the ozone regime has clearly coordinated action on the basis of explicitly formulated standards (in this case, protocols and amendments), but national action has been implemented solely on a unilateral basis. Moreover, national reports have formed the basis for some sort of centralized appraisal of effectiveness.

Turning to *entrepreneurial leadership*, this was strong in the crucial first regime-development phase from 1985 to 1990 (see Benedick 1991; Gehring 1994; Oberthur 1997b). Several types of leadership can be discerned, both on the intellectual and political sides. Intellectually, the American leadership in the ozone-depletion issue had its roots back in the mid-1970s. NASA played an important part in the Coordinating Committee on the Ozone Layer (CCOL) established in 1977. In 1984, a large international research program was launched by WMO and NASA, including 150 scientists from eleven countries. From the mid-1980s, several sources point to Robert Watson as a leading American figure within the scientific community engaged in ozone research. He had a keen awareness of the need to balance complicated requirements to the scientific venture in terms of national and industrial legitimacy, with scientific integrity and not least political usability.[16] Politically, with regard to national efforts in the negotiations, again the American efforts stand out in a field that also includes the Nordic countries. The United States provided strong leadership from early 1986 on (see Benedick 1991, ch. 5; Parson 1993, 60; Gehring 1994, ch. 6). Over sixty U.S. embassies were utilized in an information and influence strategy; diplomatic initiatives were closely coordinated with like-minded countries like Canada, New Zealand, Switzerland, and the Nordics; and the chief U.S. negotiator, Benedick, led several missions to European capitals (Benedick 1991, 55–59).

Moreover, the role of UNEP and its leader at the time, Mostafa Tolba, warrants some special attention. UNEP played an important part as arena for ozone negotiations from the early 1980s on. Tolba entered the negotiating scene in spring 1987, soon exerting considerable influence (see Benedick 1991, ch. 7). He could draw on at least three power sources: first, he was the head of the host organization; second, given the scientific and technological complexities of the ozone issue, he was himself a scientist; and third, he was clearly a charismatic person. Tolba utilized various leadership and negotiation strategies, institutional changes through small, shielded meetings of key delegations, as well as direct confrontations and "midnight theatrics" (Benedick 1991, 75). After Montreal, he continued to influence the process at least until the 1990 London meeting. In the 1990s, the leadership question has become more diffuse. There was probably a less pronounced need for strong leadership as the regulatory process acquired a stronger momentum of its own and, at least temporarily, entered more cooperative waters (see Skjærseth 1992, 46).

Let us then move on to some potentially important *institutional issues*. For instance, with regard to the issue of *decision-making rules*, the ozone regime contains some interesting elements. Although considerable weight is given to the principle of consensus, decision-making rules contain the formal possibility of decisions to be made by a three-fourths (Vienna Convention) or two-thirds (Montreal Protocol) majority and even to be binding for possible outvoted minorities. However, in practice, a kind of flexible consensus has been utilized at meetings within the regime, not allowing one single party to block the whole process. This must be seen in the light of the more general ozone decision-making context, characterized by the balancing of different interests and concerns both within the group of developed countries and between North and South. Hence, the practical effects of the majority voting clauses could stem from their symbolic and integrative functions or their "hidden stick if consensus fails" functions. Both effects need further investigation.

Turning to the *role of the Secretariat*, as indicated earlier UNEP was clearly one of the leading international organizations that focused on the depletion of the ozone-layer issue in the late 1970s and early 1980s. This was natural, given the global character of the issue. Hence, UNEP carried out the secretarial functions de facto from the early 1980s on. However, at the 1985 Vienna conference, disagreement arose on the secretarial issue, with scientists largely favoring WMO and diplomats and administrators favoring UNEP. Hence, the formal assignment was deferred to the first Meeting of the Parties (Parson 1993, 40). At this meeting, the secretarial question was formally settled in favor of UNEP. In terms of resources, UNEP as an organization is in a totally different league than what is common for international environmental regimes. However, the size of the Ozone Secretariat itself has not been so remarkable. The number of staff by March 1998 was five professionals and eight support staff. Turning to the financial situation, with regard to the Vienna Convention, the annual budget increased from around $0.3 million in 1994 to around $1.3 million for 1999; regarding the Montreal Protocol, the annual budget for the most recent years has been of the order of $3 million; and the administrative budget for the Multilateral Fund has been of the order of around $9 million (*Green Globe Yearbook* 1997, 93–97). Altogether, despite limited resources and staffing, the Ozone Secretariat has in several ways contributed to the progress within the regime and perhaps more so than envisioned in the regime-creation phase. The Secretariat has handled informal and formal meetings of the parties, provided treaty text drafts, convened review panels, and played an increasing role in matters of reporting and compliance issues. More generally, the UNEP connection has been a strength, especially in the early phases of regime development. However, we have heard much about the entrepreneurial leadership carried out by UNEP's leader, Tolba, in the initial phases of the regime. We know less about the interplay and relative importance of organizational and personal factors in this initial phase, and not least, we also have only limited knowledge about the more recent functioning of the Secretariat. A reasonable interpretation is that the main functions and role of the Secretariat have changed somewhat—from the more entrepreneurial role in the crucial, initial Montreal days to a still important, though more low-key and technical role in more recent years.

Given the high scientific uncertainty and complexity and the strong technological dimension characterizing this issue area in the mid-1980s, the regime could clearly play a potentially important role through the more systematic coordination and organization of scientific and technological activities. Hence, moving on to the *organization of the scientific and technological input*, as mentioned, the initial scientific process was dominated by U.S. scientific agencies. The organization history of the science and politics interface in this issue area goes back to the late 1970s, with the establishment of the Coordinating Committee on the Ozone Layer as an important step in 1977. Generally, before 1988 the ozone regime had a rather loose scientific-political structure, more in the form of a network.[17] However, as noted by Skodvin (1994, 85), the network was quite firmly connected. Throughout the process, many of the same scientific institutions and even the same scientists were mainly responsible for initiating and coordinating research, with the United States a dominant actor. Pursuant to article 6 of the Montreal Protocol, in 1988 the scientific-political complex was institutionalized in four panels of experts—Scientific, Environmental, Technological, and Economic Assessment Panels—and an integrated synthesis report was presented in 1989. In later rounds, combining the technological and economic aspects, the number of panels has been reduced to three main bodies. In addition, there are ad hoc groups on data reporting and destruction technologies (*Green Globe Yearbook* 1997, 97). The establishment of the panels implied broader participation and served to reduce the earlier marked American bias and increase the general geopolitical representativeness of the process (Skodvin 1994, 87). Moreover, the panels also served to broaden the area of research by including issue areas that had not been covered in a systematic manner earlier— for instance, the economic issues.

So how should we assess the ozone science and politics model? First, the impression is definitely that the ozone scientific and political complex has achieved a rather effective way of functioning in a short period of time. Second, it is not easy to say how much of the apparently well-functioning science-politics dialogue can be credited to the organizational model per se. As the panels have been regarded as forums "with the stamp of international objectivity and authoritativeness" (Parson 1993, 61), international institutional design seems to have mattered at least a bit. Moreover, effective two-way communication between the panels and the meetings of the parties seems to have been achieved, partly with the Open-Ended Working Group established in 1988 as mediator. Hence, it has been indicated that, after 1987 with institutionalization of scientists' roles in the assessment panels, scientists' influence over the negotiations has advanced their prior agenda-setting role to the exercise of substantial influence over certain aspects of the negotiated decisions (Parson 1993, 61). Referring to the process leading up to the London 1990 meeting, Gehring (1994, 277) maintains that "never before in the decade-long process of development of the international regime for the protection of the ozone-layer had the scientific and technical knowledge necessary as a foundation for political negotiations been prepared with similar care." Hence, and in conclusion, the ozone model has developed over time, with the formal establishment of the review-panel process in 1988 as an important watershed event. The organizational model has functioned well and may be an important institutional contributing factor to shedding light over the rather high effectiveness achieved so far.

Finally, there have been several *specific institutional mechanisms to increase participation*. First, there is the issue of *restrictions on trade with nonparties* to prevent nonparticipating countries from gaining competitive advantages. Such trade sanctions were to be based on article 4 of the Montreal Protocol (Control of Trade with Nonparties) and apply to both imports and exports. According to several sources, these clauses have worked effectively in terms of increasing state participation.[18] Moreover, there is the establishment of the *Multilateral Fund*, financed by developed countries to pay for developing countries' incremental costs inherent in meeting control obligations and some technology-transfer provisions.[19] The idea of a fund arose in the process leading up to the 1990 London Conference, related to the stricter phaseout control measures in the pipeline. After complicated negotiations, characterized as the most difficult issue in the entire 1990 treaty revision process (Benedick 1991, 152), a compromise financial mechanism was laid out in article 10 in the London Revisions. The Multilateral Fund, financed by the industrialized countries and based on the principle of additionality, was to be overseen by an Executive Committee with a balanced fourteen-member North-South representation, in cooperation with the World Bank, UNEP, and the United Nations Development Program (UNDP). The Executive Committee has a secretariat in Montreal, with a three-year operating budget of $7.6 million (drawn from the fund) and a nine-member professional staff. The first three-year budget of the interim multilateral fund was stipulated to be "up to" $240 million, with a 25 percent contribution from the United States and nearly 35 percent from the European Community (based on the U.N. assessment scale) (Skjærseth 1992, 43–44; Parson 1993, 50–51). At the eighth meeting of the parties, a replenishment of the Multilateral Fund at a level of $540 million was determined for the period 1997 to 1999 (Green Globe Yearbook 1997, 95).

In terms of practice, the initial phase of the Fund was slow and encountered problems like staffing delays and turf battles between involved agencies. The voluntary contributions from the industrialized countries have so far been inadequate, and

Table 6.2a
Summary scores: Chapter 6 (OZONE)

		Effectiveness		
Regime	Component or Phase	Improvement	Functional Optimum	Level of Collaboration
OZONE	1982–1985	NA	NA	NA
OZONE	1985–1987	Very low	Low	Intermediate
OZONE	1987–1996	Major	High	Intermediate

this goes especially for the CEITs. Still, by 1996, seventy-six country programs had been approved, and the coverage of the Fund's activities had been expanded to over ninety-nine countries. Moreover, by October 1996, 12,612 tons of CFCs of a projected 75,300 tons had been phased out (*Green Globe Yearbook* 1997, 95). As already noted, this has in no way been enough to stem the tide of rising ODS production and consumption in the developing countries. This touches on what has been termed a fundamental "contractual difficulty" within the regime, as "Article 5 countries are allowed to increase their consumption of ODS during the Period that the Fund is paying for projects to phase out their ODS use" (DeSombre and Kaufman 1996, 125). Hence, some progress has been made, but important bureaucratic and strategic problems remain. For instance, it has been indicated that the World Bank is ill-suited for this type of project, and as indicated above, the job may in itself be characterized as ambiguous and lacking focus and direction.[20] Lacking progress in this regard may lead to reduced willingness from the developing countries to take on new obligations and implement existing ones. Hence, there is a complicated interplay of factors, and both "vicious circles" and "positive circles" are imaginable.

Conclusion: What Are the Principal Determinants of the Success Witnessed So Far?

Despite reporting problems and lacking national knowledge, the overall impression is that the international regulatory development has been impressive and behaviorally influential (see table 6.2). Available national data indicate that the required behavioral changes are being implemented, although more steadfastly in the North than in the East and South—and clearly influenced by the international process.

Table 6.2b

Problem Type		Problem-Solving Capacity			
Malignancy	Uncertainty	Institutional Capacity	Power (basic)	Informal Leadership	Total
Moder	Intermediate	—	—	—	—
Mixed	Intermediate	Intermediate	Intermediate	Strong	High
Mixed	Low	Intermediate	Intermediate	Strong	High

Although assessing the distance to the collective optimum is complicated, both a diachronic and more synchronic assessment support the impression of a quite effective regime so far—but with uncertainties related to the future role of the South and the East.

My impression is clearly that the interim development with regard to problems, preferences, and capabilities—from moderately malign to relatively benign characteristics—can throw substantial light over the increasing and quite high effectiveness of the ozone regime so far. Although initially an intellectually complex problem with high reliance on computer models and scarce and long-term empirical effects, consensus and certainty have increased quite rapidly, due to a broad-based international research effort. Due to the global, largely indivisible character of the radiation-shielding ozone layer, affectedness has been quite symmetrical. Moreover and just as important, this affectedness has been brought into everyday life through dramatic ozone-hole reports and related increased skin-cancer risk. Another important problem characteristic is that the roots of the problem lie in a relatively small (but increasing) group of substances produced mainly by the United States and some European countries and utilized in certain domestic and commercial activities like refrigeration and cooling. In an interesting (but analytically complicating) interplay with international regulatory development, technological development has been steady but has gradually uncovered the latent fundamental asymmetry in every global regulatory effort—between the rich countries of the North that can afford greener, more expensive alternative substances and the poorer countries of the South that would like to keep present and future development costs to a minimum. This development indicates that the political character of the ozone-depletion problems has turned more malign again.

Hence, as indicated above, there has been an important but intricate interplay between problem-solving capacity, the underlying problems, and regime decisions. With regard to problem-solving capacity, the issue of entrepreneurial leadership stands out. Both scientific and several types of political leadership contributed crucial direction to the initial regime-development phase between 1985 and 1990. Strong leadership shaped the form of the regulatory ball and made it roll faster. Why strong leadership? Two key concepts here are (1) skin-cancer risks and public concern and (2) technological substitution options—further substantiating the interplay between problem-solving capacity and basic problem characteristics. More specifically on the institutional side, the organization of the science and politics relationship has been successful and contributed to the continuing evolvement of the process in the 1990s. Moreover, two specific institutional mechanisms are salient.

First, there is the issue of restrictions on trade with nonparties to prevent non-participating countries from gaining competitive advantages. According to several sources, these clauses have worked effectively in terms of increasing state participation. Second, the establishment of the Multilateral Fund was crucial for bringing the developing countries firmly into the process. The Fund has also enhanced processes in developing countries, although so far less than could be hoped for. Given the important future role of developing country action, making the Fund work is probably the most important institutional key to continuing regime progress. This issue indicates that success so far does not ensure smooth sailing with regard to complete problem solving.

Notes

1. In this case, interviews were conducted in 1995 with Per M. Bakken, Norwegian Ministry of Environment (December 21) and Ivar S. Isaksen, University of Oslo (December 12). I have drawn on a written response to a selected list of institutional questions provided by G. M. Bankobeza, program officer and lawyer with the Ozone Secretariat, dated November 6, 1995. In addition to the project group and the MIT reviewers, I thank Georg Børsting, the Norwegian Ministry of Environment, and Sebastian Oberthur, Ecologic, for helpful comments.

2. The work of Stolarski and Cicerone suggests that chlorine atoms have destroyed thousands of ozone molecules in the stratosphere, while the work of Molina and Rowlands has focused on CFCs and the release of large quantities of chlorine in the upper atmosphere. See, for instance, Morisette (1989).

3. For further information on the regime-creation process, see, for instance, Gehring (1994, 221–234) and Parson (1993, 34–40).

4. This table is based on Gehring (1994) and Oberthur (1997b).

5. This section is primarily based on Oberthur (1997a, 1997b).

6. However, the Montreal Protocol's article 5 stated that developing countries' consumption should "not exceed an annual calculated level of consumption of 0.3 kilograms per capita."

7. This is further discussed on pages 164 to 165.

8. This section draws heavily on Skjærseth (1992) and Parson (1993).

9. Parson (1993, 70–71) identifies at least three crucial stages where U.S. support for the international process was threatened by a domestic backlash—1985, 1987, and 1990. He maintains: "For the United States, and indeed for all activists, a continuing international process provided the momentum needed to smooth over uneven commitment and attention at the national level."

10. For instance, according to Skjærseth (1992), the correspondence was "remarkable." Oberthur (1997b, 6) states that "the political decisions of the Parties have always been very close to the options that appeared to be preferred in the reports of the assessment panels." See also Skodvin (1994, 2000).

11. See, for instance, Skjærseth (1992), Gehring (1994), and Rowlands (1995) for general overviews of these issues.

12. The work of Stolarski and Cicerone suggested that chlorine atoms destroyed thousands of ozone molecules in the stratosphere, while the work of Molina and Rowlands focused on CFCs and the release of large quantities of chlorine in the upper atmosphere. See, for instance, Morisette (1989).

13. It should be noted here that the broad overview provided in the following brief sections cover a period of ten years of complex knowledge and political development. Readers interested in specifics about scientific development, national positions, international negotiation processes, and so on are advised to consult sources like Benedick (1991), Gehring (1994), Litfin (1994), and Parson and Greene (1995).

14. See, for instance, *International Environment Reporter*, Sept. 20, 1995; *Environmental Policy and Law* 26/2/3, 1996; *International Environment Reporter*, Sept. 17 and Oct. 1, 1997; *ENDS-Report*, 271, 272, 1997.

15. For a more comprehensive discussion of the institutional issues, see Wettestad (1999).

16. See, for instance, Skjærseth (1992, 46–47), building on an interview with Bob Watson in *New Scientist* (April 29, 1989), 69–70.

17. According to Peter Haas, the ozone regime is a clear example of a case where an epistemic community operated and influenced the development of the regime. See, for instance, Haas (1990).

18. See Ulfstein (1996, 106), based on information from the Secretariat, and Brack (1996).

19. For more general discussions of the functioning of the Fund, see, for instance, Parson and Greene (1995), DeSombre and Kaufman (1996), and Biermann (1997).

20. In addition to DeSombre and Kaufman (1996), see also Parson and Greene (1995, 38–41) and Parson (1996).

References

Benedick, R. E. 1991. *Ozone Diplomacy*. Cambridge, MA: Harvard University Press.

Biermann, F. 1997. "Financing Environmental Policies in the South: Experiences from the Multilateral Ozone Fund." *International Environmental Affairs* 9(3) (Summer): 179–219.

Brack, D. 1996. *International Trade and the Montreal Protocol*. London: Royal Institute of International Affairs.

Brown, L., et al. 1996. *Vital Signs*. New York: Worldwatch Institute.

"The Copenhagen Meeting." *Environmental Poling and Law* 23(1): 6–12.

DeSombre, E., and Kaufman, J. 1996. "The Montreal Protocol Multilateral Fund: Partial Success" In R. Keohane and M. Levy, eds., *Institutions for Environmental Aid: Pitfalls and Promise* (pp. 89–126). Cambridge, MA: MIT Press.

"Effectiveness of Montreal Protocol in Stemming Ozone Loss Hailed by Scientists." 1996. *International Environment Reporter* (October 2), 867–868.

"The Vienna Meeting." *Environmental Policy and Law* 26(2–3): 66–71.

Gehring, T. 1994. *Dynamic International Regimes: Institutions for International Environmental Governance*. Berlin: Peter Lang Verlag.

"Funding and MDIs Addressed at Ozone Summit." 1996. *Global Environmental Change Report* (December 13), 1–3.

Greene, O. 1996. "The Montreal Protocol: Implementation and Development in 1995." In J. Poole and R. Guthrie, eds., *Verification 1996: Arms Control, Environment and Peacekeeping* (pp. 407–426). Bouldes, CO.: Westview Press.

Green Globe Yearbook. 1996, 1997. Oxford: Oxford University Press and Fridtjof Nansen Institute.

Haas, P. 1990. "Obtaining International Environmental Protection Through Epistemic Consensus." *Millennium* 19(3): 347–363.

Jachtenfuchs, M. 1990. "The European Community and the Protection of the Ozone Layer." *Journal of Common Market Studies* 28(3): 261–277.

Litfin, K. 1994. *Ozone Discourses: Science and Politics in Global Environmental Cooperation*. New York: Columbia University Press.

"Methyl Bromide Among Issues Unresolved as Montreal Protocol Meeting Nears End." 1997. *International Environment Reporter* (September 17), 861–862.

"Methyl Bromide Phase-Out Plans, CFC Licensing System Only Gains in Montreal." 1997. *International Environment Reporter* (October 1), 903–904.

"Ministers Agree to New Curbs on Ozone Depleters." 1997. *ENDS Report* 272 (September): 44–45.

"Montreal Protocol Successful in Reducing Methyl Chloroform Levels, Scientists Say." 1995. *International Environment Reporter* (July 26), 566–567.

"Montreal Protocol Working Group Puts Off New Proposals Until November." 1995. *International Environment Reporter* (May 17), 361–362.

Morisette, P. 1989. "The Evolution of Policy Responses to Stratospheric Ozone Depletion." *Natural Resources Journal* 29: 793–820.

NOAA, NASA, UNEP, and WMO. 1994. *Scientific Assessment of Ozone Depletion: Executive Summary*, World Meteorological Organization, Global Ozone Research and Monitoring Project, Report no. 37, Geneva.

"Non-Compliance with Ozone Agreement." 1989. *Environmental Policy and Law* 19(5): 147–148.

Oberthur, S. 1997a. *Production and Consumption of Ozone-Depleting Substances 1986–1995*. Berlin: GTZ, Deutsche Gesellschaft fur Technische Zusammenarbeit.

Oberthur, S. 1997b. "The Role of Europe in the International Co-operation for the Protection of the Stratospheric Ozone Layer." Paper prepared for the Strategies for European Leadership of International Climate and Sustainability Regimes, October 14.

"Ozone Layer Left at Risk as Talks Stumble on Funding." 1995. *ENDS Report* 251 (December): 35–37.

Parson, E. A. 1993. "Protecting the Ozone Layer." In P. Haas, R. O. Keohane, and M. Levy, eds., *Institutions for the Earth* (pp. 27–75). Cambridge, MA: MIT Press.

Parson, E. A. 1996. "International Protection of the Ozone Layer." *Green Globe Yearbook* (pp. 19–28). Oxford: Oxford University Press.

Parson, E. A., and O. Greene. 1995. "The Complex Chemistry of the International Ozone Agreements." *Environment* 37(2): 16–20, 35–43.

Rowlands, I. 1995. *The Politics of Global Atmospheric Change.* Manchester: Manchester University Press.

Skjærseth, J. B. 1992. *From Regime Formation to Regime Functioning "Effectiveness": Coping with the Problem of Ozone Depletion.* Oslo: Fridtjof Nansen Institute.

Skodvin, T. 1994. "The Ozone Regim." In S. Andresen, T. Skodvin, A. Underdal, and J. Wettestad, eds., *"Scientific" Management of the Environment? Science, Politics and Institutional Design.* Oslo: Fridtjof Nansen Institute.

Skodvin, T. 2000. "The Ozone Regime." In A. Underdal, S. Andresen, T. Skodvin, and J. Wettestad, eds., *Science and International Environmental Regimes: Combining Integrity with Involvement.* Manchester: Manchester University Press.

"Tough Talks at Ozone Layer Meeting on Methyl Bromide, HCFCs." 1997. *ENDS Report* 271 (August): 42–43.

Ulfstein, G. 1996. "The Vienna Convention for the Protection of the Ozone Layer and the Montreal Protocol." *The Effectiveness of Multilateral Environmental Agreements* (pp. 105–117). Copenhagen: Nordic Council of Ministers Report.

Wettestad, J. 1999. "A Triumph for Institutional Incentives and Flexible Design? The Vienna Convention and Montreal Protocol on Ozone Layer Depletion." In *Designing Effective Environmental Regimes: The Key Conditions* (pp. 125–165). Aldershot: Elgar.

III

Mixed-Performance Regimes

Introduction: Common Features of Mixed-Performance Regimes

The line of reasoning that we developed in chapter 1 suggests that there are two main roads to mixed performance. One goes through a set of intermediate scores—a combination of problems combining benign and malign features, intermediate problem-solving capacity, and a context that is largely neutral. The other goes through some combination of positive and negative scores—for example, malignant problems and high capacity, or benign problems and low capacity. We refer the reader to tables II.1 and V.1 for further specifics.

7

Cleaning Up the North Sea: The Case of Land-Based Pollution Control

Jon Birger Skjærseth

Significant Achievements, But Still a Long Way to Go

Today, the main challenge of cleaning up the North Sea lies more in domestic implementation than in the making of stringent joint international commitments.[1] Surrounded by densely populated areas, the North Sea is an area of intense human activity causing considerable pressure on the marine environment. In response to this problem, the Convention for the Prevention of Marine Pollution from Land-Based Sources—the Paris Convention—was signed by the West European maritime states in 1974 and allowed the European Union to join as a contracting member. The twelve signatory states bordering the Northeast Atlantic including the North Sea agreed to take all possible steps to prevent the pollution of the sea from land-based sources.[2] These sources account for about 70 percent of all contaminants entering the marine environment. To transform this general goal into more specific joint commitments, an Interim Commission was set up the same year as the Convention was signed and before the Paris Commission was officially established in 1978 when the Convention entered into force. Although the parties met annually in the Commissions, they were not able to adopt any joint commitments with strings attached until the latter part of the 1980s. The joint commitments adopted from the 1970s to the mid-1980s required only very limited behavioral change. In short, discretion was high, and the constraints placed on the parties correspondingly low.

This pattern of lenient commitments changed at the 1987 and 1990 International North Sea Conferences (INSCs), which adopted Ministerial Declarations aimed at significant reductions in inputs of hazardous substances and nutrients within specific time frames (see chapter 3). The Paris Commission responded to this development by adopting a wide range of joint commitments on best available technology (BAT) to be applied in specific industry sectors as well as best environmental practices (BEP) to be applied at diffuse sources. These joint efforts also led to significant domestic implementation activities, and the controlled substances were actually reduced quite significantly. However, stringent and specific joint international commitments were, in many cases, not sufficient to secure behavioral changes at the domestic level in accordance with international obligations. In contrast to dumping at sea, for example, lack of domestic implementation has been a significant problem concerning land-based sources, even in those cases where governments were determined to follow through and implement measures to achieve international goals.

The Complexity of Controlling Land-Based Sources of Pollution Problems

There was a general recognition at the beginning of the 1970s of the need to regulate land-based sources of marine pollution in the Northeast Atlantic. After the Oslo Convention on dumping was signed in 1972, the next logical step was to draw up a similar document on land-based sources and watercourses (see chapter 3). The global United Nations Conference on the Human Environment took place in Stockholm in 1972. This conference identified land-based sources, such as rivers, as the principal route by which most pollutants reach the seas. The U.N. conference led to the establishment of the United Nations Environment Program (UNEP), which became instrumental in preparing the Regional Seas Program and the Barcelona Convention for the protection of the Mediterranean. However, global efforts to control marine pollution have since been on the wane compared to atmospheric problems, for example. The Law of the Sea Convention of 1982 regulates in principle all uses of the oceans and the ocean floor. Nevertheless, provisions relevant for land-based emissions are no more than suggestions to the parties to consider the problem (Koers and Oxman 1984). Marine pollution was also an issue under the United Nations Conference on Environment and Development in Rio de Janeiro in 1992. However, the question of establishing a (new) comprehensive global convention on marine pollution was never really put on the agenda. Marine pollution was dealt with in Agenda 21, which outlined general aims and recommendations without any strings attached.

Land-based sources of marine pollution have always been a more complex and malign problem as compared to dumping at sea, for example. While the Oslo Convention on dumping was drafted and adopted within a few months, three conferences had to be held between 1972 and 1974 before the Paris Convention was open for signature (Oslo and Paris Commissions 1984a, 6). Various features made the problem of land-based pollution malign both as an international collective-action problem and as a unilateral problem causing implementation problems for each state. The anticlockwise direction of the North Sea currents places the states in a chain order according to import and export of pollution. In contrast to dumping at sea, the actual exchange of land-based pollution is much larger since land-based emissions account for some 70 percent of all marine pollution. Thus, the problem is very significant in terms of externalities, and these externalities are asymmetrically distributed.

While the current patterns were roughly known from the 1970s, their potential to transport contaminants was largely unknown until the latter part of the 1980s. A map on residual currents was presented in the first 1984 quality status report (QSR) on the North Sea representing the first attempt to jointly visualize the transport routes in the North Sea. As a consequence, the report also visualized the potential net importers and net exporters of marine pollution. A more sophisticated map was included in the 1987 QSR, and the report concluded that (*QSR* 1987, 5):

Observations indicate that the coastal margins of the southern North Sea, the Skagerak, the northern Kattegat, and the southern part of the Norwegian trench are the most important areas for transport and deposition of suspended matter. These areas show the highest concentrations of suspended matter and highest deposition rates and are therefore likely to receive any contaminants absorbed onto particulate matter.

The 1993 QSR emphasized the complex process of contaminant transport (*QSR*, 1993, 102). Nevertheless, it explicitly placed the North Sea states in a chain order (ibid.):

Water circulation may transport—via counterclockwise circulation—inputs from coastal areas of the United Kingdom, Belgium, the Netherlands, and Germany to those of Norway, Denmark, and Germany, including the biologically important Wadden Sea.

The information produced by the QSRs made it evidently clear that the currents were potentially capable of ranking the countries in a chainwise relationship with net exporters at the top of the chain and net importers at the end. More certainty as to transport patterns was also reflected in how the parties perceived the problem. In Norway, for example, there was a growing recognition of its role as a net importer of marine pollution. While Norway's general vulnerability was emphasized in the mid-1980s, the government tried to estimate the relative share of import in the beginning of the 1990s (*QSR* 1993, 102):

The pollution status along the Norwegian coast is thus affected both by inputs from other countries via the coastal current and by discharges from Norway itself. In general, the inputs carried by ocean currents from southern parts of the North Sea will determine the status of pollution in the coastal current and will to some extent affect the more open fjords. However,

Norwegian discharges will also affect the coastal current in several places. (Ministry of the Environment 1992, 12)

Since land-based pollution stems from a wide range of sources covering a large share of the territory of the participating states, abatement costs are high, and regulations may affect the international competitive situation. According to the British government in 1990, the then newly privatized water industry in England and Wales will invest £28 billion by 2000 including £13.7 billion to improve sewerage works (*This Common Inheritance* 1990). In Norway, total investment costs for implementing the North Sea Declarations were estimated at NOK 10 billion between 1985 and 1995 (Ministry of the Environment 1992, 84). In the Netherlands, the aggregated abatement costs of the public and private sectors have increased significantly from 1980, and implementation of the 1985 Water Action Plan was estimated at about 0.5 percent of the gross national product (Water Action Program, 1984, 36).

The heavy abatement costs involved in curbing land-based emissions contributed in making the uniform emission standard (UES) versus the environmental quality objectives (EQO) discourse a core topic within the Paris Commission. The majority of the parties supported the UES approach and stressed that it had a preventive aspect, could be based on best available technology, and opened equal international standards for all parties (Oslo and Paris Commissions, 1984a, 24). In the eyes of the UES supporters, the latter would also have a stabilizing effect on the competitive situation between the states since UES was based on equal standards for all actors. Conversely, quality objectives and standards may distort competition since the quality of one region can differ from a neighboring region. Supporters of the EQO approach (Portugal and the United Kingdom) emphasized that the final objective of the Paris Convention is the protection of the marine environment and of humans as consumers of seafood. They quoted article 6 of the Convention, which provides that the contracting parties should take account of various criteria including the quality and *absorptive capacity* of the receiving waters of the maritime area (ibid.). In short, the UES advocates emphasized that discharges of substances that were known to be toxic, persistent, and bioaccumulative and were listed in the Convention's blacklists should be limited as far as possible at source. Conversely, the EQO defenders claimed that standards set should be determined by observable negative effects in the marine environment for each substance. Although this discrepancy was closely related to economic interests, it was also related to a normative question of the basic philosophy behind pollution control. For example, it is hard to explain in material terms why Portugal preferred EQOs while Spain did not.

The problem of land-based emissions was also complex as a unilateral implementation problem for at least three reasons. First, the causes of land-based discharges are due to legitimate activities within a number of domestic sectors such as municipal services and production of food and goods. This implies that subnational actors are liable to be affected differently by the problem in terms of abatement costs and damage costs. However, a typical situation is that abatement costs are concentrated while benefits are widespread throughout society. Thus, support for governmental initiatives by those subnational actors causing pollution (target groups) cannot be taken for granted. Second, transsectoral problems require concerted action among relevant public authorities both horizontally, and vertically. Different regulatory agencies and ministries have to coordinate their activities horizontally, and governments have to integrate environmental concerns into a wide range of sectors to cope with the problem. In addition, regulatory competence is frequently distributed vertically between the central, regional, and local levels. Vertical distribution of competence varies not only between countries but also between sectors within countries. Whenever decentralization of competence overlaps with local opposition to governmental initiatives, implementation cannot be taken for granted. Third, the sources of land-based emissions are sometimes diffuse and invisible and hard to identify and control. For example, inputs of hazardous substances and nutrients stem both from point sources and diffuse sources. The most important implication of these three features is that domestic implementation cannot be taken for granted even though there is political will to take action at the level of governments.

The Process of Regime Implementation: From Environmental Quality Toward Curbing Emissions at Source

The main objective of the Paris Convention is to take all possible steps to prevent pollution of the sea by adopting individually and jointly measures to combat marine pollution and by harmonizing the parties' policies in this regard. Twelve parties, including the European Union, have ratified the Convention.[3] Like the Oslo Convention (see chapter 3), the Paris Convention is structured around a black list (and greylist) system. The parties are required to implement measures to eliminate pollution by substances in the blacklist, and to reduce or eliminate pollution by substances in the greylist. The Commission exercises overall supervision of implementation of the convention. The Paris Commission (jointly with the Oslo Commission) is assisted by a small secretariat in London. The Paris Convention originally applied to watercourses from the coast. In 1986, the Convention was amended to include atmospheric sources.

The First Phase: Lenient Joint Commitments

According to the Paris Convention, various formal cooperative fora were to be established to promote a *dynamic* structure of the cooperation consisting of four elements:

· Identification and diagnosis of problems,
· Action in terms of joint decision making and implementation,
· Verification and assessment of the measures taken, and
· Reappraisal.

To diagnose problems, a Joint Monitoring Group (JMG) was established by the Oslo and Paris Commissions and a Joint Monitoring Program (JMP) became operational in accordance with articles 13 and 11 of the Oslo and Paris Conventions, respectively. At the political decision-making level, the Paris Commission was established as the permanent decision-making body. Duties related to verification and assessment were placed at the scientific and technical level, and the Paris Commission established a standing scientific group—the Technical Working Group (TWG)—set up at the first meeting of the Interim Commission. The TWG was assisted by more or less permanent working groups. For example, the Paris Commission established a working group on oil pollution in 1978 that has existed up to the present. Even though a warm-up period is required for most complex international-cooperation agreements, this system did not function as expected and intended from the mid-1970s to the mid-1980s. In general, the parties did not succeed in adopting commitments aimed at significant behavioral change at any level. As previously noted, one important reason was a profound disagreement on whether goals should be based on uniform emission standards (UES) or environmental-quality objectives (EQO).

The UES-EQO dispute appeared immediately at the agenda in the Paris Commission. According to article 4 of the Paris Convention, the Commission may adopt both UESs and EQOs. At the first meeting of the Interim Paris Commission in 1974, the parties discussed the approach that should serve as the basic principle for the cooperation (Oslo and Paris Commissions 1984a, 25). Although the first discussion was linked to mercury pollution, it was clear that the approaches exceeded the framework for the study of a single substance.

This discussion continued the first four years, and the parties did not succeed in adopting any common basis for action. At the first formal meeting of the Paris Commission in 1978, the parties agreed to disagree and to continue with the dual approach for the next five years. By 1983, the Commission expected that the dispute could be settled "scientifically" by collecting sufficient data to determine the approach with the most positive impact on the marine environment. Since the problem also concerned aspects other than science, this strategy was likely to fail. At the Commission meeting in 1983, it became clear that it was impossible to settle the dispute due to insufficient data collected by the Technical Working Group and different interpretations of the data actually collected. Thus, the parties decided to continue with the dual approach. In effect, each party could choose to follow its preferred approach (Paris Commission 1983). Since not everybody agreed that inputs of blacklisted substances were potentially harmful, a basis for concerted action did not exist. The dual approach was further confirmed at the 1984 Commission meeting and included in the 1984 document, *A Strategy for the Future* (Oslo and Paris Commission 1984b) adopted at the sixth meeting. According to Sætevik (1988, 43), the parties failed to agree on a solution that could take care of both environmental and competitive aspects. As we shall see below, the two-track approach also contributed in watering down potential strings attached to the joint commitments actually adopted.

The parties to the Paris Commission are required to implement measures for the elimination of pollution by substances listed in the blacklist and for the reduction or elimination of pollution by substances listed in the greylist. The parties are also required to inform the Commission of the legislative and administrative measures they have taken to ensure compliance with the provisions of the Convention. The Commission exercises overall supervision of implementation of the Convention. Level of collaboration was thus medium, including centralized appraisal of effectiveness. More specifically, the Commission adopts decisions and recommendations for the prevention of pollution (see table 7.1).

The obligations included in the Paris Convention and the commitments adopted in the Paris Commission during the first phase centered first and foremost on quality standards and emission standards. Most of the recommendations and decisions adopted during the first phase focused on the blacklist substances mercury and cadmium. Concerning these substances, the parties were free to choose the type of commitments that should apply. Quality standards define the minimum quality of water that states have to preserve, by indicating the permissible presence of certain

Table 7.1
Decisions and recommendations adopted within PARCOM 1978 to 1987

Year	Decisions	Recommendations
1978	1	2
1980	2	2
1981	3	2
1982	1	1
1983	—	2
1984	1	3
1985	4	2
1986	2	1
1987	—	4
Total	*14*	*19*

Source: *Paris Commission Procedures and Decision Manual* (1987).
Note: Joint commitments other than Decisions and Recommendations are not included. There was no PARCOM meeting in 1979.

substances in the water. In practice, however, quality standards leave the parties with a considerable amount of discretion, resting as they do on a high level of scientific uncertainty. It is extremely difficult to assess the level of pollution against quality standards. States favoring such standards had, in practice, virtually complete freedom to judge independently whether a certain activity was permissible. Emission standards define the maximum admissible concentrations of substances in effluent entering the waters. Such standards allow states less room for calculating whether discharges are permissible since their measurement is less subject to scientific interpretation and evaluation. However, states that failed to comply with emission standards could argue that they complied with the quality standards.

While prohibition may eliminate discretion completely, the commitments adopted aimed at phasing out substances included loopholes that, in effect, placed no clear obligations on the states. In 1978, the parties agreed that the "drins" (endrin, aldrin dieldrin) should be phased out of use as soon as ecologically less harmful substitutes were available for each of the present uses of these substances. In 1985, the Commission decided that all contracting parties will phase out the use of such substances *as soon as practicable*. It was up to the parties to determine whether less harmful substitutes were available and what were considered as practicable. A similar set of wording was adopted for polychlorinated biphenyls (PCBs) and polychlorinated triphenyls (PCTs) that should be phased out in new equipment *as soon as possible*. The elasticity of these concepts means that they hardly contain any substantive obligations (Skjærseth 2000).

Although article 11 of the Paris Convention places a clear duty on the contracting parties to assess the effectiveness of measures, monitoring performance was almost absent in the first decade. With regard to the assessment of measures, the parties had not developed any common and mandatory implementation report routines. The monitoring practice developed by the Paris Commission, appears as ad hoc and based essentially on each party's freedom to report on whatever it wished. As late as in 1987, this practice was still applied by the Paris Commission, and the Commission recognized that there were shortcomings in contracting parties' reporting performance, even in those cases where mandatory obligations existed. The recommendation adopted in 1978 on the phasing out of the "drins" was changed to a legally binding decision in 1985 (Paris Commission 1987, 16):

The Commission agreed . . . that further discussion should be deferred to the TWG meeting in 1989 at which time all contracting parties concerned would be *invited* to submit updated papers for discussion.

Concerning mercury, which had been regulated by various recommendations and decisions since 1980, the Commission agreed in 1983 that TWG would review all aspects of mercury pollution in 1987. However, only the Netherlands had submitted a review of the programs and measures it had taken. With regard to cadmium, only two countries had submitted information on the measures they had taken (ibid.). The Paris Commission also recommended phasing out PCBs and PCTs in 1983. The Commission had received information from three countries, but "TWG had been unable to discuss progress on the phasing out of existing use of PCBs and PCTs" (ibid.).

The Second Phase: Toward a Solution to the UES-EQO Discourse

While the contracting parties continued to agree to disagree within the Paris Commission, the UES-EQO controversy became a major topic at the first International North Sea Conference (INSC) in Bremen 1984. According to the declaration, the ministers recognized (Bremen Declaration sec. A16):

that the approaches according to uniform emission standards (UES) or according to environmental quality objectives (EQO) aim at the protection, conservation, and improvement of the marine environment.

Moreover, the declaration hinted at the precautionary principle that in effect would imply a common recognition of the UES approach since it implies that action is taken prior to observable damaging effects. The core scientific argument of the UES advocates was linked to the fact that present scientific knowledge would not allow for establishing indisputable causal links between inputs, concentrations, and effects.

In the wake of the Bremen Conference, an ad hoc working group was established by the Paris Commission in 1985 to evaluate the EQO and UES approaches. At the first meeting of the group the same year, it became clear that the United Kingdom was left alone as the most dedicated defender of EQOs. Moreover, the parties disagreed about the scientific, economic, and environmental bases of the approaches. For example, the United Kingdom did not accept that the use of UES prevents distortion of competition. The Policy Working Group preparing the 1987 Conference in London set out to find a solution. The chair of the working group prepared a paper in which he emphasized the *complementary* nature of the EQO and UES approaches. The paper did not include any new arguments, but the different approaches' strengths and weaknesses were systematically discussed. The paper concludes that uncertainty concerning cause-and-effect relationships is the principal limitation to EQS, but both approaches have something to contribute (EQUSa 1988, 9):

> The simple conclusion to which this essay points is that the tide of thought has flowed beyond the highwater mark it reached in Bremen in 1984. To argue about the EQO/EQS and UES approaches today as if they were self-sufficient and conflicting alternatives is to miss the point. There is now general acceptance of the need for a precautionary approach that draws on the best available science in order to define which substances are most dangerous and why, and deploys the best available technology to curb them at tolerable costs.

Although the paper indicates a growing recognition of the complementary nature of the two approaches, acceptance by all parties would nevertheless imply a change in the position of the United Kingdom. Many other states actually applied a combination of the two approaches and did not oppose quality standards and objectives as such. The main limitation to an agreement was that the United Kingdom did not accept the UES approach.

The breakthrough came at the 1987 London Conference. The ministers agreed on the precautionary principle that emissions should normally be limited at source taking into account the best technical means and further that quality objectives should be fixed on the basis of latest environmental data. This agreement paved the way for the uniform percentage-reduction targets on hazardous substances and nutrients. Subsequently, a second meeting of the ad hoc working group on the EQO and UES approaches was convened in 1988 under the auspices of the Paris

Commission. At this meeting, the United Kingdom explicitly accepted the UES approach by the following statement (EQUSb 1988, 2):

Since 1985, however, the UK has modified its position in relation to the most dangerous substances, those which are very toxic, persistent and liable to bioaccumulate. It is now recognized that, particularly in view of the degree of uncertainty about the possible long-term effects of these substances, the precautionary principle dictates that their input from all sources should be minimized.

This change in U.K. policy had far-reaching implications for North Sea and Northeast Atlantic cooperation. For the first time, joint decisions were adopted to impose significant targets on land-based sources within fixed time limits. Three key elements of the sixteen sections of the London Declaration related to land-based sources:

· A substantial reduction (of the order of 50 percent) between 1985 and 1995 in the total inputs to the North Sea via rivers and estuaries of substances that are persistent, toxic, and liable to bioaccumulate (hazardous substances);
· A substantial reduction (of the order of 50 percent) between 1985 and 1995 in inputs of phosphorus and nitrogen to those areas of the North Sea where such inputs are likely, directly or indirectly, to cause pollution; and
· The preparation of national action plans to achieve both these goals.

The 1990 Hague Conference clarified and strengthened the London Declaration particularly concerning land-based sources. A list of thirty-six hazardous substances was adopted and directly linked to the 50 percent reduction target concerning hazardous substances. Targets aimed at 70 percent reduction of land-based and atmospheric inputs were adopted for the most dangerous substances—dioxines, cadmium, mercury, and lead. Some new measures were also adopted at the Hague Conference. Agreement was reached to phase out and destroy PCBs and hazardous PCB substitutes by 1999 at the latest.

At the Hague Conference, it became evidently clear that most countries would face problems in reaching the goals on nutrients and pesticides. One major problem was the agricultural sector, which had traditionally been weakly regulated by the authorities in most relevant states and slow to implement required measures due to opposition among farmer's organizations. These concerns led, among other things, to an Intermediate Ministerial Meeting (IMM-93) as a preparatory ministerial conference to the 1995 Esbjerg Conference. For the first time, ministers of agriculture were invited to discuss common measures on nutrients and pesticides. The IMM-93 tightened up the commitments on nutrients by aiming at balanced fertilization in agricultural production by 2002 at the latest.[4]

The commitments adopted on hazardous substances in 1995 may stand as a symbol of the significant change in the amount of behavioral change required from the 1970s to the present. In the mid-1980s, only a few substances were under international regulation, and even fewer were made subject to elimination. Ten years later, the ministers agreed to prevent the pollution of the North Sea by phasing out all hazardous substances (Danish EPA 1995, art. 17):

by continuously reducing discharges, emissions, and losses of hazardous substances thereby moving towards the target of their cessation within one generation (25 years) with the ultimate aim of concentrations in the environment near background values for naturally occurring substances and close to zero concentrations for man-made synthetic substances.

The soft-law ministerial declarations were not legally binding for the parties, and they could be vulnerable to changes in political leadership. Ministers come and go, and policies can change as governments change. Thus, the INSCs clearly carried a potential to end up as paper tigers. However, this was avoided through a process of *international implementation* that gave the political declarations a firm footing in international law. While the London Declaration made its strongest impact on the Oslo Commission and dumping, the Hague Declaration made its strongest impact on the Paris Commission. Based on the list of hazardous substances adopted in the Hague, the Paris Commission (PARCOM) started to address systematically discharges from specific industrial sectors. With regard to hazardous substances, PARCOM took action on industrial sectors, including best environmental practices and best available technology on specific industrial activities. In addition to the two recommendations adopted in 1988 and 1989 on nutrients, PARCOM adopted Recommendation 92/7 on the Reduction of Nutrient Inputs from Agriculture. In September 1992, the ministers representing the parties to the OSPAR Convention as well as Luxembourg and Switzerland met in Paris to sign a new Convention for the Protection of the Marine Environment of the Northeast Atlantic. The new Convention mainly codifies achievements of the Oslo and Paris Conventions and to a more limited extent the declarations of the INSCs. The OSPAR Convention entered into force in March 1998 and replaced the Oslo and Paris Conventions.

Compared to the first phase, we see from table 7.2 that the joint commitments adopted increased not only in specificity but also in number. The total amount of joint commitments adopted almost doubled in the second phase although the number of decisions remained quite stable. Three institutional changes undertaken after 1987 contribute in explaining the increased joint decision effectiveness. First, the principle of regionalization adopted in 1988 in direct response to the INSCs allowed the parties to the Paris Convention to adopt commitments that differenti-

Table 7.2
Decisions and recommendations adopted within PARCOM, 1988 to 1995

Year	Decisions	Recommendations
1988	1	5
1989	—	5
1990	4	3
1991	—	5
1992	4	8
1993	1	5
1994	1	9
1995	4	1
Total	*15*	*41*

Source: *Paris Commission Procedures Manual* (1995).
Note: Joint commitments adopted until June 1995.

ated between the North Sea and the non-North Sea states. Thus, countries like Spain and Portugal could be allowed longer time frames or more lenient obligations than the North Sea states. Second, the parties changed decision rules in use. While the Paris Convention opened up for a qualified majority, the parties did not utilize this opportunity during the first decade. In the second decade, the parties started to utilize this practice. The number of reservations filed by individual parties increased significantly from 1988. Third, the OSPAR copied the idea of convening high-level ministerial conferences. The change in the level of representation was instrumental in increasing the authority needed to cut through deadlock and thus increasing the system's aggregation capacity. In addition to a number of the joint obligations adopted, the ministers adopted an action plan including objectives for the future work of the Commissions that affected the course of actions from 1992 to 1995.

The INSC declarations and PARCOM activities also affected EU marine policy. Two important EU directives were adopted in 1991 based on the initiatives in the latter part of the 1980s. Besides being important for the North Sea, the nitrates and waste-water directives reflected a slightly different approach compared to previous directives. Like the North Sea commitments, the new directives attack the sources of pollution, describe clear goals within time frames, and rely less on quality objectives (Richardson 1994, 150). The Urban Waste Water Directive (91/271/EEC) set specific requirements on waste-water-collecting systems to be implemented by the year 2000 or 2005 concerning nutrient discharges. According to the EU Commission, investments in the order of somewhere between 100 and 200 billion ECU are needed thus making this directive the most far-reaching piece of environmental legislation of the Community in economic terms (Oslo and Paris Commissions 1993, 21). The nitrate directive (91/676/EEC) aims at supplementing the above directive by specifically addressing nutrient emissions from the agricultural sector. These directives overlap both the INSC declarations as well as PARCOM commitments. In short, the process of international implementation has led to a comprehensive international regime for controlling land-based inputs to the North Sea. This process also led to a higher level of collaboration as compared to the first phase.

Three Cases and Causes of Domestic Implementation Failure

The core model underlying this project explains outcomes and impacts by conceiving of states more as unitary actors than as a constellation of subnational actors. This case deviates from the main focus since lacking ability to follow through constitutes an important explanation for implementation failure in addition to low willingness among the relevant governments.

Since the mid-1980s, collective decisions within the European Union, OSPAR, and the North Sea Conferences have increased in number, stringency, and scope. At the domestic level, there has been a stream of new legislation and administrative directives. Moreover, target groups have actually changed their behavior significantly, and emissions of regulated substances have decreased substantially in most of the North Sea states (Danish EPA 1995). However, there are clear variations in achievements not only between countries but also between affected sectors within countries. As previously noted, land-based sources of marine pollution is not only malign as a collective-action problem but also very complex to manage unilaterally. In the following, I briefly illustrate three different pathways that may lead to domestic implementation failure concerning the joint international commitments on nutrients.

The first case is concerned with implementation or rather compliance failure due to deliberate choice. As pointed out by Weale (1992, 48), governments sometimes only reluctantly embark on policies concerning international environmental obligations. Since North Sea pollution is an asymmetrical problem in the sense that actors are affected differently by the actions of others, a given commitment may be less favorable than an actor would have chosen by itself. For example, a net importer of pollution would probably prefer that others (notably the net exporters) contribute more, while the actor itself contributes less. Whenever joint commitments do not represent the arrangement maximizing utility for each and every party involved, compliance cannot be taken for granted (Underdal 1995, 5).

The United Kingdom is clearly a net exporter of marine pollution. Water circulation may transport inputs from the coastal areas of the United Kingdom to those of Norway. Due to the counterclockwise circulation, inputs from other North Sea states are not likely to affect U.K. coastal waters. This suggests first that both inland pollution and coastal pollution are caused by domestic sources in the United Kingdom. Consequently, the United Kingdom has minor incentives to advocate joint action. Since U.K. damage costs are caused by its own emissions, which are partly exported, the United Kingdom has actually incentives to avoid joint action because victims of U.K. exports may try to force the United Kingdom to internalize these externalities. Thus, the United Kingdom had virtually no interest in supporting the joint commitments on nutrient reductions. However, the 1987 Conference took place in London, and the United Kingdom was under strong political pressure to sign up to the agreement. This pressure was generated both internally by the green movement and externally by other North Sea states. As a consequence, the U.K. government signed up to the 50 percent reduction target on nutrients to be achieved between 1985 and 1995 since the political costs of blocking an agreement would have been too high. The 50 percent reduction target was reaffirmed at the 1990 INSC in the Hague and consequently by the United Kingdom. Between the 1987 and the 1990 conferences, the exceptional algal blooms in 1988 and 1989 gave the joint commitments more political weight since the blooms were directly related to eutrophication causing problems for fishery, aquaculture, tourism, and other recreational interests (Netherlands Ministry of Transport 1990). Moreover, public opinion changed significantly toward green values between 1987 and 1990 in most North Sea states (Hofrichter 1991). The growth of the green movement was particularly strong in the United Kingdom (McCormick 1991).

However, the United Kingdom never had serious intentions to implement measures to change the behavior of target groups like farmers—one major source of inputs of nutrients. The 50 percent reduction target on nutrients that was adopted in 1987 and reaffirmed in 1990 was qualified by the phrase "into areas where these inputs are likely, directly or indirectly, to cause pollution." The U.K. government has argued that it is not obliged to take action since such areas do not exist around the U.K. coast. The United Kingdom has not reported on nutrient inputs since 1992. Between 1985 and 1990, U.K. inputs of phosphorus substances decreased by 14 percent, while nitrogen inputs had increased by 2 percent (Oslo and Paris

Commission 1992). Opinions clearly vary as to whether the United Kingdom was formally committed by the agreement. On the one hand, the official view is that the United Kingdom has a formal right not to take action since it is up to each state individually to identify sensitive areas (Danish EPA 1995). On the other, it has been argued that (MacGarvin 1995)

while the situation regarding marine eutrophication in the UK may not be as bad as on the continental coast, it seems perverse to argue that there are no possible examples of actual or potential eutrophication areas, particularly with regard to the estuaries.[5]

Nevertheless, most other North Sea states did not appreciate the interpretation of the United Kingdom due to its position as a net exporter. In any case, the first pathway shows that the United Kingdom has not succeeded in reducing nutrients significantly due to lacking willingness at the state level.

The second pathway is concerned with horizontal implementation failure—the ability to integrate environmental concerns into various sectors of society. To a large extent the Netherlands suffers the consequences of its own national discharges since water circulation is restricted along other parts of Dutch coastal waters (QSR 1993, 108). Thus, as could be expected, the Dutch government rapidly and ambitiously incorporated the international commitments on nutrients into domestic programs. The three ministries involved—Transport and Public Works, Agriculture, and Environment—agreed in principle that a 50 percent reduction was the basic aim of policy. In fact, they aimed at a balance between input and output of nutrients before 2000. This would require a 70 to 90 percent reduction in emissions of eutrophying substances (Netherlands Ministry of Housing 1989). The main cause of nutrient emissions in the Netherlands is the agricultural sector. Given the ambitious goals concerning these substances, it is interesting to note that actual reductions in this sector between 1985 and 1995 were in fact zero (Danish EPA 1995). Thus, implementation problems have been severe in the agricultural sector. One important reason is lack of integration of environmental concerns in practice. A necessary condition for implementation is the ability to integrate environmental concerns into the Ministry of Agriculture, its subordinated bodies, as well as farmer organizations. Public and private organizations with competence in agricultural pollution have traditionally formed a social block in the Netherlands. The difference between this block and other sectors of society has traditionally been more significant than the difference between the public and private sectors. In effect, it has been difficult to integrate environmental concerns, and the agricultural block has been very slow to react on early warnings of environmental problems (Hegeman et al. 1992). In short, the Dutch government failed to change the behavior of farmers due to implementation problems in the agricultural sector.

The third example is concerned with vertical implementation failure. Whenever competence in regulation is decentralized to local levels, implementation cannot be taken for granted. As a net importer of marine pollution, Norway would generally prefer that other North Sea states contribute more than itself. However, the algae blooms hit Norway hard, the 1988 bloom being particularly serious. The less severe 1989 bloom caused losses for fish farms amounting to NOK 30 million (Netherlands Ministry of Transport 1990, 14). Moreover, mass deaths of marine organisms occurred with serious implications for recreation interests and fisheries, especially along the south coast. The algae blooms were taken very seriously by the Norwegian government, which brought the case to the 1988 Paris Commission meeting. Norway proposed the PARCOM Recommendation on the reduction of inputs of nutrients, and the Norwegian delegation (Paris Commission 1988, 12, emphasis added)

stressed the need to reduce the input of nutrients, not only from countries in the immediate vicinity of the area affected, but also from other riparian States to the North Sea. In the opinion of the Norwegian delegation, it is necessary and urgent that all North Sea States *strictly follow* the agreement reached at the North Sea Conference in 1987

Thus, the Norwegian government was determined to implement the commitments on nutrients, and it developed a comprehensive and detailed domestic action program on how to achieve the percentage reduction goals on nutrients (Ministry of the Environment 1992). However, actual reductions of nitrogen lag far behind both national goals and international obligations (Danish EPA 1995). One important reason is decentralized distribution of regulatory competence concerning waste-water treatment combined with strong opposition at the local level. Investments in waste-water treatment and operation of treatment plants are mainly in the hands of local authorities. Such plants are a main means of removing nitrogen and other nutrients from waste waters. A law from 1974 provides the municipalities with the possibility of financing all investments in waste-water treatment through local taxes. The municipal nitrogen-removal program has partly been cancelled due to local opposition. To cut a long story short on how great expectations in Oslo have been dashed at the municipal level, the main reason is the unwillingness of the affected municipalities to bear their part of the financial burden because they do not see the local benefits of complying with international commitments. Even though the Ministry of Environment has promised to cover 50 percent of the investment costs, the municipalities in alliance with various marine researchers and research institutions claim that the required investments represent wasted money. At the opposite pole, the State Pollution Control Authority in alliance with other marine research institutions claim that the investments are needed to combat eutrophication. The Minister of the Environment decided in 1996 to cancel the nitrogen obligations for twenty-one out of twenty-seven municipal treatment plants, due to local opposition. This case shows that whenever local perceptions of interests deviate from what the government has defined as the national interest in terms of international commitments, significant opposition at local levels can be expected. If local opposition goes hand in hand with decentralization of competence in the relevant sector, there is a risk of defection from international commitments.

Conclusions

Northeast Atlantic and North Sea cooperation has in many ways solved collective challenges by means of adopting stringent joint commitments (see table 7.3). The main challenge facing the participating states today is related to the implementation of measures to meet their obligations. Let us start with collective challenges. The commitments adopted aimed at phasing out hazardous substances and cutting nutrient emissions in half are two cases in point. Since effective collective action frequently is hard to bring about by voluntary means, the conditions for this collaborative success may provide us with important lessons. We should first note that the problem itself has remained roughly unchanged as a very malign and asymmetrical externality problem involving not only aspects of competition but also normative questions related to the basic philosophy behind pollution control in general. Most problems hampering swift collective action in the first phase of the

Table 7.3a
Summary scores: Chapter 7 (PARCOM)

| Regime | Component or Phase | Effectiveness | | Level of Collaboration |
		Improvement	Functional Optimum	
PARCOM	1974–1987	Very low	Low	Intermediate
PARCOM	1987–1995	Significant	Intermediate	High

cooperation were in fact solved by changing the rules of the game. Of particular importance was the creation of the INSC process representing a truly dynamic instrument that was able to generate political input to an administrative and incremental way of cooperation. Equally important was the existence of a legally binding convention and commission that were instrumental in creating stable mutual expectations and in implementing the political declarations. Also the European Union has been important in this respect, and the relevant E.U. directives adopted will increase in importance when the parties are to implement measures. In short, there has been a symbiotic relationship between hard and soft international law in the sense that different institutional arrangements have fulfilled complementary functions in the same issue area.

As political scientists, we are accustomed to thinking about the existence of a national government that can enforce regulations and solve disputes among affected actors as one main advantage compared to the international system. Against this backdrop, it is somewhat surprising that the main challenge today is related to a lack of domestic implementation and not the making of joint international commitments. This diagnosis may suggest that stepping up the problem-solving capacity at the national level may be more important than reforming international institutions. This chapter has been able to offer only some examples of why the process of implementing adequate measures may fail to materialize. What these measures have in common, however, is a mismatch in participation between those actors participating in making joint commitments and those whose task it is to implement measures. As noted, the agricultural ministers participated for the first time in 1993—that is, six years after the decision to cut nutrients by 50 percent was originally taken. Since the agricultural sector is a major source of nutrient inputs, representatives from this sectors should have been included earlier. Closing the mismatch between those authorized to make and implement joint commitments may

Table 7.3b

Problem Type		Problem-Solving Capacity			
Malignancy	Uncertainty	Institutional Capacity	Power (basic)	Informal Leadership	Total
Strong	High	Intermediate	Laggard	Weak	Intermediate
Strong	Intermediate	High	Laggard	Weak	Intermediate

lead to less stringent joint commitments in the short run but increase the probabilities of subsequent implementation.

Notes

1. For a comprehensive analysis of the making and implementation of North Sea pollution commitments, see Skjærseth (2000).

2. The parties to the Paris Convention are Belgium, Denmark, France, Germany, Iceland, Ireland, the Netherlands, Norway, Portugal, Spain, Sweden, the United Kingdom, Northern Ireland, and the European Community.

3. The adoption and implementation of relevant E.U. Directives are not the main focus of this chapter, although E.U. developments will be included when such developments are particularly relevant.

4. France and the United Kingdom did not accept this goal.

5. Malcolm MacGarvin works for the department of Zoology, University of Aberdeen, Scotland.

References

Danish Environmental Protection Agency (EPA), Ministry of the Environment and Energy. 1995a. Esbjerg Declaration. Copenhagen.

Danish Environmental Protection Agency (EPA), Ministry of the Environment and Energy. 1995b. *Progress Report*. Copenhagen.

Danish Environmental Protection Agency (EPA), Ministry of the Environment and Energy. *Ministerial Declarations* Bremen 1984, London 1987 and the Hague 1990. 1995. Various years.

EQUSa, 1988. Second Meeting of the Ad Hoc Working Group to evaluate the EQO and UES Approaches, London 22–23 November 1988. Environmental Quality Objectives and Standards and Uniform Emission Standards: Complementary Approaches. EQUS 2/Info.I-E.

EQUSb. 1988. Second Meeting of the Ad Hoc Working Group to evaluate the EQO and UES Approaches, London 22–23 November 1988. Up-Date of UK Position as regards the use of Environmental Quality Objectives and Uniform Emission Standards. EQUS 2/3/3-E.

Hegeman, J. H. F., L. Korver-Alzerda, J. J. Bouma, F. J. Duijnhouwer, and R. W. Hommes. 1992. *Management of North Sea Eutrophication. Erasmus Center for Environmental Studies.* Rotterdam: Erasmus University.

Hofrichter, J. 1991. *Evolution of Attitudes Towards Environmental Issues 1974–1991.* Mannheim: Zentrum für Europäische Umfrageanalysen und Studien, Universität Mannheim.

Koers, A., and B. Oxman, eds. 1984. *The 1982 Convention on the Law of the Sea.* Law of the Sea Institute, University of Hawaii.

McCormick, J. 1991. *British Politics and the Environment.* London: Earthscan.

MacGarvin, M. 1995. "Marine Eutrophication in the UK: A Discussion Document." Paper prepared for the WWF (UK). Department of Zoology, University of Aberdeen.

Ministry of the Environment. 1992. *Report No. 64 to the Storting (1991–92). Concerning Norway's Implementation of the North Sea Declarations.* Oslo.

Netherlands Ministry of Housing, Physical Planning, and Environment. 1989. *National Environmental Policy Plan*: To Choose or to Lose. The Hague.

Netherlands Ministry of Transport and Public Works. 1990. *Interim Report on the Quality Status of the North Sea.* The Hague.

Netherlands Ministry of Transport and Public Works. 1980–1984. *Water Action Program.* The Hague.

Oslo and Paris Commissions. 1984a. *International Co-operation in Protecting Our Marine Environment.* London: Oslo and Paris Commissions.

Oslo and Paris Commissions. 1984b. *A Strategy for the Future.* Adopted by the Commissions at their Sixth Joint Meeting (Oslo). 6/13/1, § 2.2.2 and Annex 5.

Oslo and Paris Commissions. 1992. *Nutrients in the Convention Area.* London.

Oslo and Paris Commissions. 1993. *Ministerial Meeting of the Oslo and Paris Commissions.* London.

Oslo and Paris Commissions. 1995. *Activities of the Oslo and Paris Commissions September 1992–June 1995.* London.

Paris Commission. 1983. *Fifth Annual Report.* London.

Paris Commission. 1987. *Ninth Annual Report.* London.

Paris Commission. 1988. *Summary Record.* London.

Paris Commission Procedures and Decision Manual. Published and continuously updated by the Secretariat of the Oslo and Paris Commissions, London.

Quality Status Report of the North Sea (QSR). 1987. London: Department of the Environment.

Quality Status Report of the North Sea (QSR). 1993. Concerning Norway's implementation of the North Sea Declarations. 1991/92. Ministry of Environment, Oslo. *Parliamentary Report* No. 64. London: Oslo and Paris Commissions.

Richardson, J. 1994. "EU Water Policy: Uncertain Agendas, Shifting Networks and Complex Coalitions." *Environmental Politics* 3(4) (Winter): 139–167.

Sætevik, S. 1988. *Environmental Cooperation between the North Sea States.* London: Belhaven Press.

Skjærseth, J. B. 2000. *North Sea Cooperation: Linking International and Domestic Pollution Control.* Manchester: Manchester University Press.

This Common Inheritance (1990). London: HMSO.

Underdal, A. 1995. "Implementing International Environmental Accords: Explaining 'Success' and 'Failure.'" Paper prepared for presentation at the thirty-sixth Annual Convention of the International Studies Association, February 21–25, Chicago.

Weale, A. 1992. "Implementation Failure: A Suitable Case for Review?" In E. Lykke, ed., *Achieving Environmental Goals: The Concept and Practice of Environmental Performance Review.* London: Belhaven Press.

8

The Convention on Long-Range Transboundary Air Pollution (CLRTAP)[1]

Jørgen Wettestad

Overall Performance: Substantial Regime Achievements But a Mixed Success

In 1968 the Swedish scientist Svante Oden published a paper in which he argued that precipitation over Scandinavia was becoming increasingly acidic, thus inflicting damage on fish and lakes (Oden 1968). Moreover, it was maintained that the acidic precipitation was to a large extent caused by sulfur compounds from British and Central European industrial emissions. This development aroused broader Scandinavian concern and diplomatic activity related to acid pollution. The specific background for formal negotiations on an air pollution convention was the East-West détente process in the mid-1970s, in which the environment was identified as one potential area for cooperation. Due to the East-West dimension, the United Nations Economic Commission for Europe (UNECE) was chosen as the institutional setting for the negotiations. The ECE Convention on Long-Range Transboundary Air Pollution (CLRTAP) was signed in Geneva in November 1979.

So why is CLRTAP a classic mixed-success regime? On the one hand, a substantial international regulatory progress has taken place. There has been a steady development of protocols, covering more substances with regulations gradually becoming both binding, specific, and fine-tuned to ecological and economic variations between the countries. Moreover, national compliance with these protocols must be characterized overall as high. In addition, this process has clearly led to a marked shift in the perceptions of the air pollution problems in most of the countries involved. Hence, the regime has clearly contributed to the reductions in emissions witnessed and the promising steps taken toward problem-solving. On the other hand, a closer scrutiny of available national and subnational knowledge indicates that many forces other than the CLRTAP regime have been involved in bringing about policy changes and emissions reductions in this field. Much behavioral change would probably have happened anyway—due to energy policy changes, more fundamental economic and industrial changes, European Community processes, and so on,[2]—even if the picture may vary between countries.[3] Hence, relative improvement brought about specifically by the regime seems only moderate so far. In terms of environmental optimality and problem solving, there was a widespread feeling in the scientific community that the targets in the first rounds of protocols were ecologically ineffective but were, however, steps in the right direction. Increasing knowledge about critical limits in the environment has revealed that the gap to healthy, sustainable conditions continues to be considerable in vulnerable areas. Nevertheless, the development of this knowledge and more active use of critical loads in decision making are important contributions to the process of narrowing the gap to such critical levels in the environment.

Brief Institutional Background

The ECE Convention on Long-Range Transboundary Air Pollution (CLRTAP) was signed by thirty-three contracting parties (thirty-two countries and the EC Commission) in Geneva in November 1979. Four main aspects of the 1979 Convention may be discerned: first, the recognition that airborne pollutants were a major problem; second, the declaration that the parties would "endeavor to limit and, as far as possible, gradually reduce and prevent air pollution, including long-range transboundary air pollution" (article 2); third, the commitment of contracting parties "by means of exchange of information, consultation, research, and monitoring, to develop without undue delay policies and strategies which should serve as a means of combating the discharge of air pollutants, taking into account efforts already made at the national and international levels" (article 3); and fourth, the intention to use "the best available technology which is economically feasible" to meet the objectives of the Convention (Nordberg 1993). The Convention did not specify any pollutants but stated that monitoring activity and information exchange should start with sulfur dioxide (SO_2). The Convention has been in force since 1983 with a current membership of forty-eight parties. Moreover, the Convention was to be overseen by an executive body, which included representatives of all the parties to the Convention as well as the European Community. Furthermore, the ECE secretariat was given a coordinating function. The institutional structure has also included several working groups, task forces, and international cooperative programs (further described below in the section on the development of problem-solving capacity). Rooted in the Convention's strong initial focus on knowledge improvement and monitoring, a specific financing protocol for the Cooperative

Program for Monitoring and Evaluation of Long-Range Transmissions of Air Pollutants in Europe (EMEP) was established in 1984.

Substantiating the Mixed Success So Far Assessment

As indicated, a substantial international regulatory development has taken place in the wake of the 1979 framework convention. Briefly summing up, the first main regulatory step in the cooperation was the 1985 Protocol on the Reduction of Sulfur Emissions. In Helsinki in July 1985, twenty-one countries and the European Community signed this legally binding protocol. The Helsinki Protocol stipulated a reduction of emissions and transboundary fluxes of *sulfur dioxide* (SO_2) by at least 30 percent as soon as possible and by 1993 at the latest, with 1980 levels as baseline. However, some major emitter states failed to join the agreement, among them the United Kingdom, the United States, and Poland. The Helsinki Protocol entered into force in September 1987 and has been ratified by twenty-one parties. In the 1988 Sofia Protocol on Nitrogen Oxides, the signatories pledged to freeze *nitrogen oxide* (NO_x) emissions at the 1987 level from 1994 onward and to negotiate subsequent reductions. Twenty-five countries signed the Sofia Protocol, including the United Kingdom and the United States. Moreover, twelve European signatories went a step further and signed an additional (and separate) joint declaration committing them to a 30 percent reduction of emissions by 1998. The Sofia Protocol entered into force in February 1991 and has been ratified by twenty-six parties.

The next step was the 1991 Geneva Protocol on Volatile Organic Compounds. *Volatile organic compounds* (VOCs) are chemicals that are precursors of ground-level ozone. The Geneva Protocol called for a reduction of 30 percent in VOC emissions between 1988 and 1999, based on 1988 levels—either at national levels or within specific tropospheric ozone-management areas. Some countries were allowed to opt for a freeze of 1988 emissions by 1999.[4] Twenty-one parties signed the Geneva Protocol in 1991; Portugal and the European Community joined in 1992. Russia and Poland chose not to sign. The Geneva Protocol entered into force in September 1987 and has been ratified by seventeen parties.

A new 1994 Sulfur Protocol was then signed in Oslo in June by twenty-eight parties. This Sulfur Protocol was based on the critical-loads approach, which negotiates emissions reductions by assessing the (varying) effects of air pollutant (here, *sulfur*) rather than by choosing an equal percentage reduction target for all countries involved.[5] Hence, the Sulfur Protocol sets out individual and varying national reduction targets for the year 2000 for half of the countries and additional 2005 and 2010 targets for the other half—with 1980 as base year. The Sulfur

Protocol entered into force in August 1998 and has been ratified by eighteen parties.

The most recent regulatory steps include two new protocols on transboundary air pollution by *heavy metals* and *persistent organic pollutants* (POPs), which were signed by thirty-four parties in Århus in June 1998. Finally, negotiations on a new *multipollutant and multieffects* protocol were concluded in September 1999, and the protocol was signed in Gothenburg in December. The Gothenburg Protocol is by far the most advanced within the regime so far and covers four substances (NO_x, VOCs, NH_3 [ammonium], and SO_2) and three environmental effects (acidification, ground-level ozone, and eutrophication).

In terms of relative improvement, have regime activity and decisions led to greener national policies in this issue area than would have been the case without this regime activity? Turning first to the formal compliance picture, CLRTAP must be characterized as an overall high-compliance regime. For instance, with regard to the 1985 Sulfur Protocol, twenty-one parties had overall reduced emissions by around 48 percent by 1993, and all parties had reached the 30 percent reduction target. Regarding the 1988 Sofia Protocol, the NO_x stabilization target had been reached by a clear majority of the parties in 1994, but the data situation was not optimal. In the years following 1994, nineteen of the twenty-six parties to the Sofia Protocol have managed to keep their emissions at or below the 1987 base level (ECE/CLRTAP 1998, 119). Turning to the Geneva Protocol, data from 1998 indicate that six parties had already achieved the VOC target levels; six more parties "appeared to be on course to do so"; three parties had experienced (for two of them, substantial) increases in emissions; and lacking data complicated assessments for the remaining two parties (ibid., 122–123).

However, even if compliance is generally quite good, to what extent can the CLRTAP regime be credited. Would the reductions and policy changes have happened anyway? To answer such questions, we have to enter the territory of implementation and regime-induced national behavioral changes. First of all, available evidence indicates that major causal factors for initial compliance levels are not at all found within the sphere of environmental politics. On summarizing and discussing the sulfur and NO_x implementation processes in the United Kingdom, Germany, the Netherlands, and Norway in another context, I concluded that the majority of the initial reductions and compliance levels achieved in three of the four countries (the United Kingdom, the Netherlands, and Norway) are apparently explained by processes that are not primarily related to environmental protection, at least with regard to sulfur reductions (Wettestad 1996). In the United Kingdom, industrial recession and reduced energy demand in the 1980s were important factors. The NO_x picture is more varied, but privatization and the switch from coal to gas were important factors. In the Netherlands, a gradual conversion to domestic natural gas and domestic political and financial reasons was clearly important. In Norway, much was achieved by reducing consumption of heavy fuel oil on land. The exception is Germany, where environmental regulations were—at least initially—the main driving forces.

Focusing more closely on the CLRTAP contribution to these processes, the direct, easily detectable influence has been moderate, at least for these particular countries. In a situation without the CLRTAP, significant initial reductions would probably have taken place anyway, due to other economic and political processes and domestic political pressure motivated by environmental damage. Take, for instance, Germany. It is important to remember that Germany was almost as reluctant as the United Kingdom at the Convention negotiations in the late 1970s. Without the rapidly increasing concern about forest damage (*Waldsterben*) in the early 1980s, both German and European acid-rain politics would have looked very different today. Moreover, with regard to German and British acid-rain politics, the European Community decision-making arena has possibly been more important than the CLRTAP. However, there is still a possibility that the CLRTAP has been more important for the other West European and East European countries. This is indicated by Levy (1993, 118–121), for example, who suggests that countries like Austria, Finland, the Netherlands, and Switzerland were influenced by the CLRTAP through increased awareness of domestic acid-rain damage. Moreover, countries like Denmark, the United Kingdom, and the Soviet Union were influenced by the CLRTAP through various types of linkage effects. In addition, the CLRTAP process has been important as an arena for creating and maintaining intergovernmental confidence and learning. As has been pointed out by one experienced regime analyst, "It is hard to imagine being where we are today regarding transboundary air pollution in the absence of the LRTAP process."[6] In summary, a preliminary causal analysis of the regime indicates that regime activity and decisions have led to somewhat greener acid-rain policies—to varying degrees in the various countries—than would have been the case without this regime activity. Regime impact increases over time, not least being related to important knowledge improvement and learning effects. However, more in-depth studies of countries such as Germany, the Netherlands, the United Kingdom, and Norway also indicate that factors not directly related to the regime seem more important in central countries for shedding light on the compliance levels achieved by the mid-1990s.

Let us then turn to the question of *distance to collective optimum* and first to the evolving match between scientific knowledge and advice and the international policy measures taken in response.[7] With regard to the sulfur process, although scientific uncertainty was reduced and consensus grew in the early 1980s (as is further specified later), there seems currently to be general agreement on regarding various aspects of the 30 percent Protocol as only remotely related to specific scientific evidence. The baseline of 1980, the target date of 1993, and the 30 percent reduction target have been characterized as quite arbitrary (Haigh 1989); 20 percent was too little, and 40 percent seemed too much. However, my own interviews have confirmed the impression that most scientists definitely saw the 30 percent reduction as a significant step in the right direction. Hence, many scientists (at least in Germany and the Nordic countries) would probably have preferred more substantial reductions but were evidently reluctant to rally around specific, higher reduction targets. In several ways the NO_x science and politics story is different from the sulfur story. First, the scientific understanding of the more complex NO_x problems, at least for emissions, was much less developed than in the case of the sulfur problems, which had been studied longer. The start of the negotiation process can be characterized as a spinoff of the successful establishment of the Sulfur Protocol (Bakken 1989, 202). Hence, initial targets discussed in the negotiations were borrowed first and foremost from the sulfur context and based more on political ambitions among pusher states than on well-developed scientific evidence. A knowledge-improving process was organized together with the negotiations. There is little doubt that improved knowledge influenced several participants' perceptions and positions and the course of the negotiations, generally in a quite sobering direction (Wettestad 1998). The initial widespread support for a 30 percent reduction target declined in the course of time, largely due to increasing scientific and technological knowledge. Still, in terms of distance to an optimal environmental situation, there is little doubt that the stabilization target finally agreed on was not a substantial contribution to narrowing this gap.

Let us then briefly turn to the science and politics relationship within the VOC process. Generally, VOC problems are more similar to the nitrogen oxide problems than to the sulfur problems—with, for instance, a multitude of emissions sources. It is not surprising, then, that the VOC process entailed several similarities with the NO_x process. Hence, a rapid knowledge production took place in a rather short time span and primarily within the regime context. The general impression is an overall acceptance of the scientific evidence at hand but also a general sobering recognition of the considerable uncertainty and complexity involved. This recognition was reflected in the unprecedented flexibility and complexity of the Protocol's commitments: the basic 30 percent reduction commitment was sweetened by elements like a flexible base year (between 1984 and 1990), a freeze option for small polluters, and for some countries only a reduction by 30 percent within specific tropospheric ozone-management areas (TOMAs). Hence, in terms of narrowing the gap to the environmental optimum, the VOC protocol must also be regarded as a moderate contribution. Turning finally to the 1994 Sulfur Protocol and the question of critical loads, this concept was already introduced in the NO_x negotiations. However, it was not before the negotiations on a new and revised sulfur protocol took place that this approach really moved to center stage. A *critical load* is defined as "a quantitative estimate of an exposure to one or more pollutants below which significantly harmful effects on specified sensitive elements of the environment do not occur according to present knowledge."[8] Early in the negotiations, it was recognized that it would be impracticable to reduce sulfur depositions below critical loads by the end of the century. It was instead agreed to reduce the gap (first by 50 percent and then in 1993 by 60 percent) between current levels of sulfur deposition and critical loads in most of Europe, except for the most acid-sensitive areas. The final Protocol was decided on in Oslo in June 1994, setting out individual and varying national reduction targets for the parties.

Given the patchy, though improving match between scientific notions of ecological requirements and the regime regulations adopted, what is the current picture concerning actual solutions to the problems? If problem solutions are regarded as synonymous with the achievement of critical loads in Europe, then it is clear that most countries have some way to go before these critical targets are achieved. With regard to the issue area of acidification, the targets in the 1994 Sulfur Protocol involve as a minimum a 60 percent reduction of the gap between 1990 sulfur depositions and the critical levels in the environment. Hence, even if the United Kingdom, for instance, achieves its 80 percent emissions cuts target before 2010 (with 1980 as baseline year), then the country's emissions will still be some 40 percent above ideal levels. Moreover, there is definitely no direct, linear relationship between emissions cuts and environmental improvements. We are talking about gradual, slow processes in which both positive and negative effects may not be seen for several years or decades after emissions changes have been effected. Take, for instance, the Norwegian situation: at the same time that emissions in Europe were cut by around 40 percent, acidification damage in Norway gradually increased—up to a 1995 peak, when an affected area of over 110,000 square kilometers received more sulfur than nature could handle. Although rainfall gradually became less acid, acid deposition did not decrease accordingly, due to increased rainfall patterns. However, more recent scientific reports indicate improvements and possibly a reversed trend with regard to acidification. Hence, overall, the transboundary air pollution problems have been reduced somewhat, though an ultimate solution to the problems, getting close to critical loads in both the rural and urban environment, is a matter both of implementing recently adopted regulations and of revising regulations. In reality, this is a venture extending well into the next century.

The Background for the 1979 Framework Convention: High Malignity and Overall Low Concern

Thinking back, how can the transboundary air pollution problems be characterized in the mid- to late 1970s in terms of knowledge and main interests? And how can these initial problem characteristics throw light on the regime-creation phase and the establishment of the 1979 Framework Convention? The overall picture was one of high uncertainty and complexity. In the initial regime-formation phase, both scientific and political attention was primarily directed toward sulfur emissions and related acid precipitation. Important questions included transboundary transport of pollutants and their subsequent impact. In 1978 and 1979 only limited knowledge existed in this field. Concerning the transboundary question, results from an OECD project commenced in 1972 and including the eleven European member countries were published in 1977. Here, countries were classified either as net exporters or net importers, although all countries were also recipients of their own emissions. For example, Denmark and the United Kingdom were identified as net exporters, while countries such as Austria, Finland, Norway, Sweden, and Switzerland were identified as net importers. The report indicated that the United Kingdom was the major exporter, producing more sulfur pollution than any other nation. However, the Organization for Economic Cooperation and Development (OECD) itself emphasized the degree of uncertainty involved in the measurement of emissions. Moreover, as the deposition and impact questions were by their very nature more complicated than the transport question, overall uncertainty was high—despite major efforts like the Norwegian acid-rain project (see Park 1987). There was some degree of scientific disagreement, which is quite natural in such a complex field of research. However, there seems to have been less pure scientific controversy than one may be led to believe from the rather heated public debate.[9] Intellectual complexity was in principle quite high, given that many types of anthropogenic emission sources were part of the picture (power stations, factories, vehicles, ships, and so on). Further, several natural sources were involved whereby a number of substances also contributed to the problem (such as SO_2, NO_x, and hydrocarbons), which through interaction combined to produce an acid cocktail. However, as indicated, attention in this early phase was particularly focused on problems associated with sulfur dioxide emissions from power stations (particularly coal-fired plants).

Key words with regard to the international regulatory situation were *asymmetry* and *complexity*. Let us first turn to the basic interdependence of the involved states. The scientific and technological uncertainty surrounding the transboundary air pollution problems at this initial stage indicated the existence of a cost-efficiency potential. A more comprehensive understanding of the regional ecosystem and a cheaper and faster route to effective abatement equipment represented the actual synergy potential. However, improved scientific and technological knowledge alone did not reduce the transboundary transmission of pollutants. To reduce these externalities, such knowledge had to be applied and transformed into practical, national programs for reduction of emissions. It should be kept in mind that emissions sources were numerous, including not only power stations and transport but also various industrial sectors. The need for such emission-reduction programs affecting vital economic sectors immediately introduced a competitive dimension to the problem. This was also largely related to the technological abatement situation at the time. Several technological abatement alternatives relating to both power stations and industry existed, these being essentially concerned with the reduction of sulfur content in fuel prior to combustion or the extraction of sulfur from the resulting emissions. However, cheap and simple technological "fixes" were simply not available at the time.[10]

With regard to environmental vulnerability, the OECD study indicated both symmetrical and asymmetrical dimensions: symmetrical, in that regionally important countries like the FRG seemed to import as much pollution as they exported. However, the asymmetrical aspects seemed more striking. Due to atmospheric currents and climatic conditions, several countries such as the United Kingdom seemed to be net exporters of pollutants. Moreover, the United Kingdom largely exported its emissions to the net importing Scandinavian countries.[11] This particular relationship was further aggravated by the fact that the soil in Norway and Sweden has a relatively poor natural tolerance of acid precipitation. With regard to vulnerability related to abatement costs, the use of coal was an important factor in the energy profiles of important emitters like the United Kingdom and West Germany. In the case of Britain, the country derived more than 90 percent of its energy needs from fossil fuels, mostly through coal-fired power stations. Thus, given the technological situation, cleaning emissions would clearly be expensive for the United Kingdom and would apparently mainly benefit other countries' environments—due to atmospheric currents and the very low concern for British acidification damage at the time (see Hajer 1995). This coal factor also had some implications for the complexity of the transboundary air pollution problems. As the United Kingdom had large domestic coal supplies of specific sulfur content, coal had supply-security implications as an alternative to imported oil. Moreover, more general economic and employment implications were involved as coal-dependent countries like the United Kingdom and the United States had experienced economic depression in the mining areas (Miller 1989).

Based on the preceding problem diagnosis, one would expect quite diverging actor preferences. This holds true. Let us briefly take a closer look at the national positions and arguments of some of the main actors during the negotiations on the LRTAP Convention.[12] The broad picture contains no surprises. Basically, there were a few, small activists and many disinterested and reluctant parties. Mainly importer countries including the Scandinavian countries insisted on binding and specified international measures. More specifically, the Scandinavians put forward a *stand-still and roll-back proposal*, within the framework of a convention. The proposal called on signatories to hold the line against further SO_2 increases (stand-still) and begin to reduce SO_2 pollution levels by fixed, across-the-board percentages (roll-back). As could be expected, major emitter and exporter countries like the FRG and the United Kingdom opposed these proposals.[13] The British delegation called for more research to firmly establish responsibility for long-range pollution. However, it should be noted that the U.K. position at this point seems to have been somewhat more open than that of the FRG. The United Kingdom accepted broad, general statements of policy and even of measures to achieve this policy, as long as no positive obligations were imposed or assumed (Wetstone 1987, 179). Furthermore, it has been maintained that the British finally decided to accept the Convention because they believed that their plans for increased reliance on nuclear power for generating electricity would bring about a net reduction in sulfur emissions. The FRG wanted the ECE agreement to include multilateral research exclusively (Wetstone and Rosencrantz 1983, 206). The Germans did not like the idea that the proposed coordinating function of the ECE Secretariat could authorize it to intervene in the internal affairs of member states. French and Scandinavian diplomatic pressure allegedly induced the Germans to sign.

In summary, the transboundary air pollution problems in this initial phase must fundamentally be characterized as malign and complex problems, shrouded in scientific uncertainty and to some degree also controversy. One of the most important aspects to keep in mind is that central roots of the problems have been—and to a large extent still are—closely associated with the energy-supply sector. This was illustrated by the significant role of coal-fired power stations for the British sulfur emissions. Industrial competitive ability is of course crucial for industrial societies, but energy supply is in addition a kind of metaindustry, being both a precondition and a basis for the development of other sectors—and an industrial sector in itself. Hence, regulatory efforts raised problems of significance to regional employment as well as to national security and defense issues. Combined with marked asymmetries with regard to environmental and regulatory impacts, the prospects were not very bright for a further development of the process.

The Background for Steady Regime Development: Less Polarization and More Concerned Actors

The empirical ground covered in this brief section is immense. We are talking about twenty years of knowledge and political development, with seven protocol negotiation processes, each with its particular intellectual and political characteristics.[14] With regard to intellectual progress, there has been a development from scientific controversy to critical loads. Uncertainty about the character of transboundary pollution was clearly reduced during the 1980s, at least concerning transport and depositions. This was largely due to the effort of the EMEP program.[15] Studies of the development of national perceptions of the acidification problems also clearly indicate shifting perceptions in countries like the Netherlands, Austria, and Switzerland in the wake of the 1982 Stockholm Conference, and the German change in perceptions and problems.[16] However, uncertainty was generally more reduced with regard to SO_2 than to NO_x and VOC emissions. This is, of course, associated with the fact that the sulfur-related problems had a much longer history in terms of scientific and political attention—and that the NO_x and VOC problems were intrinsically more complex. Regarding emission sources, a substantial portion of NO_x emissions stems in many countries from the inland and coastal transport sectors in addition to emissions from power stations and other industry. The VOC emission picture is even more complex. Overall, VOC emissions mainly emanate from three sources—motor traffic, the use of solvents in industrial and household appliances, and oil and gas industries. However, hundreds of compounds qualify as VOCs, and they are emitted in a wide range of industrial activities.[17] Still, regime-initiated knowledge-improvement processes led to significant improvements also in the knowledge of NO_x and VOCs. From the late 1980s the critical-loads tool becomes the main factor.[18] It has been indicated that this approach contributed to a change in the fundamental cooperative atmosphere (see Levy 1995, 63): "The most fundamental effect that critical loads had was to shift the nature of the public debate, both internationally and in many domestic settings, away from determining who the bad guys were, and towards determining how vulnerable each party was to acid rain." Moreover, it is an outsider's impression that knowledge of the interplay between different substances—and also between these substances and various natural factors—advanced considerably during the 1980s and early 1990s. In summary, these developments led to a more sophisticated and relaxed debate on international air pollution, in both scientific and political contexts.

Turning to the more specific political development, an overall trend toward a more symmetrical and complex picture can be noted. With regard to the general environmental vulnerability dimension, some important changes occurred in the beginning of the 1980s. The main change was the altered position of the FRG. After the increased concern over potentially extensive forest damage in 1981 and 1982, the FRG emerged both as an important emitter of polluting substances and also as a country that perceived itself as seriously affected by such emissions. More or less similar changes occurred in countries including the Netherlands, Switzerland, and Austria. In the United Kingdom, the situation was mainly unchanged, although some critical voices were making themselves heard. Gradually, however, acid damage to the U.K. countryside became more visible, although the extent of damage remained disputed throughout the 1980s. For instance, the Forestry Commission stated in 1987: "There is no sign of the type of damage seen in West Germany occurring in Britain at the moment" (Waterton 1993, 83–86).

With regard to aspects related to abatement costs and economic competition, this was an important complicating factor up through the 1980s and into the 1990s. The shifting of regulatory focus first to NO_x and then to VOC emissions added new political games to the initial sulfur, power-station-focused context. An important element in both the NO_x and VOC contexts was the regulation of the transport sector. In a way, increasing focus on this sector contributed to a more complex and less categorical picture in terms of regulatory impact. The transport sector was rather important in all the relevant countries, and most of them experienced regulatory problems in dealing with such emissions. Hence, the asymmetrical sulfur picture was gradually replaced by a more symmetrical one. In addition, it should be noted that competitive aspects may in some instances strengthen the cooperative drive, not only complicate it. This is clearly part of the picture in the development of the air pollution issue: when the FRG changed its mind and decided on various types of abatement measures and programs in the early 1980s, the competitive dimension necessitated an active international effort to harmonize international terms of competition—and the FRG was at least regionally a major actor.[19] On the technological side, there were no major breakthroughs during the 1980s. Within the sulfur context, flue-gas desulphurization was the main technological option, generally regarded as a quite expensive one (see McCormick 1989, 99).

With regard to the more specific development of preferences and processes at the beginning of the 1980s within the first sulfur context, the main change in the broad preference picture was the rather dramatic about-turn of West Germany, internationally announced at the Stockholm Conference on Acidification in 1982. Having been a somewhat reluctant signatory of the ECE Convention, the FRG now emerged in support of the Scandinavian call for international action. Interior Minister Baum pledged his government's commitment to a 60 percent reduction in SO_2 emissions from power stations and large factories by 1993. The domestic background for this change in position was multifarious, with *Waldsterben* and the emergence of the Green Party as important factors, but there was also a regional political dimension.[20] The change in policies that moved Germany from international laggard to leader has been further confirmed in the ensuing NO_x, VOC, and second sulfur negotiations.[21] In my view, the German policy change must be characterized as the most important event within the CLRTAP context—due to Germany's high score both with regard to basic game-power capabilities (being an important emitter) and policy game capabilities (due to Germany's influential position scientifically, economically, and politically among neighboring European states).

With regard to the NO_x and VOC processes, they partly reflected different problem structures compared to the sulfur context. Hence, a more balanced picture with regard to leaders and laggards—and also rather different coalitions and lines of conflict—can be noted in these processes. First, a quite stable group of leaders can be identified in both processes, including Germany, Sweden, the Netherlands, Austria, and Switzerland. Hence, the initial, Nordic "sulfur Mafia" was considerably broadened. Second, Nordic sulfur unity broke down. For instance, in the NO_x context, midway through the negotiations Norway and Finland adopted more intermediary and some would even say laggard positions, as the important role of troublesome transport emissions was gradually clarified (see Stenstadvold 1991, 104). Third, as indicated above, the position of the United Kingdom, the important acid emitter and sulfur laggard, became more intermediary and less obstructive in these processes.[22] The more recent development of preferences and positions, witnessed in connection with the negotiations concerning the 1994 Sulfur Protocol, supports the impression that the main political game has changed somewhat (see Levy 1995; Churchill, Kutting, and Warren 1995). For instance, the initial Nordic leader countries from the 1970s and early 1980s were now joined not only by Germany but also by countries including the Netherlands and Austria. Countries that have not experienced a change in general affectedness of acid rain—such as Greece, Italy, Spain, and Portugal—formed a disinterested or reluctant group. Member states of the European Union did not adopt a common position in the negotiations, and the rifts between some member states during special EU policy-coordinating sessions were allegedly difficult to bridge. The United Kingdom was not regarded as in the vanguard of supporters for the protocol and was often perceived as obstructive (Churchill, Kutting, and Warren 1995, 182).

Problem characteristics have become somewhat less malign over time, even on the intellectual side. It has been indicated that the regime has contributed significantly to knowledge improvement over time. On this background, let us turn to a broader discussion of the issue of problem-solving capacity.

The Development of Problem-Solving Capacity: More and Stronger Leaders, Fair Institutional Record[23]

In terms of *level of collaboration*, the CLRTAP regime has clearly coordinated action on the basis of explicitly formulated standards—in this case protocols—but national action has been implemented solely on a unilateral basis. As is further described below, national reports have formed the basis for some sort of centralized appraisal of effectiveness, with the appraisal element increasing in importance over time.

With regard to the related issues of *entrepreneurial leadership* and *distribution of power*, there have been some variations, but these aspects of the problem-solving capacity have generally been increasing in strength over time. As some relevant aspects of these issues were introduced in the previous section, only some important points related to the various processes within the regime are summed up here. With regard to the regime-formation stage, the broad picture is not surprising: main importing countries like the Scandinavian countries insisted on binding and specified international measures and exercised a moderately effective entrepreneurial leadership. The power balance was clearly skewed in favor of the major emitters and process laggards. Although the Nordic countries achieved the establishment of an international regulatory framework, it was weak in content. It is highly debatable whether the Scandinavian efforts would have achieved even this moderate victory if the more or less hidden agenda of East-West confidence building had not been present in important actors' calculation of interests. In the first sulfur process, Scandinavian leadership efforts continued. However, the main change in the picture was, of course, the changed position of the FRG. After the discovery of extensive forest damage in 1981 and 1982, the FRG emerged as an important emitter that also perceived itself as seriously negatively affected by these emissions. Moreover, the adoption of domestic German abatement programs created clear incentives to take on an international leadership position to avoid a related reduction of the regulated industries' international competitive situation. In sum, the power balance within the regime was somewhat altered, and entrepreneurial leadership was considerably strengthened in the process toward the first Sulfur Protocol.

The picture in the NO_x and VOC negotiations share many similarities: first, a quite stable group of entrepreneurial leaders can be identified in both processes, comprising countries like Germany, Sweden, the Netherlands, Austria, and Switzerland. Hence, the initial, Nordic "sulfur Mafia" was considerably broadened. On the other hand, Nordic sulfur unity broke down and changed the main entrepreneurial picture. For instance in the NO_x context, midway through the negotiations, Norway and Finland adopted more intermediate and some would even say laggard positions, as the important role of troublesome transport emissions was gradually clarified. In the second sulfur negotiations, the initial Nordic pushers were joined not only by Germany but also by the Netherlands and Austria, indicating marked changes in the power balance within the regime.

Turning to some potentially important *institutional* issues, in terms of *decision-making rules* the CLRTAP regime has been fairly standard, although there are some interesting flexibility elements. As point of departure, article 12 in the 1979 Convention stated that amendments to the Convention should be adopted by consensus. A natural assumption is that the consensus requirement has obviously reduced the strength of the Protocols and hence the course of the cooperation. However, closer scrutiny raises questions on such a thesis. First, consensus has been practiced with flexibility. Informally, a liberal norm has developed, opening up for establishing protocols without the consent of all Convention parties. For instance, in connection with the establishment of the 1985 Sulfur Protocol, the majority of the parties secured formal acceptance from the reluctant United Kingdom.[24] Second, regime participants indicate that the consensus requirement also has changed somewhat during the course of time, with greater willingness more recently to push the reluctants harder. This may then be looked on as another sign of a flexible application of the basic consensus requirement. Still, the formal consensus approach is seen as an integral element in the whole functioning of the cooperation, rooted in the initial East-West confidence-building dimension of the regime. Hence, more formal changes seem very unlikely (Norwegian interviews, fall 1995).

However, would regulatory strength and possibly regime effectiveness necessarily have been higher with majority-voting procedures? This question is perhaps not as obvious as may appear at first glance. For instance, it is questionable if in any of the negotiation processes carried out so far there has been a majority for much stronger decisions. To my knowledge, in connection with the 1985 sulfur negotiations, there were never any serious discussions about a more ambitious 50 or 60 percent cut. A 30 percent cut became a focal point quite early in the negotiations. In the NO_x negotiations, there was possibly an initial majority in support of a 30 percent cut, roughly similar to the target adopted in the preceding sulfur context. However, when the final stage of the negotiations started, this possible majority was long gone.

The *role of the Secretariat* does not add very much explanatory power. In essence, it has functioned well but within clear constraints. Related to the initial choice of ECE as the basic cooperative context, the CLRTAP Secretariat functions have been provided by the Air Pollution Section of the ECE Environment and Human Settlements Division, with technical support from the ECE Secretariat. Hence, the CLRTAP secretariat functions were added to an existing institution's agenda. In terms of resources, given the broad regional context with a substantial number of parties with poor domestic administrative resources for environmental issues, the CLRTAP secretariat staff can hardly be characterized as oversized and overendowed with resources.[25] The Air Pollution Section has long had five professional staff members and two support staff members. The costs of meetings, documentation, and secretariat services have been covered by the regular budget of the UNECE (Green Globe Yearbook 1996, 86). Hence, the Secretariat is not financed directly by the parties but is dependent on the wider ECE resource situation. As stated by participants in the cooperation, "We are entirely at the mercy of the ECE" (Norwegian interviews, fall 1995). Against the background of the general financial crisis in the United Nations, the ECE financial situation has not developed positively either. Hence, perhaps to a greater extent than many other international environmental regimes, the CLRTAP Secretariat is weak by design. Given the modest resource situation and no mandate for activism from the parties, the role of the Secretariat has become very much that of a stagehand, arranging meetings within the various scientific and political bodies within the regime and collecting an increasing amount of emission and compliance information. With the scientific and political issues having become considerably complex, the issues and tasks that can practically be addressed have been restricted. A more recent additional task is related to the servicing of the 1994 Sulfur Protocol Implementation Committee. Hence, the resource situation of the CLRTAP Secretariat seems to have developed unfavorably over the years as servicing functions related to new protocols have successively been added to the workload—and the administrative capacity has remained more or less constant. Still, the impression is that the Secretariat has done a good job, despite limited resources (Norwegian interviews, fall 1995).

In the previous sections, the role of improved knowledge and the regime's contributions in this connection has been mentioned several times. This indicates that the *organization of the scientific and technological input* should be an institutional aspect of both interest and potential relevance. Let us first briefly describe the evolution of the institutional set-up. A first thing to note is that one part of the CLRTAP scientific and political complex remained from pre-Convention days (more specifically the Working Party on Air Pollution Problems). Other parts of the scientific and political complex grew out of the aforementioned OECD monitoring program—namely, the EMEP. The rest of the complex was established in connection with the Convention, with subgroups gradually being added to the structure. The EMEP has been financed by mandatory contributions from the parties, according to a specific Protocol established in 1984. The costs related to other scientific bodies under the Convention have been covered by voluntary contributions from the parties. Generally, over time, bodies have been added to the institutional structure according to developing scientific and political needs. In recent years the organization has emerged along the following general lines: on the administrative and political side, in addition to the Executive Body (EB), there is now a Working Group on Strategies (WGS); on the scientific and technical side, in addition to the Working Group on Effects (WGE) and the EMEP Steering Body, there is now a Working Group on Technology (WGT); current International Cooperative Programs under the WGE are forests (led by Germany), freshwater (Norway), materials (Sweden), crops (United Kingdom), and integrated monitoring (Sweden).

Has this organizational model been a successful one? First, the basic flexibility of the system must be declared a success. As noted by Levy (1993), the fact that the CLRTAP has been "consistently science- and ecosystem-driven" means that working groups have progressively been organized around potential environmental damage and permitted transfrontier pollutants to enter onto the diplomatic agenda: "This accounts for the ease with which VOCs entered the agenda, as well as for the current investigations into mercury and persistent organic compounds" (Levy 1993, 111). Second, the formally advanced (with financing based on a separate, specific protocol) and well-functioning EMEP system has represented a strong scientific foundation and core in the development of the regime. A third interesting element in the CLRTAP model is the establishment of a permanent negotiating forum in the Working Group on Strategies (WGS). This body may be seen as a mediating buffer between science and politics—a not too formal meeting place for scientists and administrators, allowing the building of consensual knowledge on both scientific and political strategic matters. Regime participants emphasize the flexibility in frequency of meetings and generally much less time-consuming formalities as an advantage of the WGS style of functioning compared to the Executive Bureau meetings (Norwegian interviews, fall 1995). Finally, regime participants also emphasize that the personal component should not be disregarded in this connection, as the CLRTAP scientific and political complex has been blessed with key persons combining scientific and political perspectives (ibid.).

As a brief follow-up to the previous section, an institutional aspect that should briefly be commented on is the role of EMEP in the verification context. In terms of *reporting and verification*, CLRTAP is quite standard. Generally, monitoring of state performance has been mainly based on the parties' annual reports to the Secretariat on emissions and procedures adopted for the abatement of emissions and on the measurement of acid precipitation. In addition to the annual reports, there is a more comprehensive review of national abatement strategies and policies every four years. Furthermore, in connection with the 1994 sulfur protocol, a special Implementation Committee was established, composed of eight legal experts from the parties. However, what makes CLRTAP rather different is a limited capacity for additional, independent verification of emission figures. This capacity stems from the EMEP monitoring system, based on emission data, measurements of air and precipitation quality, and atmospheric distribution models. Referring to the periodic reviews and the EMEP system, Sand (1990, 259) maintains: "Few other international agreements can be said to come equipped with verification instruments of this calibre." However, the EMEP reports are also partly based on national reports.[26] Moreover, some countries have failed to report emissions, and monitoring coverage in Southern Europe is seen by EMEP itself as insufficient (Levy 1993, 89). On balance, the impression is that the EMEP system in theory does provide additional verification capacity, although less than could easily be envisaged (see Di Primio 1996). Nevertheless, the practical compliance control potential has so far not been utilized very much. Critical compliance discussions have so far been nonexistent as there have been no provisions in the Convention or the Protocols until quite recently that authorize the Executive Body to critically examine the data provided in the national reports (Szell 1995). At most, countries have made oral clarifying statements at EB meetings.[27] Again, this must partly be seen in light of the East-West nonconfrontational dimension prevalent in the regime. Does all this mean that the verification instruments "even of this caliber" have been without importance for state policies? My guess would be that the very existence of EMEP data has contributed somewhat to reporting soberness in the West and hence general confidence building within this group of countries but not as a very important aspect in the process.

Conclusion: What Are the Principal Determinants of Medium Success?

The CLRTAP regime was to some extent "designed for weakness" (see table 8.1). In the regime-formation process, important emitter states like the United Kingdom and Germany were quite negative, and many countries were indifferent. None of these states had experienced any significant domestic acid damages at this early stage. Eastern European countries were not very eager either. Hence, a framework convention and much weight given to knowledge improvement and general confidence building are understandable elements in the regime-formation outcome. Overall, it is important to keep in mind the general East-West confidence-building dimension underlying this regime from the very start.

My assessment is that CLRTAP's mixed success so far has much to do with an initially strongly malign case—societally important emitter activities related to energy production and consumption, industrial processes, and transport, with related powerful target groups to deal with for regulators; coal availability and cheapness in important emitter countries, and hence employment implications related to regulations; technologically complicated and quite expensive abatement options and lack of technological breakthroughs; asymmetrical transboundary flows of pollutants, with some nations being net importers and some net exporters; asymmetrical vulnerability to air pollutants, with unfortunate combinations like the cases of Norway and Sweden, being both considerable net importers of pollutants and having particularly vulnerable soil characteristics. In addition, one must not forget the East-West context that increased the need for delicate diplomatic balancing acts and consensual and not intrusive processes. However, for understanding the degree of success the CLRTAP regime has achieved, symbolized by an impressive regulatory development in the 1990s, the catalytic event is definitely Germany's about-face in 1982 related to its domestic *Waldsterben* uproar. Germany's shift from laggard to leader represented a crucial and symbolic lasting shift in the power

Table 8.1a
Summary Scores: Chapter 8 (CLRTAP)

Regime	Component or Phase	Effectiveness		Level of Collaboration
		Improvement	Functional Optimum	
LRTAP	1: 1977–1979	NA	NA	Intermediate
LRTAP	2: 1983–1993	Low	Low	Intermediate
LRTAP	3: 1985–1994	Low	Low	Intermediate
LRTAP	4: 1988–1996	Low	Low	Intermediate
LRTAP	5: 1991–1996	Significant	Low	Intermediate

balance between reluctants and pushers within the regime. Moreover, on the intellectual side, there is no doubt that knowledge about the problems gradually improved and contributed benign elements to the process, although in various degrees related to various substances. From the late 1980s onward, the development of the critical-loads approach and subsequent application in several negotiation processes has marked an important intellectual and political step forward within the cooperation.

This naturally brings us over to the question of problem-solving capacity, which overall must be characterized as moderate, although increasing over time. The moderate element is clearly reflected in institutional aspects like the limited secretarial capacity and consensual decision-making style. However, it should be noted that the consensus requirement has been practiced with some flexibility. Perhaps the most important institutional contribution to effectiveness so far has been the evolution of the scientific and political complex. There are several positive aspects here—the strong scientific core represented by the EMEP monitoring program, the flexible addition of new substances and bodies to the evolving institutional structure, and the role of the Working Group of Strategies as a coordinating and mediative body between science and politics. From the NO_x and VOC processes on, the regime contribution to increased knowledge has been substantial. The strength of entrepreneurial leadership has increased over time, primarily related to the catalytic change in German acid policies as described above. German leadership has added considerable political weight to these processes and leadership continuity at the point in the regime development process where the interests of several Nordic countries got much more complicated and the initial Nordic leadership coalition broke down. This breakdown symbolizes the fact that even if processual leadership has been

Table 8.1b

Problem Type		Problem-Solving Capacity			
Malignancy	Uncertainty	Institutional Capacity	Power (basic)	Informal Leadership	Total
Strong	High	Low	Laggard	Intermediate	Intermediate
Strong	Intermediate	Low	Balanced	Intermediate	Intermediate
Strong	High	Low	Balanced	Intermediate	Intermediate
Strong	High	Intermediate	Balanced	Intermediate	Intermediate
Strong	Low	Low	Balanced	Intermediate	Intermediate

strengthened over time and general problem-solving capacity has improved, the basic interests of many countries have continued to be complicated in this issue area, both with regard to domestic regulatory capacities and international competitive aspects.

Notes

1. This chapter has benefited from a number of interviews over the years. Several interviews with Per M. Bakken, Norwegian Ministry of the Environment (1991, 1995); Harald Dovland, formerly with the Norwegian Institute of Air Research and now with the Norwegian Ministry of the Environment (1991, 1995, 1996); and Erik Lykke, Norwegian Ministry of the Environment (1991) have been conducted. Moreover, I have had talks with Anton Eliassen, Norwegian Meteorological Institute (1991), Jan Thompson, Norwegian Ministry of the Environment (1991), and Lars Nordberg, CLRTAP Secretariat in Geneva (1996). In addition, I have received very helpful comments from the project group and from three MIT Press reviewers. Although the chapter has been revised and updated several times, protocol negotiation processes after 1995 are not analyzed in detail.

2. Regarding EC air pollution policies and the relationship to CLRTAP, see, for instance, Haigh (1987, 1989), Boehmer-Christiansen and Skea (1991), and Liberatore (1993).

3. According to Levy (1993, 126), "the sulfur protocol *probably* had significant effects on the emission reductions in seven countries, including the largest and fourth-largest emitters in Europe (USSR and United Kingdom). A protocol that affects only these seven *probably* counts as a success" (emphasis added).

4. Among the signatories, fifteen countries and the European Community committed themselves to the regular 30 percent reduction, four chose the freeze option, and three chose the TOMA option (see Gehring 1994, 180).

5. For an analysis of the negotiations and content of the 1994 sulfur protocol, see, e.g., Gehring (1994, 185–193) and Churchill, Kutting, and Warren (1995).

6. Communication with Oran Young, October 1997.

7. The science and politics relationship within CLRTAP is more fully discussed in Wettestad (1995, 2000).

8. This approach was launched by Swedish scientists and government officials in the early 1980s (see, for instance, Levy 1995, 61–64).

9. Interview with Harald Dovland, then director of the Norwegian Institute for Air Research, May 1989.

10. As, for instance, stated by McCormick (1989, 69) "pollution control is expensive—no one seriously questions this basic proposition."

11. As stated by Boehmer-Christiansen and Skea (1991, 4): "While Germany, at the heart of Europe, shares its atmosphere with many neighbors, some relatively clean, others to the east decidedly not, Britain is a wind-swept offshore island little affected by atmospheric pollution originating abroad. However, the U.K. and F.R. Germany are themselves the largest polluters in Western Europe and winds spread the SO_2 and NO_x originating from German chimneys over a wide continental area."

12. See, for instance, Wetstone and Rosencrantz (1983), Chossudovsky (1989), and Gehring (1994) for detailed accounts of the negotiation process.

13. According for instance to Wetstone and Rosencrantz (1983, 142), "British scientists recognized Scandinavia's acidification problems, but questioned whether sulfur dioxide discharged by Britain's power plants was to blame."

14. Readers more interested in the details of the various processes up to the mid-1990s are advised to consult sources like Gehring (1994).

15. As stated by Sand (1990, 247): "[EMEP] has produced voluminous and increasingly reliable evidence that sulfur and nitrogen compounds emitted by a wide range of stationary and mobile pollution sources are dispersed through the atmosphere over thousands of miles. . . . EMEP results—annually reviewed, updated, and approved by an intergovernmental steering body—make it possible to quantify the pollutant depositions in each country that can be attributed to emissions in other countries."

16. See, for instance, Hajer (1995) on the shift in Dutch perceptions.

17. Levy (1995, 68) further notes that "their life span in the atmosphere, the distances they travel, and the amount of ozone they are responsible for are all subjects of scientific controversy."

18. As indicated, a *critical load* can be defined as a "quantitative estimate of an exposure to one or more pollutants below which significant harmful effects on specified sensitive elements of the environment do not occur according to present knowledge" (see, for instance, Boehmer-Christiansen and Skea 1991, 29).

19. This dynamic became even clearer in the more EC-related context of vehicle emissions regulation: "For the car industry, German policy priorities were to protect export markets, contain Japanese imports and maintain a reputation of well-engineered, high-performance cars. These challenges were perceived to be best met by moving over as rapidly as possible to clean cars equipped with three-way catalytic convertors. Trade and investment implications favored higher environmental standards" (Boehmer-Christiansen and Skea 1991, 279).

20. Commenting on the *Waldsterben* and Green Party thesis, Boehmer-Christiansen and Skea (1987, 3) maintain: "However, this explanation, although not incorrect, is insufficient. From a German perspective, broader social movements had succeeded in affecting party political as well as regional considerations which in turn helped to tip the balance of interest in a long-standing debate in favor of strict, federally mandated emission controls."

21. As documented by Gehring's (1994) detailed account of the negotiation processes (see chapter 4).

22. See Wettestad (1998) for a discussion of the different political game for the United Kingdom in the NO_x context, compared to the sulfur context.

23. The institutional issues are more fully discussed in Wettestad (1999, ch. 4).

24. Interview with Per Bakken, Norwegian Ministry of Environment, 1991.

25. For instance, Levy (1993, 84) has characterized the Secretariat as "small, overworked, and underfunded."

26. Harald Dovland, personal communication, 1991.

27. As stated by Levy (1993, 91): "Strategy and policy reviews are not interpreted; they are simply collated and published. There is no effort to ascertain whose measures place them in compliance with either specific protocols or broader norms.... Although no one is ever 'cross-examined,' states frequently make oral statements offering clarifications and emphasizing major points at EB meetings."

References

Bakken, P. 1989. "Science and Politics in the Protection of the Ozone Layer." In S. Andresen and W. Østreng eds., *International Resource Management* (pp. 198–205). London: Pinter.

Boehmer-Christiansen, S., and J. Skea. 1987. "The Development of the Acid Rain Issue in West Germany and the UK." Paper presented at the ENER meeting, Brussels, June.

Boehmer-Christiansen, S., and J. Skea. 1991. *Acid Politics: Environmental and Energy Policies in Britain and Germany.* London: Belhaven Press.

Chossudovsky, E. 1989. *East-West Diplomacy for Environment in the United Nations.* New York: UNITAR.

Churchill, R., G. Kutting, and L. M. Warren. 1995. "The 1994 UN ECE Sulphur Protocol." *Journal of Environmental Law* 7(2): 169–197.

Di Primio, J. C. 1996. *Monitoring and Verification in the European Air Pollution Regime.* WP-96-47. Laxenburg: IIASA.

Economic Commission for Europe/Convention on Long-Range Transboundary Air Pollution (ECE/CLRTAP). 1998. *1998. Major Review of Strategies and Policies for Air Pollution Abatement.* Draft report, October 2.

Gehring, T. 1994. *Dynamic International Regimes: Institutions for International Environmental Governance.* Berlin: Peter Lang Verlag.

Green Globe Yearbook. 1996. Oxford: Oxford University Press and Fridtjof Nansen Institute. After 1998 *Yearbook on International Co-operation on Environment and Development.* Oslo: Fridtjof Nansen Institute and Earthscan.

Haigh, N. 1987. *EEC Environmental Policy and Britain*. Essex: Longman.

Haigh, N. 1989. "New Tools for European Air Pollution Control." In *International Environmental Affairs* 1(Winter): 26–38.

Hajer, M. A. 1995. *The Politics of Environmental Discourse: Ecological Modernization and the Policy Process*. New York: Oxford University Press.

Levy, M. 1993. "European Acid Rain: The Power of Tote Board Diplomacy." In P. Haas, R. O. Keohane, and M. Levy, eds., *Institutions for the Earth* (pp. 75–133). Cambridge, MA: MIT Press.

Levy, M. 1995. "International Co-operation to Combat Acid Rain." In *Green Globe Yearbook 1995*. Oxford: Oxford University Press.

Liberatore, A. 1993. *The European Community's Acid Rain Policy*. Draft for the project on Social Learning in the Management of Global Environmental Risks. European University Institute.

McCormick, J. 1989. *Acid Earth*. London: Earthscan.

Miller, K. 1989. "The Greenhouse Effect: Potential for Effective Regional Response." Thesis, Institute for Marine Studies, Seattle.

Nordberg, L. 1993. "Combatting Air Pollution." LRTAP nonpaper, March. Geneva.

Oden, S. 1968. "The Acidification of Air and Precipitation and Its Consequences in the Natural Environment." *Ecology Committee Bulletin* No. 1, Swedish National Science Research Council, Stockholm.

Park, C. 1987. *Acid Rain: Rhetoric and Reality*. London: Methuen.

Sand, P. 1990. "Regional Approaches to Transboundary Air Pollution." In J. Helm, ed., *Energy: Production, Consumption and Consequences*. Washington, DC: National Academy Press.

Stenstadvold, M. 1991. "The Evolution of Cooperation: A Case Study of the NO_x-Protocol." Thesis, University of Oslo, Fall. (In Norwegian).

Szell, P. 1995. "The Development of Multilateral Mechanisms for Monitoring Compliance." In W. Lang, ed., *Sustainable Development and International Law* (pp. 97–109). London: Graham and Trotman.

Waterton, C. 1993. *The UK Case Study for Acidification and Transboundary Air Pollution: A Preliminary Survey*. Draft for the project on Social Learning in the Management of Global Environmental Risks.

Wetstone, G. 1987. "A History of the Acid Rain Issue." In H. Brooks and C. L. Cooper, *Science for Public Policy* (pp. 163–195). Oxford: Pergamon Press.

Wetstone, G., and A. Rosencrantz. 1983. *Acid Rain in Europe and North America*. Washington, DC: Environmental Law Institute.

Wettestad, J. 1995. "Science, Politics and Institutional Design: Some Initial Notes on the Long-Range Transboundary Air Pollution Regime." *Journal of Environment and Development* 4(2)(Summer). pp. 381–431.

Wettestad, J. 1996. *Acid Lessons? Assessing and Explaining LRTAP Implementation and Effectiveness*. IIASA Working Paper WP-96-18, March.

Wettestad, J. 1998. "Participation in NO$_x$ Policy-Making and Implementation in the Netherlands, UK, and Norway: Different Approaches, but Similar Results?" In David Victor, Kal Raustiala, and Euqene Skolnikoff, eds., *The Implementation and Effectiveness of International Environmental Commitments* (pp. 381–431). Cambridge, MA: MIT Press.

Wettestad, J. 1999. "More 'Discursive Diplomacy' Than 'Dashing Design'? The Convention on Long-Range Transboundary Air Pollution (LRTAP)." In *Designing Effective Environmental Regimes: The Key Conditions* (pp. 85–125). Cheltenham: Elgar.

Wettestad, J. 2000. "From Common Cuts to Critical Loads: The ECE Convention on Long-Range Transboundary Air Pollution (CLRTAP)." In A. Underdal, et al., *Science and International Environmental Regimes: Combining Integrity with Involvement.* Manchester: Manchester University Press.

9

Satellite Telecommunication

Edward L. Miles

Introduction and Reasons for Success and Failure

This chapter presents an evaluation of the effectiveness of the international regime governing satellite telecommunication since 1957 and particularly since 1963.[1] The regime has two components: one dealing with regulating the use of the radio-frequency spectrum and the other making use of orbiting satellites through the radio-frequency spectrum for purposes of national and international communication. Both subregimes form around primarily benign problems, but each is affected to varying degrees by malign (competitive) elements. The performance of the regulatory subregime demonstrates only low to intermediate effectiveness, while that of the operational subregime demonstrates high effectiveness.

It is necessary to make clear that this case is included not as a control but as representative of an open-access international resource, just like fisheries, with a certain amount of international coordination (an externality) to avoid harmful interference in the use of particular radio frequencies. The difference between this case and fisheries in the Grotian regime is that the system of allocation of radio frequencies conveys a property right (first come, first served, with seniority rights), whereas fisheries allocations in open-access regimes do not.

On October 4, 1957, the Soviet Union launched both the first manned orbital earth satellite, *Sputnik 1*, and the space age. By the end of 1961 the outlines of a global regime for outer space had been negotiated within the General Assembly of the United Nations (Resolution 1721 (XVI)). However, since human activities in outer space occurred in one of the global commons, the regime outlined by the General Assembly was to consist of a treaty to be negotiated on the status of outer space, the moon and other celestial bodies, and a series of subregimes dealing with satellite telecommunication, weather forecasting, cooperative space science research, and the like. These subregimes were still to be worked out, but the General Assembly delegated authority to elaborate them to several existing intergovernmental organizations (IGOs) (Miles 1970–1971).

Not surprisingly the International Telecommunication Union (ITU) was delegated authority to elaborate the subregime applicable to satellite telecommunication, and in 1963 an Extraordinary Administrative Radio Conference was convened to establish the first set of regulations applicable to radio communication with satellites in and from outer space.

The problem to be solved was not creation of a regime *ab initio*. It was to adapt a regime created in 1865 to the demands of a revolutionary new technology. The radio-frequency spectrum is a common property resource. Consequently, since no state by itself can control the medium, a certain amount of international cooperation (coordination, in fact) is required if harmful interference is to be avoided or minimized. Avoiding harmful interference is the common interest underlying coordination. This stimulus produces a willingness to devise collective procedures to facilitate technical coordination of frequency utilization and to make recommendations for technical performance standards to which all equipment must conform.

The radio-frequency spectrum—measured in terms of the number of cycles per second characterizing the sound wave (that is, frequency)—in conventional designations goes from 30 cycles or herz per second through thousands (kiloherz), millions (megaherz), billions (gigaherz), and trillions (teraherz) of cycles per second. It is important to understand that the spectrum is not a finite resource in the ordinary meaning of the word because, at least theoretically, one could increase transmissions up to the speed of light. The real constraint here, which does indeed create scarcity aspects, is the cost of transmitting and receiving technology as the frequency is increased.

Consequently, one finds that as technology advances, the spectrum becomes crowded in certain ranges from the high-frequency (HF) range (3 to 30 megaherz), through the very high-frequency (VHF) range (30 to 300 megaherz), the ultra high-frequency (UHF) range (300 to 3,000 megaherz), and the super high-frequency (SHF) range (3 to 30 gigaherz). Crowding occurs because frequencies must be physically suitable for the communications link (between two points on the earth's surface or between outer space and the earth) on which they are used, and harmful interference with other users of the same frequencies must be avoided (ITU 1968).

Moreover, the earth's atmosphere is frequency selective, thereby creating windows through which the sound waves readily pass, escaping severe attenuation. One major window exists between 10 Mhz and 20 Ghz (ITU 1968). Note that this window includes the HF, VHF, UHF, and SHF ranges, but it sets an upper technical limit of about 20 Ghz beyond which the sound waves tend to be absorbed by rainfall and atmospheric gases (ITU 1968). The other window exists at the very low or long wave end in the combined visual and infrared ranges.

The regime created to govern international utilization of the radio-frequency spectrum since the invention of long-distance telegraphy and telephony in the last four decades of the nineteenth century has been based on a first-come, first-served criterion with seniority rights. This kind of regime, which is characteristic of frontier thinking in mining and water resources in the western United States, effectively encourages development in the early stages of a process but usually runs into trouble on both equity and efficiency dimensions in more mature stages when many more participants are present and many more conflicting uses are evident.

The International Telecommunication Union (ITU) was itself a creature of the telegraph. The ITU was the result of an attempt by states to establish the framework within which this invention would be exploited since it dealt with a medium no single state could control by itself. Institutionalization was required at the international level for agreement to be reached on the technical systems to be used, methods of handling messages, establishment of rates, and accounting procedures. The process of international institutionalization arose out of bilateral interactions between neighboring states and then spread to the European continent. It was globalized by another technological advance—the invention of the radio in 1927.

Member states saw to it that organizational tasks would be restricted to providing a forum within which decisions would be made, providing for technical coordination of frequency utilization, and making recommendations for technical performance standards to which equipment must conform.[2] This was the implementation strategy chosen. An *operational* role for the ITU was never considered, and when the idea was proposed much later with respect to the need for supervisory and enforcement capabilities, it was never welcome.

At one level the basic problem to be solved was straightforward—adapt the global communications requirements to the space age by formulating new rules to govern telecommunication by satellite. At another level the problem to be solved was not so technical and straightforward because satellite telecommunication was always a template for the working out of preexisting organizational issues and conflicts in the ITU at the same time that it represented a new set of demands to which the organization had to respond. Moreover, given the nature of the good and the divisibility of the spectrum in particular, technical cooperation went hand in hand with conflict. These are constant themes in the organization's history; the political spillover effect of technical cooperation is consequently always constrained. Whether there will be cooperation or conflict is a function of the kind of issue defined in terms of the payoff matrix. Whenever the aim of any party is to redistribute benefits or whenever the effect of any proposal is to threaten existing large-scale investment in system design and operation, conflict arises.

Because protagonists for major organizational change—the need for global planning in frequency allocation versus the first-come, first-served traditional approach—sought to use the advent of satellite telecommunication as the lever to induce the change they desired, this immediately raised the stakes of the struggle. Given the pattern of political control within the ITU between 1958 and 1965, there was little chance that the attempt would succeed. On the regulatory dimension, therefore, choices made with respect to organizational arrangements policies and implementation strategies were all the traditional ones. Levels of international cooperation with the ITU did not therefore increase noticeably.

Where the conflict developed and, indeed, became intense was at the conjunction between issues related to organizational arrangements, types of policies, and implementation strategies. These conflicts occurred on two dimensions—within the ITU with respect to regulatory approaches and between the ITU and a new organization, INTELSAT, created to exploit the commercial opportunities offered by the new technology. This latter organization was deliberately pushed by the United States as an alternative to the ITU for reasons that will be explained. The point to be made here is that regime adaptation occurred in the context of routine implementation of the larger global regime for all communications and that the two processes were decidedly connected.

We would therefore characterize the problem to be solved as mixed. On one level, the job of elaborating a global subregime to govern satellite telecommunication is relatively benign because synergy and contingency relationships predominate. At the same time, on some issues, the problem becomes relatively malign since incongruities emerge in the form of externalities and competition. We would therefore expect that where the problem set exhibits mixed characteristics in this fashion, the level of cooperation achieved would be less than what one would expect if they were wholly benign but greater than what one would expect if they were wholly malign.

The Regime-Adaptation and -Implementation Phase, 1958 to 1971

We characterize the period immediately following the orbiting of *Sputnik 1* in October 1957 to the conclusion of the negotiations for the Definitive Arrangements for the International Telecommunications Satellite Organization (INTELSAT) in 1971 as comprising the phase in which the subregime for satellite telecommunications was fully adapted. Three second-order effects are detectable over the course of this phase. The first is an enormous expansion in the scope of the ITU's regulatory tasks. The second is a significant increase in interorganizational conflict and cooperation within the components of the ITU's task environment. And the third is a loss of regulatory initiative to INTELSAT both with respect to assigning positions in the geostationary orbit and the setting of technical standards for satellite telecommunication systems. Each deserves some elaboration.

Some appreciation for the scope of expansion in the ITU's tasks can be gained by a glance at the demands that were placed on the organization between 1958 and 1968 relating solely to the advent of satellite telecommunication:

· The initial demands for the allocation of frequencies for satellites, including bands for telemetry, command, and control facilities;

· The necessity for determining technical standards for simultaneous channel sharing (a subsidiary part of this includes requirements for the automatic or ground-command shut-off of transmitters whose batteries are recharged with solar energy);

· The control and limitation of microwave communication over land necessitated by the requirements of transoceanic satellite communication;

· The registration of call letters for the identification of signals;

· The avoidance of the saturation of sections of the equatorial orbit for geostationary satellites given limitations in the resolving power of antennae;

· The coordination between existing and planned satellite communication systems (this also includes those regional and national systems then under study);

· The regulation of direct broadcasting by satellite;

· The better policing of the use of radio transmitters by nationals of member countries to guard against harmful interference between ground stations and orbiting satellites;

· The expansion of technical assistance programs in the form of education and training of necessary personnel.

One would have assumed from this list of demands commensurate growth in budget and professional staff, but this was not the result. As the organization grew, the professional to general staff ratio in fact increased from 64.15 percent in 1958 to a high point of 99.27 percent in 1960, then decreased to 71.07 percent in 1963, slid to 46.46 percent in 1967, and finally increased slightly to 50.87 percent in 1969. Particularly after 1963, as the technical demands grew, advanced member states would second their own staff to the organization for limited periods of time. No doubt this was designed to restrict budgetary growth, but it also had the effect of constraining organizational autonomy since it made the ITU more dependent on its most advanced member states.

With respect to interorganizational conflict and cooperation, as ITU's role expanded, the organization experienced a commensurate expansion of its domain. However, since other organizations were also being affected by satellite telecommunications, the number of other organizations with which ITU had to maintain relations increased as well. Some, like the World Meteorological Organization (WMO) and the Committee on Space Research (COSPAR) of the International Council of Scientific Unions (ICSU) were allies. Others, like the United Nations Committee on the Peaceful Uses of Outer Space (COPUOUS) and INTELSAT were seen as antagonists.[3] The problem with COPUOUS was accentuated by the creation of the Working Group on Direct Broadcasting by Satellites within COPUOUS as a major focus of international regulation. The difficulty arose because this Working Group was driven by lawyer-diplomats from Ministries of Foreign Affairs who knew little or nothing about ITU and national coordination appeared to be weak. Early actions of this group were therefore seen to be troublesome by ITU officials.

As a contrast, the ITU and WMO connection was a happy one for several reasons. WMO found that it needed ITU as a result of the emergence of the weather satellite, which brought with it a corresponding need for special frequency bands for transmission of data. Since there was no conflict of jurisdictions, the relationship was symbiotic. The fact that the two organizations were then located across the street from each other, which facilitated interpersonal relations, was also a help.

Demands for interaction were heavy and continuous so that the two executive heads agreed to institutionalize the relationship from 1960. The content of interaction ranged over issues like avoidance and elimination of harmful interference, allocation of frequencies, organization and operation of data transmission systems, and technical studies and recommendations concerning radio communication. But at the same time we should note that there are effective barriers to intersecretariat interaction concerning the allocation of frequencies. For instance, in both the WMO and the International Civil Aviation Organization (ICAO) national members decide which frequencies are particularly appropriate for transmission of meteorological and navigational data. To obtain these frequency allocations from the ITU, the secretariats of other organizations may not approach ITU officials directly. Of course, occasionally there are informal discussions by secretariat officials as they chance to meet, but these do not carry any weight.

The process by which IGOs decide which frequencies are needed and then lobby for them has two stages. In the first or internal stage, an organizational position is developed by appropriate groups within the organization participating depending on the technical problem involved. In the WMO, for example, this stage calls for interaction between the secretariat and national representatives so that each national meteorological service is consulted about its particular needs. But even this is a delicate operation because each country jealously guards its frequency-allocation rights. Once the organization chooses a position, the second stage begins. At this point each national meteorological service takes up the question with its own postal, telephone, and telegraph administration and tries to convince it to seek these frequencies for the WMO in the ITU. Final allocations are then worked out among national representatives in ITU meetings.

As a result of these conditions, coordinated interorganizational planning cannot go beyond a very low level (sending observers to meetings) because the effective decision-making capacity rests with governments, not secretariats. Admittedly, joint working groups sometimes go beyond the observer level but not sufficiently beyond it to have a major impact on governmental control.

Finally, ITU and INTELSAT tension and conflict were an inevitable consequence of two decisions taken by ITU member states. The first decision, not surprisingly, was to separate the regulatory from the developmental function by vesting responsibility for exploiting the commercial promise of satellite telecommunication in a novel entity (mixed public and private) called INTELSAT. The second decision was to restrict the nature of change within the ITU, particularly to constrain the International Frequency Registration Board (IFRB) and to retain the seniority-rights system of allocation against demands for global planning. Therefore, given the fact that the technology would be driven by the manager of INTELSAT, the U.S. public corporation COMSAT, there would at the least be rivalry over technical standard setting and choice of positions in the geostationary orbit. This competition occurred consistently in the early years but then declined from the mid-1970s.

The emergence of scarcity conflicts in assigning positions in the geostationary orbit requires some explanation. If a satellite is placed in orbit at an altitude of 22,375 miles (or 35,800 kilometers) in the plane of the earth's equator, its rotation around the earth is synchronous with a defined location on the earth's surface. It therefore appears to be stationary. But since the circumference of the earth is a finite distance (about 150,000 miles) and satellites tend to wander in orbit over a range of about 100 miles, the total number of satellites in geostationary orbit at any point in time is also potentially finite (Wihlborg and Wijkman 1981).

Given these characteristics of the geostationary orbit, scarcity is imposed by a combination of the frequencies assigned to particular positions in that orbit. The resource being regulated thereby becomes two dimensional: it is an orbit and spectrum resource in which each orbit and spectrum combination generates different economic benefits and costs (Wihlborg and Wijkman 1981). However, we should not assume that this is a resource in fixed supply since we can expect that advancing technology can reduce the minimum spacing requirements between satellites placed in geostationary orbits.

In this first phase of regime adaptation and implementation, the relationship between the independent, intervening, and dependent variables we are investigating is very curious. Type of problem (mixed) is a powerful variable in its own right, but it is followed by the merging of institutional setting and the distribution of capabilities as the second combined independent variable. The configuration of actor interests and positions becomes in this case an intervening and not an independent variable that is tightly merged with coalitions and negotiating strategies. There is no uncertainty in the knowledge base that restricts the significance of the epistemic communities that do exist. The latter are forced into taking positions on the major conflicts related to ITU's role. In doing so, they become only one more player, without overwhelming influence, on a much larger canvas of political conflict. Let us see how the relationships described above actually worked out in practice.

Institutional setting is critical, as we have already pointed out, because the new responsibility was given to an organization with a history of almost 100 years. It was therefore inevitable that preexisting issues and conflicts would form the backdrop against which a new organization role would be elaborated. Tradition initially both determined the *modus operandi* to be employed and channeled conflict along preexisting lines. This means that the effect of institutional setting cannot be fully understood without first understanding the effects of distribution of capabilities within the organization and the nature of the regime.

Why are these two variables so closely related in this case? We suggest that since the regime was based on a first-come, first-served rule with seniority rights, it placed the premium on the capability to exploit the radio frequency spectrum and later the orbit-spectrum resource. The structure of control within the ITU was therefore based primarily on the technological capability to exploit the medium so that a rigid hierarchy of the most advanced states in radio engineering dominated the ITU—by exercising basic game *and* negotiating game power. At the top of the hierarchy in the 1960s sat the United States, the United Kingdom, France, the Netherlands, the Federal Republic of Germany (FRG), and the USSR. Japan did not really join this group until the late 1960s.

The major interests at stake among the players included (1) access to and control over markets, (2) protection of existing investment, (3) prevention and control of harmful interference, (4) the approach to the system of allocating the orbit and spectrum resource, and (5) enhancing national communication capabilities in a cost-effective way.

The major issues over which conflict arose were of two types: (1) issues relating to the nature of the regime and demands for a shift from seniority rights to global planning in frequency allocation and award of frequencies irrespective of use and (2) technical proposals about standards or frequency assignments that threatened existing, large-scale investment in system design and operation.

The first type of conflict shaped the coalitions and vice-versa. It pitted the developing countries, the developed smaller states, the IFRB of the ITU, and the Secretary-General against the United States, the United Kingdom, France, the Netherlands, the FRG, the USSR, the Executive Council (dominated by the traditional leadership), and the International Radio and Telephone and Telegraph Consultative Committees (CCIR and CCITT) (dominated by the industry). The challengers sought to use the advent of satellite telecommunications as a lever to induce the change they desired. But given the traditional structure of control, there was little chance they would succeed.

The second type of conflict was largely confined within the most developed group because it arose out of the fierce competition that existed among them over market share. Internally, this conflict was fought out within the two Consultative Committees tasked with the responsibility of recommending performance standards to which the design of new equipment had to conform.

It is interesting to chart the evolution of conflict and the proliferation of issues since one would expect that both would multiply with time as more applications emerged. At the Extraordinary Administrative Radio Conference held in 1963, the major issues were (1) superpower priorities in protecting their national security interests in radio frequencies, (2) advanced industrial country attempts to safeguard their investments in equipment, (3) the scope and location of control of the global operational system, and (4) the fear of the rest of the world that the choicest frequencies would be preempted, irrespective of their future needs (Miles 1970–1971).

Not surprisingly, the security issues fell into an East-West coalition mode (in which the issue of representation of China was brought into play by the USSR); the market share issue resolved itself into a fierce conflict *within* the advanced industrial country group as suppliers of electronic equipment; a challenge of the seniority rights regime by small countries with advanced electronic capabilities, including the Scandinavians and Switzerland; and the rest of the world coalition led by Israel and some of the developing countries like Nigeria and Mexico (Miles 1970–1971).

By the late 1960s to early 1970s, the issues had evolved into a host of questions concerning the design of INTELSAT—scope, U.S. dominance, awarding of R&D contracts, services and pricing, rates of technological advance, and the like; INTELSAT versus INTERSPUTNIK;[4] INTELSAT versus ITU; the growing multiplicity of planned systems (national systems plus the European *Projet Symphonie*); control over direct broadcasting by satellite; and resource sensing by satellite. On INTELSAT issues, the coalitions were the United States against the Western Europeans and Japan over U.S. dominance and the U.S. and developing-country alliance combined against the Western Europeans and Japan on awards of R&D contracts. Everything having to do with INTERSPUTNIK fell into the East-West conflict mode, and the developing-country coalition (soon to become known generally as the Group of 77) was developing its own positions on all issues in an increasingly coordinated way led informally by India, Morocco, Mexico, Algeria, and Nigeria (Levy 1975).

Over the whole of this first phase, actor interests and negotiation positions were clear and well-known, but negotiating strategies were limited since technological capability was almost the sole basis for political influence. The outcome, therefore, was predictable—expansion in the scope of the ITU's tasks without significantly increasing organizational autonomy, resources, and role. Within the ITU, the outcome (amount of cooperation achieved) never exceeded level 2.[5] Since the radio-frequency spectrum is an impure collective good exhibiting the characteristics of divisibility and nonappropriability and since the nature of the regime put a premium on technological capability as the currency of influence, it is not surprising that competition over the considerable commercial stakes involved would constrain provision of the collective good. But at the same time, since benefits are widely shared and states do not have a more effective alternative for achieving the same benefits at less cost, the level of conflict will be somewhat muted and cooperation will be unbroken.

The Regime-Implementation Phase, 1972 to 1992

This second phase is clearly dominated by INTELSAT, and INTELSAT's experience differs in several important respects from that of ITU. The most important difference was in the nature of the problem to be solved because, at this point, regime adaptation was secondary to a determination by the U.S. government and industry to exploit the communication satellite for commercial purposes. This objective dictated both the approach to organizational design and the types of policies to be pursued. That this strategy was both unique and successful is to be explained by the compelling stream of benefits relative to costs to be derived over time by *all* participants in the system, not just the dominant player. This was possible as a result of the type of technology and its implications for the nature of the collective good.

Three characteristics of satellites offered significant promise of benefits over time. Relative to all other forms of communication, especially cables, the factor of distance is of no relevance when a satellite is placed in a geostationary orbit at an altitude of 35,800 km in the plane of the earth's equator. In one leap, therefore, the problem of remoteness, whether faced by a single country or a region, could disappear. Moreover, satellites possess multiaccess capabilities, whereas cables are only point-to-point. Over time, significant economies of scale could be achieved that would affect the design life not only of satellites but also of receiving ground stations because the more powerful the satellite, the simpler and cheaper the ground station required.

Given these characteristics and given the fact that only one of the two superpowers was then willing to exploit this technology on a commercial basis globally, a combined executive, congressional, private-industry coalition in the United States moved quickly to realize the promise of satellite communications. Congress passed the Communications Satellite Act in 1962, which established the Communications Satellite Corporation (COMSAT) as a profit-making entity governed by public and private shareholders with no public investment (Colino 1984). The COMSAT team at the same time foresaw the possibility of INTELSAT because the full promise of the technology could not be realized without global application.

Accordingly, the criteria derived for the implementation strategy emphasized (Colino 1984)

· Urgency,
· The commercial aspects of the venture,
· Global sharing of the technology,
· Avoidance of the usual shortcomings of intergovernmental organizations (fractionated authority, lack of effective control, protracted decision time, entrenched bureaucracy, and so on),
· Achievement of the broadest participation possible while avoidance of purely political compromises,
· Commitment to establish a single global entity to avoid duplication and waste, and
· The need to achieve economies of scale.

This implementation strategy dictated as a first step the convening of a global conference to create a regime (decision rules plus organizational arrangements). It was expected that definitive arrangements would take some time to negotiate and should have as a basis a period of experience. This called for negotiation of interim arrangements as a first step. The Interim Arrangements for a Global Commercial Communications Satellite System were concluded within six months in 1964, a historic achievement indeed. The first INTELSAT satellite was launched in 1965, and the Definitive Arrangements were concluded in 1971 and entered into force in 1973 (Colino 1984). The USSR chose not to participate and dissuaded other member states of the Soviet bloc from doing so.

The Implications and Consequences of Type of Problem for the Decision Process
At first glance the type of problem can be characterized as benign with compelling benefits to all parties. Looking at in more detail, however, we would have to characterize type of problem as mixed but predominantly benign with compelling benefits. The difference is subtle but important because it touches on the expected responses of other parties to U.S. political and technological dominance. Remember that the United States was then so far in the technological lead that one of the design criteria for the new system, in U.S. eyes, would be crafting decision rules to preserve this dominance. On this basis alone, we could expect that the major competitors of the United States in the global communications market would have an interest to limit U.S. dominance even to the point of slowing down the overall rate of technological advance and ensuring themselves a large enough share of R&D contracts to build up their own capabilities.

U.S. dominance called for global applications to generate the necessary economies of scale to make this whole effort a going (and paying) concern and the most rapid rate of technological advance to bring down unit costs in the shortest possible time. This interest resonated with the rest of the world, which did not have significant commercial competitors in the communications market. This group of players certainly had an interest in limiting U.S. domination but not to the point where it would slow down the rate of technological advance and therefore narrow the stream of benefits to be derived. And they certainly had no interest in seeing the Western Europeans get a larger share of the R&D pie because such a development would slow things down and increase costs. The political result of all of this was that the United States was clearly the dominant player in an overwhelming global coalition, but even so, it could be expected to make a minimal deal with Western Europe and

Japan, which were the principal allies of the United States in the larger cold-war world.

Looking at the problem in this fashion, we note that type of problem merges fully with the variable of distribution of capabilities. These jointly determine actor interests and positions relative to the creation of an institutional setting (INTELSAT) that does not yet exist. Of the intervening variables, uncertainty in the knowledge base and presence or absence of epistemic communities both have a zero loading, and only coalitions and negotiating strategies remain to be accounted for in explaining both outcomes and impacts.

The INTELSAT Experience

In terms of organizational design, as a *global* intergovernmental agency INTELSAT was then unique. Since that time, only a successor organization, INMARSAT, designed to provide maritime satellite communication services to member states of the International Maritime Organization (IMP) has been created. Even at that time, only two other *regional* intergovernmental organizations in Europe were of a similar design: the European Company for the Chemical Processing of Irradiated Fuels (EUROCHEMIC) and the European Company for the Financing of Railway Rolling Stock (EUROFIMA).

The structure of INTELSAT under the Interim Arrangements was very simple and reflected the clear dominance of the United States. The governing body was called the Interim Communications Satellite Committee (ICSC), and the manager was COMSAT: "Investment shares in the organization were determined on the basis of projections of long-distance traffic likely to be carried by satellite" (Colino 1984, 62). And shares determined voting power so that the United States had a share or vote of 61 percent, Western Europe jointly 30 percent, and the combined shares of Australia, Canada, and Japan then accounted for only 9 percent. But the overwhelming dominance of COMSAT, in addition to its potential conflict of interest, presented disadvantages as well as advantages.

As we have expected from the configuration of actor interests, the dominance of COMSAT ensured rapid development of the technology and was a major drawing card for participation by developing countries, but it stimulated concerted opposition in the European camp to the point where this issue became a very sticky point in the negotiations over the Definitive Arrangements. The result was that no single state was permitted to exercise more than 40 percent of the vote. It is worth noting that the analog for organizational arrangements under the Interim regime was the approach utilized by multinational consortia operating networks of transoceanic cables. These networks were based on two fundamental principles: undivided ownership of intercontinental facilities and investment in cable systems in proportion to use. Control was vested in the carriers, not in governments (Levy 1975).

Membership in INTELSAT is open to any state that is at the same time a member of ITU. The structure of governance includes the Assembly of Parties where states participate directly as the principal organ. Each party has one vote, and the Assembly meets biennially. In addition, another plenary body, this one of all shareholders (including therefore corporate entities), is called the Meeting of Signatories (Colino 1984). This body meets annually, and each representative has one vote. The executive oversight function is performed by the Board of Governors, consisting of twenty-seven members from any signatory or group of signatories meeting the minimum-share requirement annually established by the Meeting of Signatories (Colino 1984). In addition, up to five seats are chosen on the basis of state representation in an ITU-defined region. This accommodates the interests of developing countries irrespective of the share requirement. Voting is on a weighted basis determined by shares with a cap of 40 percent placed on all signatories. Finally, an executive organ, directed by a secretary-general responsible to the Board of Governors, provides the management function.

In addition to the issue of U.S. domination and the voting majorities to be required in the Board of Governors, there were several other major issues of conflict in the negotiations for the Definitive Arrangements. The solutions arrived at generated very important long-term consequences for the organization. These consequences include quite significant second- and third-order effects. The issues in conflict were (1) R&D, procurement, and patent policies and (2) principles for avoiding competition.

The R&D issue generated the expected coalitions, and Western Europe, Canada, and Japan pointed out that COMSAT would remain the manager for six years after the entry into force of the Definitive Arrangements, then expected for 1973. Not only would this mean that COMSAT would have done all the R&D for INTELSAT III and IV, but it would also have written the specifications for the contracts to be let for INTELSAT V and even VI. This would mean that in the early years of the internationalized manager, operations would still be based on decisions taken by COMSAT (Colino 1984).

The Europeans, on the other hand, were intent on developing their own technological capabilities in the field and were very much opposed to the policy of conducting R&D in-house by COMSAT. They insisted that contracts be let and that a ceiling be placed on in-house R&D by COMSAT (Levy 1975). This was opposed not only by COMSAT but the developing countries whose representatives refused to subsidize the technological development of Western Europe, Canada, and Japan. They had no objection to a U.S. monopoly on R&D because this was the most efficient solution.[6] To them, U.S. control guaranteed faster, more reliable technological development at the lowest costs over the shortest time horizon. Conversely, because the Europeans were behind the United States technologically, their products would take longer to appear, be less reliable initially, and cost more, and the time horizon for planned cost reductions would be longer. This differential in cost and time streams between the United States and Europe was regarded by the developing countries as a potential subsidy to the Europeans. As such, they were against it.

The Europeans were therefore unsuccessful in their main objective, but they did succeed in getting a larger share of contracts being let. But even though bidders quickly formed international teams to respond to invitations by INTELSAT, and even though INTELSAT insisted that a lot of effort went into decisions to allocate work among team members from different countries, the result through INTELSAT VI was still that U.S. companies were the prime bidders (Colino 1984).

The issue of avoidance of competition had two parts. First, draft article 14 of the Definitive Arrangements on the Rights and Obligations of Members required any member that intended to establish, acquire, or utilize space segment facilities separate from INTELSAT for domestic purposes to consult first with the Board of Governors on the issue of technical compatibility. If, however, a member intended to participate in an *international system* other than INTELSAT, the issue would again have to be considered by the Board, but in this case the Board's findings would include judgments of economic as well as technical incompatibility. Europe, Canada, and Japan saw in this an implied threat, since a U.S. agency (NASA) was the sole launcher of satellites for INTELSAT. The threat was that should the Board, which was still dominated by the United States, find any system either technically or economically incompatible, NASA could and would refuse to launch the satellite.

The second part of this issue had to do with patents. COMSAT as the manager sought to prohibit the use of INTELSAT-funded inventions and data by competing systems. Consequently, COMSAT sought to have INTELSAT retain title to all inventions and data (Levy 1975). The Europeans, on the other hand, wanted INTELSAT to sell licenses with the contractors retaining title. The Europeans lost this fight again. Both of these losses had long-term effects.

In the short run, the COMSAT position was seen to be inconsistent, if not hypocritical. COMSAT pressed very strongly for a global organization with universal membership, but at the same time their insistence on control prevented this and provided incentive for competing systems to be created. France, the FRG, Italy, Canada, and Japan found the technological dependency uncomfortable and therefore redoubled their efforts to develop their capabilities outside the consortium, focusing on national systems. The French, in particular, in 1970 saw article 14 as putting at risk their plans for a joint development with the FRG of their own regional system, *Projet Symphonie*. Since NASA could refuse to provide a launcher, this threat added substantially to the French incentive to develop the *Ariane*.[7]

The longer-term effects of the COMSAT policy were twofold. First, it stimulated the proliferation of the very same competing systems they feared. Second, the more COMSAT insisted on control of INTELSAT, the more narrowly the scope of the organization was defined by others. In design, therefore, it became no more than an international common carrier, and it lost the chance to develop INMARSAT, which then went to the Intergovernmental Maritime Consultative Organization (IMCO, later IMO). It also lost the chance of universal membership, and it stimulated the creation of the competing Soviet-sponsored organization, INTERSPUTNIK. The proliferation of competitors is an issue that has haunted INTELSAT since the 1980s.

INTELSAT Performance, 1965 to 1986

Looking at the consequences of the conflict as outlined above, one might be tempted to conclude that COMSAT's policies produced a decline in the level of cooperation observed, but this would be an incorrect judgment. The fact is that *the system created is a level 5 outcome*—coordination through fully integrated planning and implementation, including centralized appraisal of effectiveness. Just as we saw in the sea dumping of low-level waste, the route to a level 5 outcome is through *compelling benefits with no alternative options for achieving them*. The effect of COMSAT's policies certainly created the seeds of many difficulties faced by INTELSAT today, but the startling successes of INTELSAT's performance under COMSAT's direction also significantly increased the level of international cooperation in utilization of the spectrum for satellite telecommunication purposes. This dimension therefore needs consideration as well.

Pollack and Weiss (1984) illustrate quite clearly the thrust of the COMSAT technological strategy for the development of satellite telecommunications on a *global scale*. They show that in the twenty-one years between the launches of INTELSAT I and INTELSAT VI, the organization produced seven generations of satellites, if we count INTELSAT IV-A and V-A as separate generations. As the thrust of the launcher increased, in-orbit satellite mass increased from 38 kilograms in INTELSAT I to 2,231 kg in INTELSAT VI. As in-orbit mass increased, satellite capability exploded. Effective band width went from 50 Mhz in INTELSAT I to 3,086 Mhz in INTELSAT VI, and satellite capacity measured in telephone circuits increased from 240 to 40,000. The speed of development shows in the very limited design-life times of INTELSAT I ((one to five years), III (five years), and IV, IV-A, V, and V-A (seven years). The first significant increase in design life since 1968 came with INTELSAT VI in 1986 (ten years).

This rapid rate of technological advance was driven by a never-ending search for economies of scale. The rate of decrease in the satellite utilization charge, measured in cost per telephone circuit per annum, is a measure of INTELSAT's spectacular success (Colino 1984). Satellite-utilization charges dropped precipitously, from $60,000 per telephone circuit per annum in 1965 to $10,000 per circuit per annum by 1981.

Another second-order effect of this process of rapid technological advance was to stimulate equally rapid advances in cable technology. Pelton (1977) compares the increases in capacity and cost efficiency of INTELSAT I and IV-A to cable systems, and the rates are remarkably similar.

Yet another second-order effect was to modify the relationship between INTELSAT and the ITU, which, as we have seen, had started out on a rather conflictual footing. If we go back to 1963 for a moment, to the Extraordinary Administrative Radio Conference called to allocate frequencies for space exploration, both the developing countries and the United States wished to broaden the scope of the agenda. The United States in particular wished to discuss the nature of the relationship between ITU and the future INTELSAT. However, the European countries, led by the United Kingdom and fearing that others might attempt to use the 1963 Conference as a vehicle to institute some radical changes in allocations and organizational structure, successfully resisted any expansion in the scope of the agenda beyond frequency allocation. The eventual result, as we have seen, was that ITU was preempted by INTELSAT in two areas: assigning positions in the geostationary orbit and setting technical standards.

The source of the problem, particularly on the standard-setting issue, was the coordination procedure stipulated in article 14 of the INTELSAT agreement. This procedure was "completely divorced from the ITU procedure, especially as regards criteria of interference" (Withers and Weiss 1984). INTELSAT's criteria were far more stringent than ITU's, which raised the issue of discrimination against member administrations. Over time it was found to be simply "impractical to maintain this separation." One major source of conflict was thereby removed.

The other reason for increased cooperation between ITU and INTELSAT has come about through technological advances affecting the capability to reuse satellites in geostationary orbit. The technical complexity and difficulty of the reuse capability is great, and the need to avoid harmful interference from inappropriate spacing of satellites is acute. These conditions facilitated an expansion in the scope of the IFRB's responsibilities with respect to application of new regulations and resolving conflicts. In turn, the increased importance of the IFRB required closer technical collaboration between INTELSAT and ITU. Close interaction also facilitated the introduction into ITU's radio regulations of many provisions that have been based primarily on INTELSAT experience. The relationship has therefore grown closer and more cooperative with time.

The frequency-reuse issue also highlights the political significance of type of technology, in particular whether technological fixes to important policy problems are available to the players. The distinguishing feature here seems to be whether the type of technology can in fact increase the benefits to be shared.

During the decade of the 1970s the structure of control within ITU had passed from the traditional leadership to the Group of 77 in the negotiation game. However, the Group of 77 remained constrained by its lack of the technological capability that was the source of political influence in the basic game, especially on operational issues. As one would have expected, the fight over the system of allocation intensified and came to a head in 1979. The surprise at that time was the outcome: *there was not a change in the system from one based on seniority rights.*

What preserved the system of allocation was a diligent lobbying effort by the U.S. delegation, which argued that the orbit and spectrum resource was not a

Table 9.1a
Summary scores: Chapter 9 (TELCOM)

| Regime | Component or Phase | Effectiveness | | Level of Collaboration |
		Improvement	Functional Optimum	
TELCOM	1: 1958–1971	Significant	Low	Intermediate
TELCOM	1: 1958–1971	Significant	Low	Intermediate
TELCOM	2: 1972–1992	Major	Intermediate	High

resource that was restricted to the degree that the challengers had claimed because a technological fix was possible. This fix was based on the newly emerging capabilities of frequency reuse and use of frequencies hitherto avoided given the degree of interference from cosmic noise (sunspots, rainfall, and so on). The United States committed itself to sharing these capabilities widely. U.S. commitments were made more credible by the fact that U.S. industry was solidly behind the U.S. proposal.

Conclusions

Let us now assess the role of coalitions and negotiation strategies in this process over time (see table 9.1). Once the definitive arrangements were settled, the pattern of coalition formation was unusual. Though to be expected given the conflicts over market share, the United States frequently allied with the developing countries against the other advanced industrial countries. The developing countries were unwilling to subsidize the technological development of advanced countries, but the United States, given larger political concerns, could not push this to the limit. The amount of R&D contracts let by INTELSAT was allowed to increase to a certain extent, but the United States succeeded in having INTELSAT retain title to all inventions and data against the wishes of the Europeans.

Negotiating strategies are also interestingly different from the usual mode in this case. In the first place, the potential benefits that could be realized by each participant were so large and important that they constituted a built-in set of economic selective incentives. The United States also wrote into the definitive arrangements, as we have seen, a sanctions procedure. This was very credible since the United States, as the only launching state, could refuse to launch any satellites for a state

Table 9.1b

Problem Type		Problem-Solving Capacity			
Malignancy	Uncertainty	Institutional Capacity	Power (basic)	Informal Leadership	Total
Benign	Low	High	Pushers	Strong	High
Mixed	Low	High	Pushers	Strong	High
Mixed	Low	High	Pushers	Intermediate	High

that was judged to be in contravention of the INTELSAT arrangements. Sanctions constituted another type of selective incentive.

The Europeans and the Japanese, on the other hand, found both the degree of U.S. dominance and the potential threat of sanctions burdensome. Their negotiating strategies were therefore twofold. First, the more COMSAT insisted on control of INTELSAT, the more narrowly the scope of the organization was defined as we have seen. In political terms, the European and Japanese response to U.S. domination was encapsulation of INTELSAT, and, much later, the organization was not allowed to develop INMARSAT, which went to the International Maritime Organization (IMO).

The other strategy was to seek to develop, as quickly as possible, their own launching capabilities, which the United States never saw as really significant for a long while. The French and Germans created a regional satellite communications system, *Projet Symphonie*, partly as a means of using this potential competition as a selective incentive for INTELSAT (the Americans) to develop the kind of technology they were interested in. But precisely because the Europeans saw NASA as a potential bottleneck in the event the sanctions procedure would be applied against them, they had an incentive to develop their own launcher (the *Ariane*) as quickly as possible.

With respect to outcomes, we have already judged the ITU to be a level 2 organization and INTELSAT to be a level 5 organization, and we have explained the difference in outcome in terms of differences in the scale and significance of benefits to be provided. Conflict over the regime rules was far more prevalent in the attempt by ITU to adapt the existing regime governing use of the radio-frequency spectrum to the coming of the communications satellite than it was in the creation of a subregime to exploit the potential of the communications satellite.

One should not, however, overemphasize the existence of conflict in the ITU even in 1963. Conflict existed on some preexisting basic issues relating to regime design and regime change. These may have been the major issues, but they were certainly not the only ones. Let us not forget our characterization of satellite communication as a mixed type of problem with a large benign component (see table 9.1b). The evidence for this characterization is the fact that most of the activity at the 1963 Extraordinary Conference consisted of cooperative problem-solving based on technical rationality (Miles 1970–1971). This behavior was facilitated by the predominance of consensual knowledge and a coherent epistemic community.

The knowledge base for regulation of satellite telecommunications was never an issue. Uncertainty played almost no observable role in the discussions and negotiations. The knowledge base had been developed historically in the twentieth-century discipline of radio engineering. Its application was overseen by radio engineers from a wide variety of countries who had been working together in the ITU for many years. And managing the radio-frequency spectrum has always involved a mix of cooperative problem solving and conflict over access to and use of the choicest frequencies. The system as a whole merely expanded to accommodate a new set of concerns generated by a new technology. The effect of consensual knowledge in the minds of this epistemic community was to contain organizational conflict and to allow the system to function relatively effectively (though not efficiently) even though challengers were seeking to overturn the existing regime based on seniority rights.

Now, the question we should ask is why this epistemic community did not play a significant role on the major issues of regime design and adaptation and why it did not support the IFRB in its push for comprehensive, systematic planning authority. The answer is that the epistemic community *on those issues* had divided organizational loyalties. The CCIR and the CCITT did not wish to be dictated to by the IFRB, and the representatives to these bodies represented a mix of industry and government officials who shared that view. Moreover, the COMSAT officials who ran INTELSAT in the early stages never hid the fact that they thought the ITU was an inefficient body that should not be allowed to get in the way of an efficient system for exploiting the potential of the communications satellite.

Given these outcomes for the ITU and therefore the subregime for satellite telecommunication and INTELSAT, what judgments can we make about impacts?

Expert Judgment: Did the Agreement (in Each Case) Solve the Problem?

The problem, in the first instance, was to adapt the global telecommunications regime to the appearance of the communications satellite. On the basis of technical rationality, the answer for the ITU in 1963 is overwhelmingly in the affirmative. However, this success was constrained by the political opposition of the dominant players to make any radical changes in the fundamental regime rule of first-come, first-served with seniority rights. This unwillingness to face up to major surgery left the ITU weakened compared for its competitor INTELSAT.

With respect to INTELSAT, the problem was to exploit the technology, and the answer is also overwhelmingly in the affirmative, as the record attests. But over time, the logic of the technology led to a muting of the rivalry between ITU and INTELSAT and an increase in the coordination authority of the IFRB.

Political Judgment: Was the Agreement the Maximum Feasible Within the Institutional Setting, and Were the Outputs the Maximum Feasible?

In the case of the ITU regime, one would have to say no. In the mid- to late 1960s the traditional dominant coalition of the ITU (the United States, the United Kingdom, France, the Netherlands, the FRG, and the USSR) had no effective challengers. The coalition could have put in place a more efficient allocation system without going to the extreme of choosing a regime based on equitable national distribution irrespective of use. There would have been difficulties with the USSR given the strong East-West conflict at the time, but Soviet interests could have been accommodated without great difficulty. The level of collaboration remained fairly low up to the late 1970s, and the dominant coalition never recognized that creation of a separate operational regime would have a corrosive effect on the traditional regulatory regime.

On the other hand, with respect to INTELSAT, on this dimension, one would have to rate the effectiveness of the operational regime as being very high. The level of collaboration achieved was the highest possible. Even the issue of domination of INTELSAT by the United States (COMSAT) was resolved by a weighted voting formula that gave the United States 40 percent of the vote and guaranteed that the most technologically advanced member would dominate decisions on the rate and type of technological advance through *six* generations of INTELSAT satellites in a relentless drive for economies of scale. Moreover, the regime included restrictions on competitive membership by states parties and effective sanctions (controls on launching). One could not realistically have asked for more.

Substantive Measurement: Were Conditions Improved as Much as Were Expected, at Costs Projected, Without Producing Involuntary Losers?

By criteria of technical rationality, the performance of the regulatory regime cannot be rated very highly. As Wihlborg and Wijkman (1981) point out, the resource to be allocated in this case consists of the radio-frequency spectrum and allocations of positions in the geostationary orbit. These two dimensions characterize the orbit and spectrum resource. This is not a single resource in fixed supply but rather a composite resource combining the spectrum itself and the electromagnetic characteristics of particular orbits. Efficiency criteria would demand that the two dimensions be freely combinable and allocated separately. Instead, the designers of both the regulatory (ITU) and operational (INTELSAT) regimes chose to treat them as units with fixed coefficients, thereby creating artificially problems of scarcity.

Admittedly, in negotiating solutions to a complex collective-action problem, it is unlikely that only efficiency criteria would determine regime design. Equity criteria would also have a very high priority. But in both cases of regime design to regulate and exploit the new technologies of satellite telecommunication, a very limited range of design alternatives was considered. In fact, only two choices were on the table—that is, maintain the existing system of seniority rights (first-come, first-served) or shift to a system of global planning that would produce equitable national distribution regardless of use. In both regimes the dominant coalition chose to continue the regime based on seniority rights at the same time that they chose a radically new regime for managing commercial operations.

On the operational dimension, COMSAT as the dominant player saw that the full promise of the new technology could not be realized without its global application, but under decision rules that would ordinarily not be achievable in the international system. That they were achieved in this case speaks to the intensity with which all parties desired the benefits such a system could produce. The pattern of control chosen ultimately satisfied the United States, the developing countries, and the challengers (Western Europeans and Japan). Thereafter, COMSAT/INTELSAT rapidly delivered the goods.

Within this pattern of *constrained technical rationality*, one would have to rate the effectiveness of INTELSAT performance as being high. However, one would have to rate the effectiveness of the ITU regime as being low on technical rationality criteria.

But it must also be admitted that INTELSAT's success has had unfortunate consequences for efficiency. In the first place, rapid technological advance in satellites stimulated rapid technological advance in cable systems to the point where all players are currently threatened by overcapacity. Second, as Pelton (1977) shows, this competition between modes of communications was aggravated by a proliferation of satellite systems competing for positions in the orbit and spectrum resource. Third, the system as a whole lacks a capacity to plan on a global basis encompassing all modes of communication.

The lack of a global capacity to plan carries with it a major threat. So far the system for exploiting the potential of satellite telecommunication thrives on a relentless search for economies of scale. This puts a premium on technological innovation for increasing and widely sharing benefits. If overcapacity is widespread, commercial rivalry will increase, and the possibility of technological fixes may decline. If this occurs, the level of conflict will intensify, and effectiveness will decrease. The threat multiplies as deregulation proceeds at the national level because deregulation fuels the rapid development of capacity in private hands. Such a large addition of new players and services threatens INTELSAT, while it increases the significance of ITU's function.

The satellite telecommunication case study shows what levels of cooperation are possible when dealing with benign problems on a global basis. Commercial rivalries constrain cooperation on regulatory questions, but even so, cooperation is continuous because the spectrum is a common property resource where all states have an equal interest in avoiding or minimizing harmful interference. This interest is supported by an inherent reciprocity and retaliation sanction, an epistemic community of considerable longevity, and consensual knowledge.

The answer to the counterfactual question is obvious. In both cases, the problems either would be worse (ITU) or could not effectively be solved (INTELSAT) without the existence of either subregime. Both subregimes therefore matter.

Notes

1. The analysis presented herein is a reformatted version of an evolutionary analysis previously offered in Miles (1989). The focus of that paper was on the long-run or second-order effects of negotiated agreements on issues related to science and technology. The focus of this version is on the relationships between type of problem and levels of international collaboration and between type of problem and regime effectiveness.

2. For full-scale evaluations of the ITU and the evolution of its role, the reader should consult Codding and Rutkowski (1982), Jacobson (1973), and Savage (1989). For full-scale treatment of the entire telecommunications regime, the reader should consult Cowhey (1990), Krasner (1991), and Zacher with Sutton (1996). The reader is reminded that the present evaluation focuses solely on that part of the global communications regime that governs satellite telecommunications.

3. During the period 1968 to 1971 the author worked closely with ITU, WMO, COSPAR, and COPUOUS on the issues treated in this chapter and systematically interviewed delegates from all organizations.

4. The Soviet-sponsored counterpart to INTELSAT.

5. Coordination of action on the basis of explicitly formulated common standards but with national action implemented solely on a unilateral basis. No centralized appraisal of effectiveness of actions taken is possible.

6. Personal interviews by the author with developing-country delegates during INTELSAT negotiations.

7. Personal interviews by the author with representatives from the Delegation of France to the INTELSAT negotiations.

References

Codding, George A., and Anthony M. Rutkowski. 1982. *The International Telecommunication Union: An Experiment in International Cooperation.* Dedham, MA: Artech House.

Colino, Richard R. 1984. "The INTELSAT System: An Overview." In Joel Alper and Joseph N. Pelton, eds., *The INTELSAT Global System.* Vol. 93, Progress in Astronautics and Aeronautics (pp. 55–94). Washington, D.C.: American Institute of Aeronautics and Astronautics.

Cowhey, Peter F. 1990. "The International Telecommunications Regime: The Political Roots of Regimes for High Technology." *International Organization* 44(2) (Spring): 169–199.

International Telecommunication Union (ITU). 1968. *ITU and Space Communication.* Geneva: ITU.

Jacobson, Harold. 1973. "The International Telecommunication Union." In Robert Cox and Harold Jacobson, eds., *Decision Making in International Organizations: The Anatomy of Influence* (pp. 59–101). New Haven: Yale University Press.

Krasner, Stephen D. 1991. "Global Communications and National Power: Life on the Pareto Frontier." *World Politics* 43(3) (April): 336–366.

Levy, Stephen A. 1975. "INTELSAT: Technology, Politics, and the Transformation of a Regime." *International Organization* 29(3) (Summer): 655–680.

Miles, Edward. 1970–1971. *International Administration of Space Exploration and Exploitation.* Monograph Series in World Affairs, Vol. 8, no. 4, Graduate School of International Studies, University of Denver.

Miles, Edward L. 1989. "Scientific and Technological Knowledge and International Cooperation in Resource Management." In Steinar Andresen and Willy Ostreng, eds., *International Resource Management: The Role of Science and Politics* (pp. 46–87). London: Belhaven Press.

Pollack, L., and H. Weiss. 1984. "Communication Satellites: Countdown for INTELSAT VI." *Science* 223 (February 10): 553–559.

Pelton, Joseph N. 1977. "Key Problems in Satellite Communications: Proliferation, Competition and Planning in an Uncertain Environment." In Joseph N. Pelton and Marcellus S. Snow, eds., *Economic and Policy Problems in Satellite Communications* (pp. 93–123). New York: Praeger.

Savage, James C. 1989. *The Politics of International Telecommunications Regulations.* Boulder: Westview Press.

Wihlborg, Clas G., and Per Magnus Wijkman. 1981. "Outer Space Resources in Efficient and Equitable Use: New Frontiers for Old Principles." *Journal of Law and Economics* 24(1) (April): 23–43.

Withers, David, and Hans Weiss. 1984. "INTELSAT and the ITU." In Joel Alper and Joseph N. Pelton, eds., *The INTELSAT Global Satellite System* (pp. 270–310). New York: American Institute of Aeronautics.

Zacher, Mark W., with Brent A. Sutton. 1996. *Global Networks: International Regimes for Transportation and Communications* (pp. 127–180). Cambridge: Cambridge University Press.

10

The Management of High-Seas Salmon in the North Pacific, 1952 to 1992

Edward L. Miles

Introduction and Reasons for Success and Failure

The focus of this chapter is the evolution of the regime for the management of high-seas salmon fishing in the central North Pacific Ocean over a forty-year period from 1952 to 1992. Five species of Pacific salmon were the targets of these fisheries: *Oncorhynchus keta* (chum salmon), *O. gorbuscha* (pink salmon), *O. nerka* (sockeye salmon), *O. kisutch* (coho or silver salmon), and *O. tshawytscha* (chinook or king salmon). The listing is in ascending order of market value. The sixth species of Pacific salmon, *O. mason* (masu salmon), is found predominantly off the Japanese and Southern Russian coasts and is of no significance in this story. The main state actors are Russia (USSR), Canada, and the United States as "states of origin" in whose waters salmon (as anadromous species) spawn. The main target of regulation is Japan, which licensed a mother-ship fishery in 1936 and a land-based gill-net fishery after 1945.

Even though only four state actors are involved here, the story is complex because until 1976 and 1977 the central North Pacific was an area of high seas as far as salmon were concerned and Japan could legitimately engage in fishing as a freedom of the high seas. Any restrictions on Japanese fishing, therefore, had to be negotiated, and Japan could not be bound without its own consent. Moreover, up to the 1960s there was considerable empirical uncertainty over the ranges of salmon migrations (all species) and the extent of intermingling of stocks in the central North Pacific involving salmon of Asian and North American origin. The International North Pacific Fisheries Commission (INPFC) presented only a level 2 organization, but it did succeed in creating a firm base of consensual knowledge about migration and intermingling of stocks within seven years. To that extent it was highly effective. With respect to its second job, however, reducing salmon interceptions, the

Commission was not very effective for most of its history until the basic regime in which it was designed itself changed in 1976 and 1977. After the basic regime change, the subregime for reducing salmon interceptions on the high seas in the central North Pacific was very effective, but this improvement still did not protect the regime from termination for political reasons.

This case is also interesting in a comparative analysis of international environmental regimes for a number of reasons. First, like the whaling, nuclear nonproliferation control, and oil pollution at sea cases, this case has a long evolutionary history. These long evolutionary histories therefore speak to the comparative life cycles of regimes. Second, even though what is at stake is a common property resource, like the radio-frequency spectrum, in this case the built-in reciprocity and retaliation sanction that exists in use of the spectrum to control harmful interference is missing, divisibility of the collective good is very high, and conflict is endemic and intense.

Third, even though conflict is endemic and intense, like all fisheries that are conducted in open-access regimes, the final outcome, as determined by the transformation in the basic regime produced by the third U.N. Conference on the Law of the Sea (UNCLOS III), is surprisingly effective as it responds to the peculiar life history of the target species. This regime transformation in essence forbids high-seas fisheries for salmon. In this respect, the salmon case adds something new to the evaluation of fisheries regimes. Almost all international fisheries *open-access* regimes have the same sorry history, which can be summarized in the following rule: "The key question in international management of fisheries is *control over the stocks*. If authority is fractionated and participation in the management system is voluntary, with serious constraints on surveillance and enforcement, then biological and economic waste will be high, conflict will be endemic, decision systems will be cumbersome with protracted decision times, and effectiveness will be low."

The high-seas salmon case is interesting because it begins as an example that fits the rule but transitions over time into a completely different set of outcomes by systematic changes in the basic conditions.[1] As such, this experience provides an effective bridge to another fisheries regime, that involving the management of tuna fisheries in the Western Central Pacific.

In the treatment that follows more weight must be placed initially on the pre-regime and regime-creation phases than on regime implementation. When the focus does shift to regime implementation, we must distinguish between three subphases. Most of the story takes place between 1952 and 1976, which corresponds to the era of open access and the heyday of high-seas fishing for salmon. This is subphase

I. The year 1976 is a convenient cut-off point because that is the year in which consensus was arrived at in UNCLOS III concerning article 66 of the Law of the Sea Convention forbidding fishing for anadromous species beyond the outer limits of the 200-mile exclusive economic zone (EEZ). This development represents the transformation of the regime. The second subphase of regime implementation, which occurs between 1977 and 1984, consists of the effort to adapt the existing arrangement (INPFC) to the new international law. The third subphase, 1985 to 1992, reflects attempts primarily by the United States (pushed by representatives of Alaska) and the USSR to terminate *all* Japanese high-seas fishing for salmon. This objective was achieved with the entry into force on February 21, 1993, of the Convention for the Conservation of Anadromous Stocks in the North Pacific Ocean involving the United States, Russia, Canada, and Japan.

The Regime-Creation Phase, 1907 to 1952

What Problem Is Being Solved?

The source of the problem was the development of a Japanese offshore fishery for salmon of Russian/Soviet origin from 1855 and the development of a distant-water Japanese mother-ship fleet fishery in the Northeast Pacific (Bristol Bay) in 1936 (Miles et al. 1982). These fisheries triggered a series of attempts to regulate them by Russia/USSR in 1855, 1907, 1925, and 1928 and pressure from the United States on Japan to refrain from operating salmon mother-ship fleets in the Northeast Pacific (Miles et al. 1982). These conflicts were interrupted by World War II, but they resumed after 1945 once hostilities had ceased.

Analytically the problem was mixed. It began as a predominantly malign problem with both direct competition and externalities dimensions for the United States and Canada and for the USSR and Russia—that is, for three of the four states of origin. It was also a problem of conservation relative to the need for protection of immature salmon of Soviet and North American (United States and Canada) origin. However, since international law clearly permitted Japanese fishing for salmon on the high seas, the United States and Canada, pursuant to the negotiation of a U.S.-Japan Peace Treaty in 1952, hit upon the idea of the principle of abstention as a means of controlling Japanese fishing efforts on the high seas (Miles et al. 1982).

According to this principle, Japan was to refrain from fishing stocks of salmon of North American origin if all of the following criteria were met (Miles 1987b):

· Evidence based on scientific research indicated that more intensive exploitation of a stock would not provide a substantial sustained increase in yield;

• The stock was being directly managed by a contracting party for the purpose of maintaining or increasing its maximum sustained productivity; or

• The stock was being studied extensively to discover whether it was being fully utilized and what the necessary conditions were for maintaining its maximum sustained productivity.

Once Japan proved willing to accept these criteria for determining the case for abstention, it became necessary to define a procedure for identifying the stocks that qualified and those that did not. At this time it was generally not known how far east salmon of Asian origin migrated; nor was it known how far west salmon of North American origin migrated. All parties to the accord, including the USSR, had symmetrical interests in answering these questions and, as an afterthought, the questions of whether there was intermingling of stocks and, if so how, much and where for each species. This was the benign part of the problem because these questions related only to defining how large the "pies" (East and West) were and how they were related to each other. As soon as the questions shifted from migration patterns to what line of meridian best separated stocks of North American origin from stocks of Asian origin, this touched on allocation or division of the pies so that the problem became malign once again.

The situation was further complicated by the fact that the Soviets claimed that Japan was taking a lot of salmon of Soviet origin as well in the Asian group. While the cold war prevented the USSR from being a part of the United States, Canada, and Japan arrangement, the western boundary of this arrangement became the eastern boundary of the Japan and USSR arrangement. Consequently, the Japanese were caught in a potential squeeze between the USSR in the west and the United States and Canada in the east. For three of the four participants—that is, the states of origin—the problem to be solved was one of allocation with heavy conservation overtones. For one participant (Japan) the problem was primarily allocation within a legitimate high-seas fishery where the rights of coastal states were constrained by international law.

But the potential squeeze on Japan between the Soviets in the west and Canada and the United States in the east was not the only complication. At the same time, when the United States and Canada were seeking to negotiate a cooperative solution to the problem with Japan, the U.S. Senate was considering ratification of the Peace Treaty between the United States and Japan. Since Japan was vitally interested in the latter, its leaders claimed that they signed the agreement establishing the cooperative arrangement under duress. It is worth emphasizing that while Japan could exercise substantial issue-specific power derived from international law, it was equally substantially constrained in the exercise of basic game power as a result of being the loser in World War II and needing peace treaties with all three former allies to regularize its status internationally. Clearly the treaty with United States was the most important in this regard.

The final complication arose out of variations in the salience of the issue to domestic constituents at national and regional levels. For the Japanese this was a national issue of great importance since it involved a traditional fishery on the high seas that was producing income, employment, and a high-value food product. In the United States and Canadian Pacific Northwest the value of the fishery, as measured by income and employment, was also very high, which gave rise to domestic pressures to protect "their resource." However, looked at from the national level, the salience of the issue was only moderate. However, national leaders chose to let their regional salmon-fishing industries set the agenda on the basis of inputs from U.S. and Canadian scientists.

Distribution of Capabilities

We must also distinguish salmon-related capabilities from generalized capabilities that could be brought to bear on the negotiations. With respect to salmon-related capabilities, the Japanese were significantly ahead of the three other states of origin, none of whom possessed an at-sea processing capability for salmon. Scientific capabilities on salmonids, however, were symmetrical as between Japan, Canada, and the United States. The Soviets were behind the other three in this respect. On the other hand, when it came to generalized capabilities, clearly the United States and the USSR overwhelmed Japan as well as Canada.

For Japan and the United States this issue was part of a generalized normalization of relations after World War II and one in a series of actions planned at the time. Japan's potential responses to pressure from the United States and Canada on the issue were consequently constrained by the fact that the U.S. Senate was considering ratification of the peace treaty at the same time. On the Soviet side, Japan was hostage to a vigorous program of enforcement, which included ship seizures and sometimes temporary incarceration of crews (Miles et al. 1982). Japan had no effective responses to these measures and, as a result, was propelled into accepting a more rapid normalization of relations with the USSR than it would have preferred.

Configuration of Actor Interests and Positions and Institutional Setting

These independent variables are merged in the high-seas salmon case because no appropriate institutional settings existed; they had to be created. In the Northeast

Pacific the protagonists (the United States, Canada, and Japan) created the International North Pacific Fisheries Commission in 1952, while in the Northwest Pacific the Northwest Pacific Fisheries Convention of 1956 established a bilateral commission between the USSR and Japan wherein these issues were to be discussed. The USSR-Japan relationship on salmon remained highly contentious into the mid-1970s. The Japan, Canada, and United States arrangement was contentious as well, but it also broke new ground.

Given that all parties had identical interests in facilitating the creation of consensual knowledge with respect to salmon migration and intermingling, the only issue remaining was that of institutional design: what set of arrangements would most efficiently and effectively yield both the answers to the questions posed and the management decisions desired by at least two of the parties to the arrangement? While U.S. leverage over Japan was most effective on the issue of the decision rules to be adopted, Japan was not helpless. Japanese interests were substantially protected by international law since these were admittedly high-seas fisheries. While the coastal-state coalition was strong and stable without any chance for Japan to play one off against the other, the coastal states could not simply demand. They had to negotiate, and Japanese consent had to be obtained. Japan's negotiating position therefore was to seek protection in the decision rules and in the constraints to be imposed in arriving at management decisions. The latter required explicit elaboration of the criteria on which these decisions would be based.

Clearly, the U.S. leverage with respect to the peace treaty provided the political incentive for Japan to agree, and the negotiators chose to design the institutional arrangements in two ways. First, the International Convention for the High Seas Fisheries of the North Pacific created the International North Pacific Fisheries Commission (INPFC). Second, the Convention stipulated that Japan would abstain from fishing stocks of halibut, herring, and salmon in areas specified in an annex to the Convention and in a protocol. The latter defined the line of meridian 175 degrees West as constituting a salmon-abstention line since there was to be no Japanese high-seas fishing for salmon east of 175°W. These provisions created what came to be known as the *abstention principle* in international law.

The Commission was given three major jobs: (1) determine annually whether the halibut, salmon, and herring stocks identified in the annex continued to qualify for abstention (the criteria for proving qualifications for abstention were also specified); (2) determine whether any additional stocks in the Convention area qualified for abstention; and (3) determine what line best divided salmon of North American origin from salmon of Asian origin. *The line of meridian 175°W was seen initially to be a provisional line*, subject to adjustment if it could be proved that another line more equitably divided the stocks.[2]

All three major jobs of the Commission depended on research that involved a large amount of field investigation. But the Commission was not given an independent staff to do the job. The Commission could only define and coordinate the research to be done by the three governments. As such, the outcome of the regime-creation phase was only level 2 cooperation. However, it is surprising that in one respect a level 2 cooperative arrangement did succeed in creating consensual knowledge. This had to do with coordinated research to determine the extent to which salmon of North American and Asian origin migrated and the extent of intermingling of stocks.

The answer to the question of what line best divided the stocks did not become consensual knowledge for almost twenty years, until a basic change in the regime occurred. The research did yield an answer—the line of meridian 175°E. But the protocol stipulated that the line could not be moved without unanimity among the three parties to the Convention. Since by the time answers were available to these questions (1954 to 1961) Japan and the USSR had made the line of meridian 175°W the easternmost boundary of their arrangement, Japan refused to agree to change the provisional line for almost two decades.

Consensual knowledge did not emerge on the issue of stocks qualifying for abstention because this issue went to the heart of the divisibility of the resource and the competitive interests of the signatories. Moreover, the institutional setting compounded the problem since underlying the Standing Committee on Biology and Research, the scientists were divided into *national* sections. Each nation did its own research, and the conclusions routinely supported national positions. The scientists were therefore too constrained to form a fully effective epistemic community.

In terms of long-run impacts, the outcome of the regime-creation phase represents what was politically feasible rather than any attempt to define what would be ecologically most efficient. Governments wanted to know the answers to the questions how far salmon of North American and Asian origin migrated, how much intermingling occurred, and where it occurred. They therefore did not interfere with this research, and the answers were available within seven years. At least two of the parties in question were also vitally interested in determining what line best divided the stocks, but the decision rules provided Japan the protection it sought that under the current regime the line would never be moved without its consent.

The First Regime-Implementation Phase, 1952 to 1976

Over the course of the next twenty-four years international law, which had provided critical protection to the Japanese, shifted. In 1952, the United States and Canada acknowledged the right of Japan to continue their high-seas salmon fishery while claiming the right to manage anadromous stocks of North American origin on the high seas, a dubious claim at that time. By 1976, however, article 66 of the emerging United Nations Convention on the Law of the Sea of 1982 clearly gave to states of origin primary interest in and responsibility for anadromous stocks that spawn in their rivers. High-seas fishing for salmon was prohibited "except in cases where this provision would result in economic dislocation for a state other than the state of origin."

That formulation was intended to accommodate Japan and the regional arrangements in the North Pacific. In those cases, consultations were required, and concessions were made for states historically fishing those stocks if they met certain criteria. With its position so clearly weakened, Japan could fight only a rear-guard action until 1992 when a new North Pacific convention terminated all high-seas fishing for salmon. By this time the USSR/Russia had formally joined the club by aligning with the United States and Canada.

Between 1954 and about 1961 most of the Commission's time was consumed by the problem of the protocol: to what extent did the six species of salmon of North American and Asian origin migrate? Was there an intermingling of stocks, and if so, where and when? If there was intermingling, what line best divided the resources?

It is clear from the protocol that the line of meridian 175°W longitude was intended to be provisional until research should indicate a better one. Information on the extent of oceanic migration of salmon was sketchy when the convention was negotiated. The dominant view was that salmon stayed within the influence of estuaries after they left the rivers. However, scientists had a hunch that salmon migrated because in some instances fish tagged on one side of the Pacific had been caught on the other side. Nonetheless, it was believed that long-range migration was the exception rather than the rule.

The setting of the specific research agenda was accomplished by the INPFC and its Standing Committee on Biology and Research (CB&R), which was created in February 1954. The CB&R included one commissioner and two scientists from each member country; the director and assistant director of the INPFC were later added ex officio, without vote (Jackson 1963). The CB&R itself created one scientific subgroup. The CB&R dealt with policy and instructions, while the subgroup was charged with implementation of the instructions. The actual investigations were conducted by the national agencies of the contracting parties.

To assess the distribution, continent of origin, and extent of intermingling of salmon on the high seas, the program of research as defined involved the following:

· The study of movements as indicated by the fishery itself;
· The use of tagging to study movements;
· The marking of seaward migrations;
· Stock differentiation using morphological, physiological, or biochemical characteristics;
· Special test fishing to study movements and factors controlling them; and
· The study of physical and biological oceanography.[3]

This work involved the coordinated investigations undertaken by fifteen or sixteen research vessels of the three countries each year for several years at an estimated cost of about $2 million per year (Jackson 1963).

The scientific subgroup of the CB&R was further divided into national sections. On the major issue of the distribution and intermingling of salmon, there were no great differences of opinion regarding research goals, and according to the testimony of those involved, communication and cooperation among the sections were easy and fruitful (Fredin n.d.). Major advances were also made in investigations of the physical oceanography of the North Pacific between 1954 and 1961, and these too provided important clues to the dynamics and extent of salmon migration.

While the cumulative results of this research settled the question of the extent of salmon migration and intermingling and were accepted as authoritative by all parties, any attempt to use this knowledge for management purposes could not succeed given the decision rules.

Let us recall that the Convention also gave the INPFC the job of determining the line that best divided salmon of North American and Asian origin. For the provisional line to be moved, proof had to be "beyond a reasonable doubt," and all three parties to the Convention had to agree. However, as Fredin points out, by 1957 field research had shown that the provisional line did not protect salmon of North American origin (particularly Bristol Bay sockeyes) from exploitation by Japanese high-seas fleets. As a result, the United States in 1957 proposed to move the line west to 175°E longitude, but Japan did not agree (Fredin n.d.). Japan could not have agreed in any event because in 1956 a bilateral Japanese and Soviet fisheries treaty made the provisional line of 175°W its eastern boundary. Any move west of that line would therefore have incurred serious losses for Japan to both the United States and the USSR. From this point on, U.S. attempts to move the line increased conflict within the INPFC to no avail because it was clear that the line would never be moved unless circumstances were radically altered.

The INPFC's third task—to determine which stocks of those identified in the annex qualified for abstention according to the three criteria stipulated by the Convention—caused even more contention than did the issue of the abstention line. The matter was first placed on the agenda in 1956, when the United States submitted its original cases for salmon, halibut, and herring (Fredin n.d.). By 1958 it was clear that no agreement would be forthcoming; the parties could agree only that those stocks identified in the annex in 1952 remained on the list and that Japan be required to abstain from fishing them (Fredin n.d.).

This is a graphic example that the veto worked both ways—in favor of Japan on the issue of the abstention line but in favor of the United States and Canada with regard to stocks qualifying for abstention. Only if the United States and Canada agreed could Japan be released from the abstention provisions.

The research on qualifying stocks was affected by the national positions of the three parties to a far greater extent than was the research on distribution and intermingling. A reasonable inference is that allocation was involved directly in the former case but not in the latter. Governments had to be concerned with calculating probable gains and losses, as well as probable political consequences. However, while most participants acknowledged the national effects on the research, they argued that the peer-review mechanisms within the CB&R acted as a brake on the extent to which politics could infuse science.[4] One participant thought that national political preferences played a role in the setting of the research agenda of each section but did not necessarily lead to corruption of research results because political effects were largely negated through the peer review mechanism.[5] In contrast, another participant argued that the national research on stocks qualifying for abstention was of lower quality than research on distribution and intermingling because there may have been a general tendency to suppress research results that did not support the national positions.[6] In addition, in cases where national positions were conflicting internally regarding priorities to be accorded salmon, pelagics, and demersal species, the research process became even more highly politicized over which species would be accorded preference. The inference therefore is clearly that the stock status and abstention-line issues presented to governments problems of allocation involving high stakes. High stakes produce high levels of government interest and therefore greater contamination of research where research is conducted on a national basis. If indeed there was systematic suppression of results that contravened national positions, this would represent serious corruption of research. However, we have no way of knowing whether such practices were widespread. Access to this documentation must be cleared by national sections and is still restricted.

Let us now summarize what this case represents to this point. Type of problem (mixed) remains the single most important variable. The benign aspects—research collaboration to answer two questions to which participating governments attached high priority—facilitate design of a mechanism that proves to be very effective in answering the questions how far salmon of Asia and North America migrate and whether they intermingle. The decision rules, which provide Japan a veto on the use of this new knowledge, effectively insulate the process of developing that knowledge from governmental meddling. But these same decision rules also guarantee that the new knowledge will not be used if it adversely affects any of the parties' priority interests relating to controls on high-seas fishing for salmon and moving the abstention line.

We note that this issue was a fight primarily between states of origin and flag states about the right to manage a high-seas fishery and about the nature of the overarching legal regime in which the issue emerges. Since the norms of the Grotian regime clearly gave authority to the flag states on the high seas (that is, Japan), the United States, Canada, and the USSR were all engaged in an attempt to transform the regime and to adapt it to the biological necessity of managing anadromous species. This transformation at the time was never intended to crack the basic norm that high-seas fishing was guaranteed as a high-seas freedom, so the compromise norm that states of origin suggested—the abstention principle—was as onerous to states of origin as to the flag state. One had to prove in each case that particular stocks qualified for abstention unless they were already protected by the abstention line.

Analytically, the independent-variable type of problem merges with an intervening-variable level of uncertainty in the knowledge base, thereby increasing significantly the weighting to be ascribed to the latter. This combination of variables is sufficient to permit a level 2 mechanism like the INPFC to produce consensual knowledge. But we note that the uncertainty in this case was empirical and not theoretical, even though scientists had to develop a series of surrogate measures for working out a very complex picture of movement by species in time.

Configuration of actor interests and positions merge completely with another intervening variable (coalitions) and yield a very stable configuration over forty years (that is, two and then three states of origin against the flag state, Japan). Distribution of capabilities merges with yet another intervening variable (negotiation strategies), but this merged variable is heavily affected by timing and capabilities outside the salmon issue and by type of regime, which gives critical protection to the flag state. Another external factor of importance is the small size of the group (only four states).

The process of regime creation, therefore, was a game between only three, and by implication four, parties in which the outcome was creation of an institutional setting to perform certain tasks. That outcome was determined essentially by four sets of merged variables: (type of problem + uncertainty in knowledge base) + (configuration of actor interests and positions + coalitions) + (distribution of capabilities + negotiation strategies + [timing + type of regime]) = Level 2 outcome. *However, the decision rules imposed on the institutional setting that was created guaranteed that an epistemic community would not form.*

Because the result ensured Japan a veto and because the U.S. government continued to let the domestic regional interests set the agenda in this case, the allocation conservation problem could not be solved. Dissatisfaction by states of origin on the management side could only feed the fire of regime transformation in which the basic rules of the game would acknowledge the exclusive right of states of origin to manage anadromous species even on the high seas. For the United States, serious internal conflict over competing regional and national interests was thereby engendered.

In direct contrast with the collaborative research process that occurred under INPFC auspices between 1952 and 1961, the management process was highly conflictual. This problem was wholly malign involving both competition between states of origin and the flag state and negative externalities for the conservation programs of states of origin as well. From the late 1960s and especially into the 1970s, these externalities also affected the salmon aquaculture plans of the states of origin.

The level of controversy was so high from 1953 to 1958 that the Commission was not permitted to make any judgments on which stocks qualified for abstention (Miles et al. 1982). In addition, from 1961 U.S. attempts to move the abstention line were blocked for almost twenty years by Japan.

On the plus side, the Commission decided between 1959 and 1962 that herring stocks off northern British Columbia and the U.S. coasts no longer qualified for abstention (Miles et al. 1982). And in 1962 this judgment was extended to halibut in the eastern Bering Sea. This judgment was a mistake, but it was done because the United States could not prove that those halibut were fully utilized by U.S.

nationals and because the United States wished to give Japan some incentive to renew the arrangement as was required after the first five years (Miles et al. 1982).

It should be pointed out that a serious design flaw in the INPFC arrangement related to new entrants. This question was not dealt with at all and was raised after 1973 by an attempt of South Korea to begin high-seas fishing for salmon (Miles et al. 1982). Luckily, this new problem arose just as the Law of the Sea negotiations were beginning. By 1976 article 66 was almost in its final form, as was the exclusive economic zone as a whole. On the basis of this regime transformation, the INPFC had to be reconfigured. Even though the exclusive authority of states of origin was acknowledged, the conflict over Japanese interceptions of North American and Soviet salmon did not end until 1992 when *all* high-seas salmon fishing was unequivocally prohibited (deReynier 1995).

The Second Regime-Implementation Subphase, 1977 to 1984

Pursuant to developments in UNCLOS III, the United States enacted the Magnuson Fisheries Conservation and Management Act (MFCMA) in 1976 with its application set for March 1977. In February 1977, the United States submitted its intention to withdraw from INPFC unless the arrangement was renegotiated to conform to the MFCMA. These negotiations began in October 1977 and concluded in April 1978 (Miles et al. 1982). Pursuant to the new conditions, Japanese high-seas salmon fishing was closely regulated by the United States to minimize Japanese interceptions of salmon of North American origin. The principal management tool was an area-time closure system (Miles et al. 1982). But in addition, the U.S. Marine Mammal Protection Act also applied to the incidental taking of Dall's porpoise, northern fur seals, and northern sea lions. Japanese fishing operations were made to conform to these regulations as well, and Japanese vessels were required to carry scientific observers at their expense. The level of Japanese interceptions of salmon of North American origin was thereby cut from an estimated average annual take of 2,545,000 fish from 1964 to 1973 to 400,000 to 670,000 after 1978 (Miles et al. 1982). Japan also had to agree to allow U.S. enforcement of its high-seas salmon regulations on the high seas after notification of and consultation with Japan about the vessels in question. This new regime extended to 175°E longitude—that is, the Provisional Line of demarcation (175°W) was finally moved to coincide with the Commission's research results (Miles 1987b).

The regime transformation that had occurred within UNCLOS III clearly put all states of origin in the driver's seat, thereby solving the authority problem. As such, the transformation of the basic regime definitively tipped the scale of issue-specific power in favor of states of origin and devalued Japanese superiority in operating capabilities. From there on, regional interests, particularly from Alaska, waged a relentless struggle to reduce all Japanese interceptions to zero. The Soviets adopted the same position. This outcome was not immediately available, but it was predictable. State-of-origin strategies therefore implied shutting down all conditions of open access. The only question left for the Japanese fleets was how long the transition period would last. The federal government of the United States sought to prolong this transition for its ally, Japan, by focusing on reducing Japanese interceptions of U.S. salmon on the high seas. This objective was successful in driving the interception rates down to a level below 1 percent but even that was unacceptable to the Alaskans (Miles 1987a, 1989). The internal political winds in the United States finally shifted in the late 1980s.

In the Northwest Pacific, after the introduction of the Soviet Exclusive Economic Zone Proclamation, Japanese salmon fishing vessels operating in the Soviet zone were cut from 2,327 in 1976 to 1,772 in 1977 (Miles et al. 1982). The 1977 salmon quota for Japan was 62,000 metric tons (mt), but the Soviets reduced this to 42,500 mt as part of a complex trade of fisheries quotas in each other's zones (Miles et al. 1982). In doing so, the Soviets rejected Japanese incentives to increase the quota, including construction of twelve salmon hatcheries or two integrated research centers or payment of a "fisheries cooperation fee" (Miles et al. 1982). This quota of 42,500 mt remained in force until 1983 but then was reduced by the Soviets to 40,000 mt in 1984, 37,600 mt in 1985, and 24,500 mt in 1986 (Miles 1987a). At that time the Soviets raised the possibility that a complete prohibition might be in place by 1989 (Miles 1987a).

The Third Regime-Implementation Subphase, 1985 to 1992

The years 1985 and 1986 were difficult for the Japanese high-seas salmon fleets because they were caught in a severe squeeze by both the USSR and the United States. At the same time that Japanese quotas were being severely curtailed in the Soviet zone, Japan was in an intense conflict with the United States over alleged Japanese interceptions of immature Western Alaska Chinook salmon. The Alaskans argued that these interceptions threatened the stability of the renegotiated INPFC, while Japan replied that the United States had already squeezed all mother-ship operations into only 13 percent of the total area available as a result of its area-time closures (Miles 1987a).

The U.S. side claimed 994,000 fish were intercepted. The Japanese estimate was 400,000 fish or less than 0.8 percent of total catches of salmon in Alaska (Miles 1987a). In the midst of an intense, public confrontation, the U.S. Department of State was seeking to find a solution that would achieve a 50 to 60 percent reduction in Japanese interception of North American salmon and still allow the Japanese high-seas fleets to catch almost the same amount of salmon of Asian origin as they had been catching (about 30 million fish). Eventually, a solution was found as a result of painful concessions made by each side. The essential elements of the compromise closed the central Bering Sea to the mother-ship fleet by 1994 in return for increased effort within the U.S. zone. Fishing in the eastern half of the Bering Sea was to be phased out by 1988 and in the western half by 1994. Total fleet days within the U.S. zone were to increase to 140 days after the high-seas area was closed. For the land-based fishery, the eastern boundary of 175°E was moved to the west by one degree (174°E) and was subject to review in three to five years in the light of joint research to be the undertaken. The agreement was denounced by Alaskan fishermen but was submitted to an extraordinary meeting of the INPFC, after consultation with Canada, and approved. The Alaskan objection was that the "solution" ignored the Japanese interceptions in the North Pacific Ocean as distinct from the Bering Sea. As such, they claimed, it did not provide adequate protection (Miles 1987a).

It is difficult to see how a reduction of Japanese salmon interceptions to around 0.04 percent of the total Alaskan catch did not provide adequate protection to Alaskan interests, but clearly they wished to terminate all Japanese high-seas fisheries for salmon. A complete phaseout from the Bering Sea could have been foreseen in 1987 by 1994, but the stability of the arrangement proposed by the Department of State for the central North Pacific was predicated on continued Japanese access within the U.S. EEZ. Such an arrangement could not have been foreseen for very long because most U.S. fisheries interests in the Pacific Northwest of the United States devoutly wished and worked hard for a complete phaseout of *all* foreign fishing within the U.S. zone. The end was not long in coming.

In 1988 a coalition of Western Alaskan fishermen and U.S. environmentalist organizations used the lever of the Marine Mammal Protection Act to block Japanese fishing for salmon within the U.S. zone as a result of the incidental catch of Dall's porpoise in the Japanese fishery (Miles 1989). In addition, illegal salmon fishing by Japanese, Taiwanese, and even North Korean vessels flying Japanese flags added to the furor, which was compounded by a major conflict over the operations of the Japanese and Korean squid drift-net fleets operating in the North Pacific from 1988 to 1990 (Miles 1989). It was therefore no surprise when the United States and the USSR began to negotiate a Convention for the Conservation of Anadromous Stocks in the North Pacific Ocean in 1990, letting the Canadians into the process only after a draft of the treaty had been completed. This Convention entered into force on February 21, 1993, and completely replaced the INPFC as it prohibited *all* high-seas fishing for salmon. Japan was finally caught in a terminal squeeze between the United States, USSR/Russia, and Canada in a formal arrangement, and the Japanese land-based gill-net and mother-ship salmon fleets passed into history.

What then was different analytically between the first implementation subphase and the second and third implementation subphases on this issue? It is the single fact that the overarching regime of the ocean, within which the law relating to high-seas fisheries is nested, had been transformed, thereby taking away from Japan its issue-specific power. Issue-specific power was transferred to states of origin, two of which also exercised the largest margin of basic game power in the international system as a whole. Moreover, while the U.S. government was disposed to be more lenient with Japan, neither the USSR nor the Alaskans shared that view. A shift in the global regime as represented by UNCLOS III was the first and major external change in conditions that stumped the Japanese. An internal shift in political power in the Reagan administration within the United States was the second significant external change because it gave to Alaskan congressional representatives, especially to Senator Ted Stevens (R), ascendancy on marine policy issues generally and issues of interest to Alaska (fisheries) in particular. Apart from this fact, the underlying dynamics of the issue remained the same.

Table 10.1a
Summary scores: Chapter 10 (INPFC)

| Regime | Component or Phase | Effectiveness | | Level of Collaboration |
		Improvement	Functional Optimum	
INPFC	1907–1952	Significant	Low	Intermediate
INPFC	1953–1976	Very low	Low	Intermediate
INPFC	1977–1984	Major	High	Intermediate
INPFC	1985–1992	Major	Overshoot	High

Conclusions

Technical Rationality: Did the Agreement Solve the Problem?

The answer to this question depends on which problem is being addressed (see table 10.1). The answer is yes if the problem concerns knowledge of the extent of salmon migration and intermingling in the North Pacific. In this connection collaboration under the INPFC produced consensual knowledge on the basis of a multiyear series of coordinated research cruises in which costs were shared by all parties.

Science benefited from a significantly improved description of the physical oceanography of the North Pacific and the environmental conditions that influence variations in distribution, abundance, and migrations of salmon. The research on salmon was driven by the need to resolve the problem of intermingling of stocks of North American and Asian origin, but the result was a qualitative, not quantitative, description of the distribution of the resources.

One major advance was made in techniques to determine continent of origin for various species of salmon. Tagging studies turned out to be less important than other methods because of the expense and logistic difficulties, though they helped to delineate areas of intermingling. On the other hand, morphological, parasitological, and meristic approaches to stock identification proved to be major successes of lasting value. In particular, it became apparent that scales and parasites could act as biological tags. Despite the successes, the terms of reference of the INPFC plus the level of internal disagreement over the abstention line and status of stocks resulted in continued research on intermingling long after it was particularly useful and foreclosed any attempt to move beyond what had been learned. Comprehensive stock forecasting, for instance, based on the entire life cycle of salmon was never pursued.

Table 10.1b

Problem Type		Problem-Solving Capacity			
Malignancy	Uncertainty	Institutional Capacity	Power (basic)	Informal Leadership	Total
Moderate	High	Low	Pushers	Intermediate	High
Strong	Low	Low	Pushers	Intermediate	High
Strong	Low	Low	Pushers	Intermediate	High
Strong	Low	Intermediate	Pushers	Intermediate	High

The INPFC framework may have lost its utility in the 1970s, but for the first decade or so after its creation it provided the opportunity for cooperative research that otherwise would not have been possible. The INPFC served as the umbrella under which competing interests communicated, coordinated large-scale research programs, and exchanged data and information.

If we shift the question to management, the answer would be no. The INPFC was indirectly connected to management concerns because it required research on stocks on the abstention list with a view to proving that these stocks qualified for abstention. No management decisions were made within the commission because unanimity among the three parties was unattainable. The multilateral contribution to management lies in the extensive advances in knowledge about North Pacific salmon and the active dissemination of that knowledge through formal and informal channels. In particular, there was a strong link between INPFC research results and later U.S. and Canadian actions on spawning escapements. Target escapements set from 1961 incorporated the knowledge that was derived through the INPFC mechanism. And we recall that while research showed by 1961 that the line of abstention should be moved west to 170°E, the decision rules provided Japan all the protection it needed until the regime was transformed by UNCLOS III and the MFCMA in 1977.

After 1977, within a changed regime, the necessary adjustments were made, the states of origin had the upper hand, and interceptions were brought down to a very low level by 1988. Access was closed, to new entrants as well, and authority was effectively centralized in the hands of states of origin. The answer to the management question at this point must be yes; the problem was solved. Was there any biological or economic justification for terminating all high-seas operations of the Japanese fleets? It is difficult to see any biological justification relative to salmon since the interception rate was so low. Even with respect to incidental catches of marine mammals, the decision of the administrative law judge did not go so far, and economically Alaskan access to the Japanese market for Western Alaska sockeye and chinook was more stable when the Japanese fleets were operating, in part because the government was involved.

On the other hand, termination of the fishery eliminated government involvement, and the industry sought substitutions for Alaskan sockeye on the Japanese market. Eventually a near perfect substitute was found in coho, which drew large amounts of Japanese investment capital to Chile and other places for salmon aquaculture. In addition, Norwegian successes in culturing Atlantic salmon, which is virtually identical to coho, facilitated their domination of the year-round market in North America and Western Europe. By 1996, during a period of high ocean productivity of all species of salmon, these Japanese- and Norwegian-produced, hatchery-reared coho flooded the Japanese and U.S. markets, driving prices down. There is therefore such a thing as being too effective.

Political Judgment: Was the Agreement the Maximum Feasible Within the Institutional Setting, and Were the Outputs the Maximum Feasible at the Time?
The answers to both questions are in the affirmative for both phases—pre- and post-MFCMA.

Substantive Measurement: Are Conditions Improved as Much as Were Expected, at Costs That Were Projected, Without Producing Involuntary Losers?
For the research problem in the first phase, the answer would be yes. For the management problem in the pre-MFCMA period, the states of origin would say no, but Japan would say yes. An impartial observer would have to agree with the states of origin. For the post-MFCMA period up to 1987, the answer would be yes; this was the optimal performance. For the period after 1988, which included complete termination of the fishery, the answer would be no *both* for the two Japanese fleets and for the U.S. (including Alaskan) industry from a marketing point of view. The U.S. industry therefore shot the Japanese fleets in the head and themselves in the foot.

Notes

1. Once more, it should be noted that the analysis presented herein is a reformatted and condensed version based on several prior publications of the author: Miles et al. (1982) and Miles (1987a, 1987b, 1989).

2. More detailed treatment of these issues can be found in Miles et al. (1982, 170–173) and in Miles (1987a, 13–21).

3. The research program and progress reports thereon can be found in the *Annual Reports* of the INPFC from 1954.

4. Between 1984 and 1985 the author interviewed as many of U.S. members of the CB&R as he could find.

5. Interview with Frank Fukuhara (1984).

6. Interview with Dayton L. Alverson (1984).

References

deReynier, Yvonne L. 1995. "International High-Seas Salmon Management in the North Pacific Ocean: From Freedom to Prohibition." M.M.A. thesis, School of Marine Affairs,

University of Washington. Later published as "Evolving Principles of International Law and the North Pacific Anadromous Fish Commission." *Ocean Development and International Law* 29 (2) (1998): 147–178.

Fredin, Reynold. n.d. "The History of INPFC." Manuscript.

Jackson, Roy I. 1963. "Salmon of the North Pacific Ocean: Introduction." *Bulletin of the International North Pacific Fisheries Commission* 12.

Miles, Edward L., et al. 1982. *The Management of Marine Regions: The North Pacific* (pp. 56–63, 90–99, 170–173). Berkeley: University of California Press.

Miles, Edward L. 1987a. "The Evolution of Fisheries Policy and Regional Commissions in the North Pacific Under the Impact of Extended Coastal State Jurisdiction." In FAO, *Essays in Memory of Jean Carroz: The Law and the Sea* (pp. 150–154). Rome: FAO.

Miles, Edward L. 1987b. *Science, Politics, and International Ocean Management*. Policy Papers in International Affairs No. 33. Berkeley: Institute of International Studies.

Miles, Edward L. 1989. *The U.S./Japan Fisheries Relationship in the Northeast Pacific: From Conflict to Cooperation*. Seattle, WA: Fisheries Management Foundation and Fisheries Research Institute, University of Washington), Doc. #FMF-FRI-D02.

IV

A Control: A High-Security Regime

Introduction: The Nuclear Nonproliferation Regime as a Control Case

Are international environmental and resource-management regimes different in kind from other types of regimes? If so, the findings of this study will have very limited applicability. If not, the findings should be generalizable across a wide range of regimes in scientific, economic, and security issue areas. While space limitations do not allow us to include a representative control sample of cases, we have chosen a high-security case—the nuclear nonproliferation treaty (NPT)—to conduct an initial test of the proposition that our findings are generalizable beyond our sample of cases, that our categories are in fact universal, and that the dynamics of regimes that we elaborate may have wide application. We realize that a single case cannot yield definitive answers to these questions, but a high-security case can be used as a critical test because it is a rebuttable presumption that the configuration of interests would be different from low-politics cases.

One other set of considerations guided our choice of the NPT regime: its relatively long life cycle as a policy issue. Fifty years provides an adequate basis for comparing this high-security case with international environmental and resource-management cases in terms of the reasons that regimes shift direction over time, generating fluctuations in scores of effectiveness. Comparative evaluation of the evolutionary aspects of regimes is provided in the concluding chapter.

11

Nuclear Nonproliferation, 1945 to 1995

Edward L. Miles

Introduction and Reasons for Success and Failure

This chapter evaluates the effectiveness of the regime for controlling proliferation of nuclear capabilities, particularly weapons, from 1945 to the present. Like the issue of sea dumping of low-level radioactive waste, the analysis is divided into three phases: the preregime phase from 1945 to 1953, the combined regime-creation and regime-implementation phase from 1953 to 1970, and the pure regime-implementation phase from 1970 to 1995. This case is complex in itself and in the fact that it has a long evolutionary history. The latter characteristic suggests that judgments of levels of regime effectiveness may shift with time, and indeed they do. We include the preregime phase in this analysis because several of the policy designs the United States preferred made their appearance in this period (1945 to 1953) and then were carried over into the combined regime-creation and regime-implementation phase (1953 to 1970). The effectiveness score here would be low. The score would still be low in the first part of the pure regime-implementation phase (1970 to 1974) and shift to moderate to high in the period 1976 to 1995, with the years 1974 to 1976 constituting a transition. However, we need to make clear that this higher score is achieved on a less ambitious regime objective—a shift from *stopping* proliferation to *slowing it down*.

What Are the Problems to Be Solved, and How Did They Change Over Time?

Over fifty years, nonproliferation policies have changed and multiplied, but the core of the problem they were designed to solve remains the limitation of the total number (and sometimes the type) of governments with the capability to launch or to threaten to launch nuclear weapons. The problem is actually multidimensional and includes the following components:

· Global diffusion of reactor technology;
· The large amount of plutonium ($^{239/240}$Pu) available in the world;
· The diffusion of the nuclear fuel cycle: uranium enrichment, fuel fabrication, and chemical reprocessing;
· The diversion of fissionable materials to weapons production; and
· The deliberate pursuit of nuclear-weapon capabilities combined with effective delivery systems.

Taken together these components make up a complex mix of benign and malign problems. As we show, this mix changes significantly over time as a consequence of endogenous as well as exogenous developments.

The evolution of *policy* exhibits the following characteristics. First, in the pre-regime phase (1948 to 1953), the United States tried to prevent the spread and use of nuclear technology to and by others, including allies, by denying them access to information and equipment (Goldschmidt 1977). When this policy of technological denial failed with the USSR, the United Kingdom, and France, the United States chose a schizophrenic policy known as the Atoms-for-Peace Plan, in which it sought rapid diffusion of nuclear technology combined with the implementation of safeguards to inhibit the diversion of fissionable materials to weapons production. This policy embodied both export controls as well as safeguards against diversion.

Export controls over time became a highly contentious issue between nuclear supplier states, while safeguards became a highly contentious issue between supplier states and those states that were intent on developing their own nuclear-weapon capability.

Burgeoning difficulties with the effects of the Atoms-for-Peace policy combined with a period of détente between the United States and USSR in the 1960s produced an increase in superpower collaboration to negotiate and bring into force the Nuclear Nonproliferation Treaty (NPT) in 1970. But while the NPT sought to close the loopholes in the safeguards system of the International Atomic Energy Agency (IAEA), it had three major sets of problems with it: it formalized the division between nuclear-weapon states (NWSs) and all others and was seen by non-nuclear-weapon states (NNWSs) to be discriminatory in its imposition of penalties; it provided no incentive for the so-called threshold powers to sign the treaty and foreswear nuclear weapons; and it did nothing to alleviate the conflict between nuclear exporters.

A series of third-order effects combined with totally new events in the mid-1970s served to usher in a crisis and to stimulate moves to increase the level of cooperation within the nonproliferation regime. Among the important catalysts were the Indian explosion of a nuclear device and the OPEC-induced oil crises in 1973, which led to wildly unrealistic projections about growth in nuclear power to satisfy electricity demand.

Under the prodding of the Ford administration, a Nuclear Suppliers Group was established in London in 1976 in the context of which the French agreed to accept the same restrictions on export policy as all the other suppliers did (Nye 1981). These new guidelines were finally in place by 1978. The Carter administration chose a suite of policies designed both to boost U.S. credibility and to have the effect of slowing down the rate of proliferation. These included, *inter alia*, a decision to defer reprocessing of spent fuel indefinitely, to slow down the U.S. fast-breeder program, and to try to convince others to do the same. In addition, Nye, then serving as Deputy Under-Secretary of State, was convinced that a device for focusing on the long-term in a global setting was necessary (Nye 1981). The device chosen—the International Fuel-Cycle Evaluation (INFCE)—was a systematic reevaluation, within the IAEA, of the whole fuel cycle over a two-year period. The net result of the exercise was significantly to slow down the pace at which the plutonium economy was being developed.

The events that were most powerful in curtailing the pace of development of the nuclear proliferation problem, however, were spurred by the structural decline of the nuclear-power industry, especially in the United States (Bupp 1981). Electricity-demand growth had shrunk from the 7 percent average annual increase that it had been for almost three decades to about 3 percent from 1974. The result was near fatal to the industry, which, all of a sudden, had a catastrophic excess-capacity problem. If that were not all, the industry was seriously affected by inflation in fuel costs, construction costs, and the real cost of money so that manufacturing costs rose rather than fell over time (Bupp 1981). In addition, there were serious management failures in the United States; the Three Mile Island incident occurred in 1979, and growing public disenchantment with and growing opposition to nuclear power developed in Western Europe, in the United States, and within the fisheries sector of Japan.

All this combined to precipitate a major long-term decline in the industry, and this removed fears about the urgent need to avoid the plutonium economy. This development has not stopped the continued growth of nuclear-weapon capabilities among the threshold states, Pakistan being the latest one to join the club, but it is slow growth. The decline of the industry, plus the lessons learned from INFCE, did serve to remove diffusion of reactor technology and the fuel cycle from the critical list of urgent policy problems.

By 1981, there were grounds for cautious optimism as far as control over proliferation of nuclear weapons was concerned. The problem had not been eliminated, but it had definitely been contained:

· A full-scope global regime was in place;
· Safeguards and inspections were routine, even though important loopholes remained;
· There was a definite norm in both customary and conventional international law against aiding and abetting the proliferation of nuclear-weapon capabilities;
· The problem related to rapid diffusion of the nuclear fuel cycle had been solved; and
· Only a few states had chosen to "go nuclear" since 1970.

However, after the relative quiescence and stability of the 1980s, the proliferation problem reemerged in the 1990s as a matter of urgency. The problem had once again changed form. In 1990, Leonard Spector (Spector 1990) published a pessimistic assessment that showed a pattern of slow, steady expansion of nuclear-weapon capabilities in developing countries supported by a pattern of rapid diffusion of ballistic missile capabilities. In addition, new patterns of nuclear trade had emerged to the point where the clandestine trade was rated as very significant for programs in Pakistan, India, Argentina, Brazil, and Iraq (Spector 1990). The sources of clandestine support were identified as industry in Germany, the Netherlands, Norway, the United States, and China.

At the same time, France announced its decision to sign the NPT (Riding 1991); South Africa foreswore nuclear weapons and stated its intention to sign the NPT and accept IAEA safeguards (Wren 1991); and Argentina and Brazil stated their willingness to accept international inspections under the regional Treaty of Tlatelolco (Christian 1990). Perhaps the most striking change of all in the 1990s was that China announced its intention to sign the NPT in 1991 and actually acceded to the treaty in March 1992 (Garret and Glaser 1995–1996).

Arguably the two most important changed conditions, however, were the disintegration of the USSR and the growing instability in the NPT regime itself resulting in complete deadlock at the 1990 review of the Treaty. The dissolution of the USSR raised several issues: questions of control of nuclear weapons and nuclear forces within the arsenal of the former Soviet Union; claims of a rapidly growing illicit trade in fissionable materials, particularly enriched uranium and plutonium;

and storage and control of weapons-grade plutonium from dismantled warheads in the former Soviet arsenal.

On the other hand, there were clear indications of a growing revolt within the NPT regime wherein nonnuclear signatories were denouncing all nuclear-weapon states signatory to the Treaty for failing to have lived up to their obligations under the treaty relative to a comprehensive test ban and effective reduction of nuclear arsenals. This revolt gained great significance since the review of the Treaty schedule for 1995 would determine whether the regime would be renewed indefinitely, for only a limited period, or denounced. These were all major changes in the nature of the problem, each of which presented major policy hurdles.

Control over nuclear proliferation is clearly a hydra-headed beast, not susceptible of any simple *Endlösung* (finite solution). Let us therefore subject each phase of the problem to more detailed analytic scrutiny.

The Preregime Phase, 1948 to 1953

Two questions really underlie the proliferation phenomenon as a policy problem: what are the relationships between nuclear power and proliferation of nuclear weapons, and what policies are likely to be effective in controlling proliferation and why? U.S. policy, as stated in the U.S. Department of State's 1946 *Report on the International Control of Atomic Energy* (known generally as the Acheson-Lilienthal Report), assumed that all nuclear technologies were inherently dual-use (for war as well as for peace) and that the diffusion of this technology *under national systems of control* would inevitably lead to the proliferation of nuclear weapons (U.S. Department of State 1946; Gilpin 1962). The United States also assumed that the two components of the nuclear fuel cycle containing the highest risk of diversion to weapons production were uranium enrichment and chemical reprocessing. All three of these assumptions were correct.

Where the United States went astray was to assume that its monopoly would permit it simultaneously to pursue a policy of *bilateral* denial combined with multilateral ownership and control of uranium resources, enrichment and reprocessing technology, and stockpiles of plutonium. In fact, even before U.S. policy had flowered into the Lilienthal-Baruch Plan in 1946, the United States, the United Kingdom, and Canada had agreed in 1945 "not to disclose any detailed information on the practical applications of atomic energy before effective enforceable safeguards against its misuse could be devised, either in the form of international inspection or otherwise. They also agreed to try to exploit or purchase

for their own use all uranium resources available in the western world"
(Goldschmidt 1977).

The assumption that U.S. nuclear hegemony could be maintained, at least for a
while, through a policy of technological denial proved wrong even for U.S. allies
(Britain and France) and even more for the USSR. The latter demonstrated its capability as early as 1949, Britain in 1952, and France in 1960. The first two events
completely shattered the basis for the initial policy initiative of the United States.
Going back to the drawing board, the Eisenhower administration then developed
a plan to create the International Atomic Energy Agency (IAEA) as a specialized
agency of the United Nations, to use the IAEA as a bank for fissionable materials,
and to facilitate the peaceful use of nuclear energy through a policy of safeguards
combined with inspections. Thus began the formal regime-creation phase in 1953.

The problem as it emerged in this phase has malign as well as benign components. It is malign insofar as we have a hegemon (the United States) and four
wartime allies (the United Kingdom, the USSR, Canada, and France) with which
the hegemon is engaged in essentially competitive relationships. Except for Canada,
the latter are determined to develop their own nuclear capabilities independently
of the United States and succeed in doing so. Facilitating development of a nuclear
industry is at first blush a benign problem characterized by synergy relationships.
But hidden beneath the surface is a high potential for competition and conflict
between the putative suppliers of nuclear technology.

The policy objective that was pursued at the time—to prevent other countries
from "going nuclear"—was inherently unachievable, especially when combined
with deliberate assistance in the development of a global nuclear industry combined
with loopholes in the safeguards and inspections system. The Atoms-for-Peace Plan
was schizophrenic because it was simultaneously a boost for and a constraint on
proliferation, but the former was stronger than the latter. Commercial development
of nuclear power in fact resulted in a *subsidy* of proliferation by the United States
through the Export-Import Bank (Long 1977). It was not recognized that nuclear
proliferation was a multidimensional suite of problems instead of a single one and
a matter of degree in which the rate of change is the crucial factor. The latter condition was not learned until the mid-1970s, and Nye (1981, 15–16) has stated this
crucial lesson very clearly:

[W]hether the policy prospects are hopeless or not depends upon the policy objective. If the
policy objective is defined as preventing another explosion of a nuclear device, then the
prospects are indeed gloomy. If the policy objective is to reduce the rate and degree of

proliferation in order to be able to cope with destabilizing effects, then the situation is by no means hopeless.

. . . From a broader perspective the policy objective is to maintain the presumption against proliferation. The great danger is the exponential curve of "speculative fever"—an accelerating change in rate. In such a situation, general restraints break down and decisions to forebear are reconsidered because "everyone is doing it."

The Combined Regime-Creation and Regime-Implementation Phase, 1953 to 1970

The two central components of the nonproliferation regime are the IAEA and its role with respect to safeguards and inspections and the NPT. The former was proposed by the United States in 1953 and formally established in 1957. The latter was negotiated in 1968 and came into force in 1970. These two events form the benchmarks in the evolution of the nonproliferation issue, which is itself metaphorically like a mushroom cloud. The preregime phase is ground zero, the combined regime-creation and regime-implementation phase is the slender neck, and the full-scale regime-implementation phase since 1970 is the head of the "mushroom cloud" in all its turbulence and complexity.

Phase II begins with the U.S. response to the defeat of its proposals derived from the Acheson-Lilienthal Report of 1946. This response was contained in President Eisenhower's Atoms-for-Peace initiative launched at the United Nations in 1953. Recognizing that proliferation was already occurring and that nuclear energy would be attractive to countries for industrial purposes, the new U.S. position rested on a basic bargain: access to U.S. technology, information, training, and even financial assistance in return for assurances of only "peaceful use" and acceptance of safeguards—such as on-site inspections by international teams under IAEA auspices and accounting of nuclear materials—to verify compliance. In the original U.S. design for the IAEA, the concept required joint contributions from stockpiles of uranium and plutonium by the principal member states, IAEA storage and protection of all fissionable materials, and development of procedures for making allocations of materials to countries as from a central bank (Goldschmidt 1977).

Eisenhower's proposal represented the first step in the formal establishment of a regime: an institutional design that combined creation of a new organization with a set of incipient norms and procedures. However, as a result of conflict between the United States, on the one hand, and the Soviet Union, France, and India, on the other, the bank concept never came into being, since it became clear that donor states wished to set conditions *bilaterally* for the transfer of material and assistance. The Soviets and India also objected strenuously to the concept of safeguards.

Let us pause for a moment to consider an obvious question. How was it possible, at the beginning of the cold war, for the United Nations to respond relatively quickly to such a major policy innovation and to implement it in no more than four years? The answer is that both donor and recipient countries, except for the Soviet Union, put a high priority on the development of nuclear capabilities for commercial reasons and, in some cases, also for military reasons. This groundswell of support produced a level 3 type organization—coordination of action on the basis of explicitly formulated common standards, with national action implemented on a unilateral basis but constrained by centralized appraisal of effectiveness (IAEA on-site inspections).

No matter how revolutionary the policy concept, real life will be the determining factor in intergovernmental negotiations on the issue. The United States for its part pursued an array of objectives besides control over proliferation, such as relieving tensions in the alliance and taking advantage of the opportunity both to create and commercially exploit a world nuclear market for the benefit of the U.S. nuclear industry (Clausen 1985). Likewise, for the IAEA the commercial-application dimensions of the bargain came to overshadow the safeguard dimensions. The latter division remained understaffed and underfunded for most of this seventeen-year period. Moreover, like all other international organizations, IAEA reflected and refracted the issues, tensions, and conflicts of the larger international system of which it was a part. This led to further diffusion of its focus from safeguards to a variety of exogenous conflicts.

Between 1958 and 1968, with only one major component of the global regime in place, the donor countries, led by the United States, implemented the rules severely hampered by the lack of full-scope safeguards and embarrassed by the determination of the Indian government to develop a nuclear facility and to use the CANDU reactor at Tarapur, which was also supported by the United States, to do so. Not surprisingly, therefore, over time support increased for elaborating a treaty that would require full-scope safeguards.

The specific trigger that released efforts leading to the NPT was an Irish proposal submitted to the General Assembly of the United Nations in 1958 calling for specific measures to prevent further spread of nuclear weapons beyond the United States, the USSR, and the United Kingdom (Lawrence and Larus 1974). However, the actual route to the NPT was a tortuous one, and it was particularly hampered by the external developments affecting the strategic balance between the United States and the USSR, particularly the emplacement of United States nuclear weapons in Europe under NATO auspices and the U.S. proposals for a multilateral nuclear naval force (the MLF) (Lawerence and Larus 1974).

But not all issues within the U.S.-USSR relationship were harmful. Prospects for the treaty improved when there was a thaw in relations during the early 1960s that facilitated the negotiation of the Limited Test Ban Treaty of 1963 (Goldschmidt 1977). This proved to be an extremely popular event for the rest of the world, except for France and China, which both rejected the Test Ban Treaty and the NPT as impermissible constraints on their own national sovereignty (Goldschmidt 1977). Eventually, however, after very difficult negotiations, the NPT was signed in 1968 and came into force in 1970 (De Gara 1970). By then we have a formal merging of the two central pillars of the regime: the NPT and the IAEA safeguards system.

What can we say about the policy implications of the nonproliferation regime as it was elaborated in 1970? First, the NPT formalized the division between NWSs and all others. NNWSs that were technically close to a nuclear capability and governments of NNWSs that were intent on becoming NWSs rejected this formalization on the basis that it was discriminatory in the imposition of penalties. Moreover, the NPT contained no incentive for these threshold states to sign since their full development could be achieved outside the scope of IAEA safeguards either through bilateral assistance or the illicit trade. A final shortcoming is that the NPT did not solve the problem of providing credible security guarantees against nuclear blackmail, which the NNWSs demanded, and it did not alleviate the rivalry and conflict between suppliers of nuclear technology. Indeed, by 1970 there were grounds for asking whether the NPT was too little too late. In an overall assessment, regime effectiveness in this stage must be rated as *low*.

On the other hand, the NPT was of great importance in three respects: it was the basic global law on the question and, as such, grounded the nonproliferation norm and facilitated its widespread acceptance (Simpson and Howlett 1994); it transferred into positive international law a general obligation to accept IAEA safeguards (Simpson and Howlett 1994); and it was the vehicle that produced a step-level change in the IAEA's capability to implement safeguards (Scheinman 1969; Quester 1970).

Even as the partial success of negotiating the NPT was in the offing, two developments occurred that portended increased competition for market share between Europe and the United States and increased conflict over the issue of control of sensitive components of the nuclear fuel cycle. The first was an agreement on December 2, 1968, between the United Kingdom, West Germany, and the Netherlands to pool their efforts to build an advanced centrifuge separation plant to produce cheap enriched uranium reactor fuel (Cook 1968). By 1968, France had already invested more than $1 billion in its gaseous diffusion separation plant to produce enriched uranium for military purposes (Cook 1968), and in 1971 the USSR changed its policy and offered enrichment services to Western Europe (Goldschmidt 1977). Thus, the United States had lost its monopoly on enriched uranium.

The United States countered with an offer to transfer enrichment technology to friendly nations on condition that the technology support only multinationally managed operations with U.S. participation and that these multinational facilities not compete commercially with U.S. production (Goldschmidt 1977). The Europeans declined the offer. Many contradictions in policies were evident. The Soviet Union was for strict control of proliferation and yet strenuously opposed the obligation of accepting IAEA on-site inspections as a means of protecting itself against U.S. intelligence-gathering probes. (France, China, and the threshold states echoed this view but were consistent in this rejection). The United States wanted to control proliferation but at the same time moved aggressively to bolster its own nuclear industry. In turn, the latter worked only to erode U.S. control. It was not easy to construct an effective regime on this basis.

In terms of our analytic categories, we would say that the problem remains mixed in this phase but that the malign components have begun to predominate as more countries seek and gain significant nuclear capabilities, whether or not they are directly weapons related. Apart from the East-West division, the following group-ings have emerged—the superpowers on some issues, three of the five NWS on others (excluding France and China), the threshold states, and the NNWS.

The Regime-Implementation Phase, 1970 to 1995

By the end of 1973, it was apparent that there were three categories of players in the nuclear game. First, there were the five nuclear-weapons states: the United States, the USSR, the United Kingdom, France, and China. Second, there were approxi-mately fifteen threshold states: India, Israel, Pakistan, South Africa, Argentina, North Korea, Brazil, the United Arab Emirates, Iran, Iraq, Venezuela, Taiwan, South Korea, Indonesia, and Yugoslavia. Third, there was the rest of the countries, some of which (like Sweden, Japan, Belgium, West Germany, and Spain) themselves

Table 11.1
Diffusion of nuclear capabilities by the mid-1970s

Natural uranium (^{238}u) supplies	United States, USSR, Australia, Canada, South Africa, Namibia
Producers of weapons-grade materials (^{235}u, ^{239}Pu, ^{240}Pu)	United States, USSR, United Kingdom, France, China, India, Israel, Pakistan
Uranium enrichment (^{238}u, ^{235}u)	United States, USSR, United Kingdom, France, and China (all by way of gas centrifuge or gaseous diffusion); potentially Israel and South Africa via laser
Chemical reprocessing of spent fuel and fuel	United States, USSR, United Kingdom, France, and China; West Germany, Belgium, Japan, India, Italy, Spain, Canada, Argentina, Sweden, Pakistan, Taiwan, and Yugoslavia
Reactor design and construction	United States, USSR, United Kingdom, France, Japan, West Germany, Sweden, Canada, India, Argentina

possessed considerable nuclear capability while foreswearing weapons. The threshold states, over the course of a decade or so, split off at both ends: India, Israel, and Pakistan became minor NWS; while Venezuela, Taiwan, and South Korea moved (sometimes under U.S. pressure) toward joining the NPT.

As the commercial dimension of nuclear-energy utilization advanced, the diffusion of capabilities grew apace. Table 11.1 shows the extent of diffusion by about the mid-1970s. Politically, therefore, one would have to say that there was a dense concentration of capabilities into a group of about twenty countries. This kind of skew in the distribution of capabilities almost always produces conflict over perceived inequities in the distribution of benefits, and the NNWSs perceived lots of inequities in the status quo, both with respect to the lack of progress among NWSs toward nuclear disarmament and to perceived inadequacies in the sharing of nuclear technology. These perceptions fed an increasing restiveness, which spilled over into public view at every quinquennial review of the NPT—1975, 1980, 1985, 1990, and 1995. The NNWS group was remarkably consistent in its support for the objectives of the NPT, its strong criticism of the NWSs for failing to live up to their treaty obligations, its demands for access to nuclear technology for peaceful purposes, and its complaints that the NPT puts all the burdens on those NNWSs that become a party to it and not on the threshold states.

At the same time, conflict within the suppliers group and beyond (the NWSs and the threshold states, plus the commercially advanced NNWSs) was fierce over market share and over access to particularly sensitive technology. To be sure, the technological and financial barriers to entry into the game were severe. Large infrastructure required large investment costs where the reactor was the largest cost component by several orders of magnitude, followed by enrichment and reprocessing plants. These costs implied that most of the NNWSs would be unable to share in the benefits of the peaceful uses of nuclear energy, no matter what the rhetoric of the regime. As a result there was initially high diffusion of reactor technology but less diffusion of the fuel cycle as a whole.

Not surprisingly, this diffusion of capabilities had considerable implications for the political, as opposed to the economic, stakes of the game. The highest-priority item remained the collective good of global security defined in terms of preventing the diversion of fissionable materials to weapons production. But the global diffusion of reactor technology and the slower diffusion of the fuel cycle ate away the capacity of the major NWSs to control the diversion problem. Moreover, it affected the superpower relationship by making more complex their calculations on deterrence. It also brought into prominence the issue of the relationship between regional conflict, proliferation, and deterrence since Israel, India, and Pakistan were all involved in intense regional conflicts. And finally it raised the problem of controlling the growing amount of $^{239/240}$Pu available in the world as a result of both civilian and military activities (Feiveson and Taylor 1977; see also Isnard 1993; Friedman 1993; Nolan 1991). The scope of the problem had thereby become appreciably greater and more malign.

Let us now shift to a developmental analysis, focusing only on critical turning points or events in the evolution of the nonproliferation issue between 1970 and 1975. Prior to 1970, the major stimuli underlying development of the issue were internal. The only serious external stimulus was superpower strategic concerns relative to deterrence. In the 1970s and 1980s, however, the major stimuli were both internal and external, and the latter loomed very large.

The first important shift of policy in the 1970s came in 1974 when the United States announced that it would not conclude any new contracts for enrichment of natural uranium, thereby leaving the field to its competitors—Western Europe and, as of 1971, the Soviet Union (Goldschmidt 1977). This loss of U.S. governmental monopoly control paralleled the rise of the Germans and the French in the export of U.S.-designed power-generating LWRs.

Competition for sales by licensees involved both American and non-American technology, and, more important, these European companies did not chafe under the U.S. restriction against the sale of enrichment and reprocessing technology to foreign states. As competition heated up, the French and the Germans therefore began offering these components of the nuclear fuel cycle as "sweeteners" for reactor sales contracts and, in so doing, dangerously intensified threats to the nonproliferation policy of the United States and others.

In the midst of this hot-house atmosphere, a combination of external and internal events triggered a major shift in proliferation policy. The external events were the 1973 oil embargo on OPEC and the Indian nuclear device in 1974. The internal events in this case were a West German contract to sell reactors to Brazil, which included enrichment and reprocessing technology as sweeteners, and French contracts to export reprocessing facilities to South Korea and Pakistan (Clausen 1985; Nye 1981; Gillette 1975a, 1975b, 1975c; Markham 1977a, 1977b; Hammond 1977).

The OPEC-engineered oil embargo in 1973, with its consequent upward pressure on oil prices, made the advanced industrial countries of North America, Western Europe, and Japan begin to look to nuclear power to make them independent of future oil embargoes. This increased priority accorded to nuclear energy had two consequences: it generated conflicting estimates about the probability of uranium shortages, and it fueled the race toward production of the liquid-metal fast-breeder reactor (LMFBR), a reactor that produces more plutonium than it consumes (Hammond 1971, 1972; Metz 1975; Kandell 1977; Hammond 1976; Miller 1977; Metz 1977a, 1977b; Walsh 1978; Krass 1977).

In such a context in which the nonproliferation issue was fed by both the OPEC oil embargo and the intensified competition and conflict among suppliers, the Indian explosion of a nuclear device in 1974 provided a severe reality check. All of a sudden, the future was on the doorstep, as it were, and this event triggered a major adjustment in French policy that facilitated a wider agreement on controls (Goldschmidt 1977; Lellouche 1981).

From the end of 1974, the United States began to shift its policy from confronting the French and the Germans to seeking establishment of a supplier cartel. The first meeting of the Nuclear Suppliers Group was convened in London in early 1975 and included as participants the USSR, the United States, the United Kingdom, Canada, West Germany, Japan, and France. (Later this group was expanded to include fifteen participants.)

At that meeting, participants were not actually united on the choice of policy options. As noted previously, the group was divided between those who preferred political and institutional options and those who preferred technological denial primarily.

An Evaluation of the Policy Options

Since choice of policy emerged as a major issue during the shift of 1974 to 1976, let us pause to assess the entire array of potential options available before we describe and evaluate what was done. We can identify the total range of potential policy options as including six types of alternatives:

1. Technological denial
   - Controls on exports
2. Dissuasion
   - A wide variety of political and institutional options as identified by Greenwood (1977)
3. Regulated transfer
   - Application of full-scope safeguards within an NPT/IAEA regime
4. Controlled uranium supply
   - Agreements among the United States, Canada, and Australia
5. Technological fixes
   - Substitution of CIVEX for PUREX in chemical reprocessing
   - Facilitation of disinvestment in the fast breeder by shifting to a thorium cycle and linkage of one fast breeder to three or four advanced converter reactors (ACRs) based on ^{233}U (the thorium cycle)
   - Running of LWRs in tandem with HWRs
   - Running of ACRs without linking them to a breeder
6. Unilateral U.S. action
   - Controls on exports
   - Controls on uranium enrichment
   - Controls on exports of other countries where U.S. technology is involved
   - Controls on assistance
   - U.S. disinvestment from fast breeder

The policy that emerged, beginning late in the Ford administration in 1977 and coming to fruition in the Carter administration, combined much of dissuasion as defined by Greenwood, technological denial, regulated transfer, and a moderate emphasis on technological fixes (that is, substitute CIVEX for PUREX). However, apart from a collective approach to securing implementation of full-scope safeguards, the United States chose a primarily unilateral, and somewhat punitive, approach. Moreover, no serious attempt was made to negotiate an agreement with Canada and Australia to control uranium supply.

The INFCE study, however, was highly innovative and very effective. It was designed by Nye himself, then serving as Deputy to the Under-Secretary for Security Assistance, Science, and Technology in the U.S. Department of State and sold to the United States and the rest of the nuclear suppliers. The intent of INFCE was to buy time and to inject the long view in what had become a "fevered situation" (Nye 1981, 1977, 1980)—"fevered" because projected energy shortages had been overestimated, which in turn produced serious overestimations of the demand for nuclear power, of predicted shortfalls in uranium supply, and of the attractiveness and economic viability of fast breeders. Thus, the rush to a plutonium economy was built on quicksand.

INFCE bought two years for a massive multilateral investigation to occur, the results of which produced an enormous amount of learning and, eventually, a turning away from reliance on the fast breeder. In Nye's (1981, 25) own evaluation:

INFCE helped to reestablish a basis for consensus on a refurbished regime for the international nuclear fuel cycle. The very process of engaging in international technology assessment helped to heighten awareness of the nonproliferation problem and the threats to the regime. In that sense, INFCE helped the United States to set the agenda for other governments. Moreover, it affected the process inside other governments. Foreign offices rather than just nuclear energy agencies became more involved. Most important, attention to the problem and to regime maintenance was spread beyond the United States.

However, the Carter administration policy in its other aspects—particularly the punitive cast of the Nuclear Nonproliferation Act (NNPA) of 1978, which elaborated the U.S. strategy of technological denial via export controls—caused much conflict (Nye 1980; Cook 1977; Hawkes 1977; Benjamin 1978; Rose and Lester 1978; Starr 1984). The objections focused in particular on the U.S. embargo on new contracts for uranium enrichment and U.S. opposition to the chemical reprocessing of spent fuel. The allies in particular objected to U.S. intentions to prevent export of fuel for reprocessing from third countries where the fuel was originally supplied by the United States, thereby directly threatening Japanese, U.K., and French plans and agreements. Even U.S. observers delivered fairly harsh judgments about the direction and effect of U.S. policy.

In our view, however, while the criticisms of the Carter administration policy are correct, most of the effects were relatively short-term in nature. This was so because another step-level change in the external environment was occurring that participants did not realize at the time. This change was the declining economic, social, and political viability of the nuclear-power industry that was set in train by responses to the OPEC oil embargo of 1973 and the Three Mile Island event of 1979. (The safety issue later skyrocketed with the explosion in Chernobyl in 1986.) The most effective response to the Organization of Petroleum Exporting Countries (OPEC) was conservation, which seriously dampened demand for electricity for the rest of the decade and made nuclear power uneconomic relative to fossil fuels, particularly coal.

From the perspective of hindsight, therefore, the most lasting and effective contribution of the Carter administration policy on nonproliferation was the process of INFCE and the learning and results it produced. These produced generally agreed on conclusions that there was no shortage of uranium in the world and that the case for the fast breeder was not compelling on either economic or national security grounds. Nothing in the intervening period has lessened the accuracy of these two conclusions, and the U.S. civilian nuclear-energy industry has virtually collapsed since then.

By the end of the 1980s, one could discern a lessening in the intensity of the threat of proliferation. There was consensus among the suppliers on the dangers of proliferation and, as a result, a more cautious approach relative to the introduction of weapons-usable fuels into the fuel cycle. There was also a realization that safe storage of spent fuel did not require reprocessing. The problem of the rapid diffusion of the entire nuclear fuel cycle had been solved.

One could definitely say that while the nuclear nonproliferation problem had not been eliminated, it had been contained. The Reagan administration had attempted a shift from the Carter emphasis on technological denial to a combined emphasis on dissuasion and regulated transfer. In addition, it sought to provide assistance to the ailing U.S. nuclear industry particularly with respect to reprocessing, but these efforts were ineffectual, and they generated growing opposition in Congress (Walsh 1981, 1982; Marshall 1981; Fox 1984). By 1988 it was plain to see that growth in the U.S. industry had virtually stopped (Spinrad 1988). Earlier economic realities had seriously curtailed growth of fast-breeder programs even in Europe (Dickson 1982; Chow 1977).

At the same time, this plateau in the evolution of the issue did not by any means represent the end of its natural life cycle. It was only a breathing space since the 1990s generated their own new and sometimes startling surprises. We analyze first the growing instability in the NPT regime by looking at shifts in the type and intensity of conflict at each quinquennial review from 1975 to 1995.

Growing Instability in the NPT Regime, 1975 to 1995

Since the NPT formalizes a world of nuclear haves and have-nots, it is no surprise that this has been a contentious issue since the first Review Conference in 1975. At that time a confrontation developed between the superpowers and the United Kingdom, on the one hand, and the developing-country NNWSs led by Mexico, Nigeria, Romania, Yugoslavia, and the Philippines (Epstein 1975b). The latter were struggling to get the NWS to implement their treaty commitments concerning general arms control, but the former refused to budge. Their aim was to have the Conference focus on control of the fuel cycle only.

On the issue of control of the fuel cycle, the NWWSs found the attitude of the NWSs to be discriminatory since they were unwilling to agree to discontinue sales to any state not willing to place all of its nuclear activities under IAEA safeguards. The Group of 77 at the Conference demanded (Epstein 1975b)

· An end to underground nuclear tests,
· A substantial reduction in nuclear arsenals,
· A pledge not to use or threaten to use nuclear weapons against nonnuclear parties to the Treaty,
· Concrete measures of substantial aid to the developing countries in the peaceful uses of nuclear energy,
· Creation of a special international regime for conducting peaceful nuclear explosions, and
· An undertaking to respect all nuclear-free zones.

This profile of conflict by and large characterized most Review Conferences over the twenty-five-year period, although the intensity fluctuated (Simpson and Howlett 1994; Epstein 1975a). In 1975, conflict was very intense because the NWSs refused to discuss many of the issues the G77 wished to discuss, and they refused to give the assurances the NNWSs demanded. As a result, the NNWSs refused to budge on the technical issues desired by the NWSs, so stalemate ensued.

By 1980, the intensity of conflict was markedly lower, reflecting the results of the INFCE study—a shift in U.S. policy away from technological denial and toward a heavier emphasis on dissuasion (Pond 1980). However, this was merely a relaxation of tensions; the basic divisive issues had not disappeared. In fact, the 1990 Review Conference again ended in deadlock with Mexico leading the charge against the NWS on the issue of a complete nuclear test ban (*New York Times*, September 16, 1990: "Deadlock Threatens Future of Nuclear Treaty").

Given this history, it was again no surprise that a major conflict was looming on the horizon of the 1995 Review Conference since the NPT stipulated that in 1995 the signatories should decide whether the Treaty should be renewed indefinitely, with specified time limits, or scrapped. The United States pushed hard for indefinite renewal, but until the end of the Conference, the outcome was very much in doubt

(Crossette 1995a; Jehl 1995; Hanley 1995; Crossette 1995c). The opposition to indefinite renewal was again led by Mexico, Venezuela, Egypt, Nigeria, and Indonesia, but it split when South Africa moved away from the G77 coalition seeking to find acceptable compromises (Crossette 1995b). This loosening of the G77 coalition allowed Sri Lanka to broker a compromise solution that facilitated acceptance of indefinite renewal (Crossette 1995d).

Developments in the 1990s

The 1990s were different from anything that went before, the major external cause of change being the political collapse of the Soviet Union and the emergence of Russia, Ukraine, and Kazakhstan as independent states with nuclear capabilities.

The end of the cold war left the United States as the only superpower and has fundamentally transformed the structure of the larger international system, permitting a far larger degree of flexibility than has existed since the 1920s. At the same time, the economic collapse of the USSR has been a major source of instability on the proliferation problem both with respect to Russian plans for export of a wide variety of nuclear technologies and to the need to safeguard increasingly large stocks of plutonium retrieved from nuclear warheads as a result of new arms-reduction agreements with the United States (Gordon and Wald 1994a, 1994b; Dufour 1994; Benjamin 1994; Whitney 1994a, 1994b; Broad 1994, 1991b). Yet another danger, as arms reductions force severe cutbacks in Russia's defense posture, is the possibility of the export of large numbers of Russian nuclear scientists and technicians to states intent on building their own nuclear weapons capabilities (Sciolino 1992c). A variety of responses have been made to lessen and control these dangers (Broad 1992a, 1992b; Sciolino 1993; Panofsky 1994; Taubes 1995a; Nunn 1995).

In a major first step the United States reached an agreement with Russia to buy the enriched uranium from dismantled nuclear arms. This uranium would be brought to the United States and diluted for sale in commercial reactors. Worth billions of U.S. dollars, the agreement shored up the Russian economy at the same time that it kept Russian nuclear technicians employed. The United States also agreed to provide $800 million to assist the Russians in dismantling their weapons, but this agreement proved to be more difficult to implement on the Russian side than had been foreseen. Finally, the United States provided funding, through the National Science Foundation, to engage Russian nuclear scientists in collaborative research.

Shifting from the external to the internal impetus for change, Spector's 1990 survey of the proliferation problem classified Israel, India, Pakistan, and South Africa as *de facto* NWSs and Libya, Brazil, Argentina, Iraq, Iran, North Korea, and Taiwan as emerging NWSs (Spector 1990). This analysis was completed prior to the Gulf War and the destruction of the Iraqi nuclear capability.

Spector found the following patterns to be especially troubling:

· A slow but steady expansion of nuclear-weapons capabilities in developing countries;

· Rapid diffusion of ballistic missile capabilities, also in developing countries, thereby increasing the relative vulnerabilities of parties to conflicts and therefore increasing the escalation potential;

· Expansions in the clandestine trade in nuclear materials, which had become very significant for programs in Pakistan, India, Argentina, Brazil, and Iraq (before the Gulf War);

· Clandestine support from industries in Germany, the Netherlands, Norway, the United States, and China (there were questions about Switzerland);

· The exporting of submarine construction technology by German shipyards to India, Brazil, Argentina, South Korea, Israel, and South Africa, and the pursuit of the development of indigenous nuclear propulsion technology by Pakistan, India, Argentina, and Brazil; and

· Attempts by India to seek to purchase a Russian SSN as a prototype, and exploratory efforts by both India and China to purchase former Soviet aircraft carriers from the Ukraine (all of these efforts were fruitless).

These developments suggested that in a post-cold-war world there would be a qualitative shift in the nature of the proliferation problem, quite apart from the effects of the collapse of the USSR. This shift raised the question of the need to rethink the approach to controlling proliferation based on the NPT-IAEA regime as presently constituted. We return to this question a little later.

The case of Iraq demonstrated all too clearly Spector's point about the new significance of the clandestine trade in nuclear materials. It also graphically pointed to holes in the NPT-IAEA regime since it showed that low-technology methods, such as electromagnetic enrichment of uranium by calutrons, can be used to circumvent export restrictions and that the clandestine trade can supply all needed high-technology equipment, including that necessary for the manufacture of hydrogen bombs (Broad 1991a; Norman 1991, 1992; Pilat 1992).

In response to the Iraq experience, the secretary-general of IAEA announced his intention to seek collaboration with the intelligence agencies of member states to report suspected cheating (Lewis 1991). This information would then be used to support aggressive spot checking of suspicions installations, whether or not the country in question had declared them as nuclear and had accepted IAEA safeguards. Spot checks would have to be authorized by the Board of Governors of the Agency, and if a country refused access, the case would be referred to the United Nations Security Council. The IAEA also wanted nuclear suppliers to create a register of sales of nuclear equipment and technology. The Board of Governors was not willing to approve specific spot checks even though the policy would apply only to signatories of the NPT.

On the other hand, it must be admitted that all is not gloom and doom and that truly significant progress was made in the 1990s. For instance, France announced in June 1991 that it would finally sign the NPT (Riding 1991); South Africa renounced nuclear weapons, decided to sign the NPT, and placed its Valindaba enrichment plant under IAEA safeguards (Wren 1991); Japan, under pressure from its neighbors, delayed for twenty years its plans to build a plutonium stockpile and to import large amounts of preprocessed plutonium annually from the United Kingdom and France (Suzuki 1991; Sanger 1991a, 1992a, 1992b, 1992c, 1994a; Swinbanks 1994a, 1994b); China acceded to the NPT; the world nuclear industry continued to decline ("Nuclear Reactors" 1989); and a decisive turn away from fast breeders was apparent everywhere but in France and Japan ("Are Fast Reactors Gone?" 1994).

The two biggest victories in support of the regime during the 1990s, however, came with respect to North Korea's plant at Youngbyon and China's proposed sale of a reactor to Iran. The road to an agreement with North Korea was long and tortuous, beginning in October 1990 and not finally resolved until June 1995 (Gordon 1990; Weisman 1991a, 1991b; Friedman 1991a, 1991b; Sterngold 1991; Gordon 1993; Greenhouse 1994; Pollack 1994; Sanger 1991a, 1991b, 1991c, 1992d, 1992e, 1992f, 1993a, 1993b, 1994b). With respect to Iran, it appears that after the Gulf War Iran stepped up its attempts to develop a nuclear capability and increased its links with China. The United States put heavy pressure on China to cancel its engagement. China, not surprisingly, resisted U.S. pressure between November 1991 and September 1995 (Sciolino 1992a, 1992b) but suddenly, on September 27, 1995, reversed itself and canceled the deal with Iran (Sciolino 1995).

While great progress has admittedly been made in slowing down the rate and scope of proliferation and in strengthening the regime, some analysts have begun to argue that the conditions and assumptions underlying control of proliferation in a cold-war world no longer apply in a post-cold-war world. Brad Roberts, for example, argues that an "antiproliferation" strategy should encompass chemical and biological, as well as nuclear weapons since the strategic value of the former to small states is very high (Roberts 1993). Moreover, the advanced state of proliferation of ballistic missile systems must be taken into account as well, along with advanced conventional systems, particularly naval. The result is that (Roberts 1993, 148; see also Isnard 1993; Friedman 1993; Nolan 1991)

the proliferation problem is one with many faces. Desegregation of the problem technologically . . . may be conceptually neat but oversimplifies the complex technical reality in which leaders of developing countries make decisions about military capabilities.

Within the nuclear proliferation problem itself, the United States is particularly concerned about rogue states and the danger of terrorism and, in this connection, the need to provide for joint U.S.-Russian monitoring and control of stockpiles of weapons plutonium that are now growing (Nuckolls 1995; Taubes 1995b; Macilwain 1994a, 1994b). So a new debate on the future of nonproliferation policy is being joined given the step-level change in the international system that has occurred.

Nye suggests that new policy must build on past accomplishments and that the categories of responses remain as security guarantees, technical restraints, unilateral measures, and multilateral institutions (Macilwain 1994a). But it seems to us that Brad Roberts's point is also well taken. As the cold war disappears, one does need to take into account chemical, biological, and advanced conventional systems as well as missile (delivery) systems. This kind of multidimensionality makes the problem even harder because the nonnuclear aspects of proliferation make more complex states' choices about policy responses.

Conclusions

Problem-Solving Effectiveness
It is by now easy to answer the counterfactual question of *behavioral change*: yes, conditions would indubitably have been worse without the regime (see Table 11.2). The world has seen only three new entrants to the nuclear club since China demonstrated its capability in November 1964—India, Israel, and Pakistan. Without the regime, all of the threshold states may have chosen to go the same route.

Did the regime *solve the problem(s)*? From 1953 to the early 1960s, as the regime was being created and implemented simultaneously, the answer is in the negative. In this period we see the Europeans, specifically France and the United Kingdom, pushing for the development of their own nuclear capabilities in resentment over the U.S. policy of technological denial. Moreover, one would have to say that the

Table 11.2a
Summary scores: Chapter 11 (NPT)

Regime	Component or Phase	Effectiveness		Level of Collaboration
		Improvement	Functional Optimum	
NPT	1: 1948–1953	NA	NA	NA
NPT	2–3: 1953–1970	Significant	Intermediate	Intermediate
NPT	2–3: 1953–1970	Significant	Intermediate	Intermediate
NPT	4–6: 1970–1995	Major	High	Intermediate
NPT	4–6: 1970–1995	Major	High	Intermediate

policy objective—prevention—was impossible for fulfill, especially given the developmental activities of the United States. At the same time the whole complex of policy intervention was not thought through systematically, and investigations of what policies were likely to work under different conditions were seriously deficient. Development and learning were based on trial and error, action and reaction, which moved the situation from an inadequate IAEA to the negotiation of the NPT. The NPT is indeed a correct step because it establishes the basic global law against proliferation and the general obligation to accept IAEA safeguards. But as we have seen, there are major objections to be made against the treaty since it formalized the division of nuclear haves and have-nots, indirectly rewarded the threshold states, and did not provide the NNWSs with credible security guarantees. One would therefore have to rate problem-solving effectiveness as *low* between 1953 and 1970.

The critical period is the one from 1974 to 1980 in which the NSG is established, new guidelines are agreed on, and INFCE is launched. The critical development of this period is the change in the *policy objective*—from prevention to slowing down the rate at which proliferation occurs. The issue then achieved a level of stability that lasted until the late 1980s. One would therefore have to rate the problem-solving effectiveness of the regime as moderate to high in this period. A range is included because the Carter administration policies, as symbolized in the NNPA of 1978, continued to generate a great deal of conflict, especially among the U.S. allies.

The period 1988 to 1995 is also a period of major new challenges and very large *dei ex machina*, but the regime has proved to be resilient. It facilitates cumulative learning; it is flexible and adaptable. The nature of the problem has intensified into an almost wholly malign one, but the regime still coped effectively. The safeguards emphasis in IAEA has been significantly increased, and organizational capabilities

Table 11.2b

Problem Type		Problem-Solving Capacity			
Malignancy	Uncertainty	Institutional Capacity	Power (basic)	Informal Leadership	Total
Strong	High	Intermediate	Balance	Weak	Intermediate
Mixed	High	High	Pushers	Intermediate	High
Mixed	High	High	Pushers	Intermediate	High
Strong	Low	High	Pushers	Strong	High
Strong	Low	High	Pushers	Strong	High

to implement full-scope safeguards have been massively expanded. Major problems remain in the Review Conference and as a result of the lack of progress toward implementation of a comprehensive test ban, but regime performance has remained at the moderate to high level. This is no mean achievement.

Finally, was the outcome the *political maximum* that could have been accomplished given the institutional setting, the configuration of interests, and the distribution of power? It is difficult to see, even in the preregime phase, how the U.S. policy of stringent technological denial was the maximum feasible outcome. Admittedly, the United States was clearly the hegemon, and attempting to delay the time when the Soviets would have the bomb, in the circumstances, was quite understandable. But the United States had engaged in intensive collaboration with the United Kingdom and Canada on the development of the atomic bomb. To think one could have cut that off at war's end with no further consequences was wishful thinking. The policy put severe strain on the Anglo-American alliance, and it reinforced the British and French desire to build bombs, independently of the United States, as quickly as they could manage.

The lesson to be learned as early as this period is that technological denial cannot prevent nuclear-weapons proliferation when the targets are states clearly determined to develop such a capability, but it can slow the rate at which proliferation occurs. On the contrary, technological denial stimulates those states to redouble their efforts to find routes independent of the normal suppliers, and it tends to exacerbate relations between those exercising the option and their allies, even if their allies do not wish to build a bomb.

The United States did not learn this lesson until the 1980s. However, as early as 1953 the United States did produce a qualitatively different and innovative policy in the design for the IAEA. This new policy could be described as a combination of technological denial and regulated transfer. This policy design was the primary output of the first half of the regime-creation and regime-implementation phase, and the primary outcome was the IAEA. Both were probably the maximum feasible in the context and for the time, but the design for the IAEA was flawed by the schizophrenia inherent in the U.S. Atoms-for-Peace Plan. The latter was stimulated by a combination of a desire to assist U.S. industry to exploit the peaceful uses of nuclear energy and an assumption that the Soviets would be able to make nuclear technology widely available on their own terms. The United States clearly misread the Soviets in this respect. The Soviets, in fact, practiced the most unrelenting form of technological denial.

Faced with major loopholes in the IAEA safeguards system, the NWSs agreed to negotiate the NPT as the second step in regime building during the second phase of the development of the nonproliferation regime. The question whether the NPT was the maximum feasible outcome at the time is a hard one to answer. As we have seen, the Treaty as it stands is demonstrably flawed and is the source of continuing conflict with the NNWSs. But the fact is that the policy design enshrined therein was probably the maximum the NWSs were prepared to agree to at the time.

Where both outputs and outcomes became the maximum feasible occurred in the middle of the third, pure implementation phase of the issue from the mid-1970s to the mid-1980s.

Explaining Effectiveness

We begin, as usual, with *the nature of the problem*. We have already noted a shift in the nature of the problem during the 1953 to 1970 period from mixed with an equal weighting of benign and malign characteristics to one in which the malign characteristics slowly begin to predominate over time as more countries acquire nuclear capabilities. The full flowering of the malign dimension of the proliferation problem does not occur until the third, pure implementation phase of the regime (after 1970). During the latter period the following offshoots of the issue grow rapidly:

· Horizontal proliferation of reactors and the fuel cycle;
· Intense competition for market share among supplier states;
· Significant expansion of the clandestine trade;
· Significant expansion of the total amount of plutonium available in the world;
· Significant connections to intense regional conflict;
· Resentment of the nuclear have-nots against the haves.

This configuration of interests generated three major coalitions—that is, the superpowers, the Western Europeans (both NWSs and NNWSs), and the NNWSs, especially the developing countries among them. The threshold states were not a formal coalition although their interests were identical with those of the regime as a whole. The NWSs were split between the superpowers, the Europeans (the United Kingdom and France), and China.

Another characteristic of the post-1970 phase of the nonproliferation issue, which was not included in the model, is the crucial importance of external events as *dei ex machina*. These external events act as wild cards in the game since they both aggravate and alleviate the problem, albeit in different ways. Three such events occurred in the time period under review:

· The 1973 OPEC oil shock and its role in giving a false sense of priority to nuclear power, presumed shortages of ^{238}U, the desire to introduce the fast breeder, and the threat of rapidly increasing the rate of change of proliferation;
· The Indian explosion of a nuclear device in 1974 and its effect as a reality check on suppliers; and
· The decline of the nuclear-power industry, beginning as early as the mid-1970s in the United States and Western Europe.

In conclusion, the increase that we have observed in regime effectiveness *cannot* be explained in terms of the development of the problem itself. In brief, regime effectiveness increased *despite* the problem gradually becoming more malign.

Our model suggests, then, that the key is to be found in the development of *problem-solving capacity* (table 11.2b). Looking first at the *institutional setting*, we can indeed observe a positive development. There were actually three arenas on the proliferation issue from the mid-1970s—the IAEA, the Nuclear Suppliers Group (NSG), and the Review Conference of the NPT. The IAEA was focused on safeguards while the Review Conference was limited to the Treaty itself. The NSG became by far the most significant policy-making group and source of critical policy innovation. In an overall assessment we conclude that institutional capacity increased significantly over time.

Also the *distribution of capabilities* changed, but here the picture is more complex. On the one hand, the hegemon lost its monopoly on reactors and reprocessing and, more slowly, on enrichment. Western Europe became a very significant source of technology transfers. The immediate effect of these developments was to weaken the basis for coercive leadership by the United States. However, as the pool of high capability grew somewhat, actor interests and negotiating positions changed, in at least two directions. In part, the situation became more difficult because

Western Europe was fully competitive with the United States and U.S. regulations and policies lost credibility with the Europeans because they were seen to give the edge to U.S. industry. This brought increased conflict. But as nuclear powers, they—and even to some extent the USSR and China—also acquired some of the interests of the previous hegemon. And the *dei ex machina* simultaneously introduced a sense of threat to *shared* interests and thus induced cooperation with newcomers (notably India).

Perhaps the most important qualitative growth occurred with respect to *epistemic communities* and uncertainty in the knowledge base. The former flowered through the coming to maturity of a variety of groups, such as the PUGWASH Conferences, the Institute of Strategic Studies in London, the Swedish Institute of Peace Research International in Stockholm, the Institute Français de Relations Internationales in Paris, the *Bulletin of the Atomic Scientists* in Washington, D.C., along with the Federation of American Scientists, the Council of Foreign Relations in New York and Washington, D.C., and the Harvard-MIT Study Group on Arms Control in Cambridge. Many of these groups had formal or informal official connections, and there was a burgeoning literature.

A division ran through the entire community, however, and was reflected in the NSG between those who favored technological denial and those who favored dissuasion combined with some form of regulated transfer. This uncertainty about choice of policy instrument was never subjected to systematic investigation, and U.S. policy came to be dominated by technological denial (to allies as well as to others). But Nye, coming out of the Harvard-MIT Study Group, used the substantial uncertainty surrounding projections about uranium shortages and the economics of the fast breeder to buy time through a systematic investigation. INFCE shifted the focus from the NSG once again to the IAEA, and the results of that study produced a prodigious amount of learning and served to reduce both uncertainty and the intensity of conflict. As a consequence, the United States and some other parties adjusted their strategies, and these adjustments seem to have facilitated regime development significantly.

While the participants and the organizational actors now have fairly sophisticated understanding of what policies are likely to be effective or ineffective in controlling the rate of proliferation, it remains to be seen what approaches will be developed for safeguarding the large amount of plutonium retrieved from the dismantling of nuclear weapons.

We think there is cause for optimism because the regime has demonstrated a capacity for cumulative learning and an institutional memory clearly exists.

Moreover, the NSG has been revived and can continue to be involved. The IAEA is also a far more capable and powerful lever than it has ever been in the past. While we agree that the scope of the proliferation problem has been immeasurably broadened to include missile-delivery systems, CBW systems, and advanced, conventional systems, we are skeptical about the advisability of adding these to the nuclear nonproliferation regime since all of the latter are more intractable problems and neither the IAEA nor the NPT is equipped to handle them. Nascent regimes are being built in each case, and we suspect that serial specialization is a potentially more effective course to follow.

What lessons have we learned about the causes, dynamics, and consequences of regime change? We think our experience with the nuclear nonproliferation issue teaches that the longer the history of a regime, the greater the probability of exogenous step-level changes in its evolution, and therefore the greater the need for a cumulative learning capability and for flexibility and adaptability in organizational design. Step-level changes can occur cumulatively over time as advancing technology is more widely diffused and changes the capability distributions and therefore the interests and utility functions of the participants. Step-level changes are also introduced by powerful *dei ex machina* that occasionally throw up major and unexpected challenges for the regime to respond to. The task of leadership is to mobilize the membership to align capabilities, interests, coalitions, and epistemic communities in such a way as to produce qualitative growth in institutional capacity to respond.

References

"Are Fast Reactors Gone for Good?" 1994. *Nature* (April 14), 571–572.

Benjamin, Daniel. 1994. "Plutonium Fund Sets Off Fears About Security." *Lexington Herald-Leader*, July 18, p. A.6.

Benjamin, Milton R. 1978. "Carter's Nuclear Policy Wins Few Converts Abroad." *Washington Post*, December 6, p. A10.

Broad, William J. 1991a. "Iraqi Atom Effort Exposes Weakness in World Controls." *New York Times*, July 15, p. A1.

Broad, William J. 1991b. "A Soviet Company Offers Nuclear Blasts for Sale to Anyone with Cash." *New York Times*, November 7, p. A5.

Broad, William J. 1992a. "Russia's A-Fuel More Plentiful, Study Says." *New York Times*, September 11, p. A5.

Broad, William J. 1992b. "U.S. to Buy Uranium Taken from Bombs Scrapped by Russia." *New York Times*, September 1, p. A1.

Broad, William J. 1994. "Experts in U.S. Call Plutonium Not Arms-Level." *New York Times*, August 17, p. A1.

Bupp, Irvin C. 1981. "The Actual Growth and Probable Future of the Worldwide Nuclear Industry." *International Organization* 35(1) (Winter): 59–76.

Chow, Brian G. 1977. "The Economic Issues of the Fast Breeder Reactor Program." *Science* 195 (February 11): 551–556.

Clausen, Peter A. 1985. "U.S. Nuclear Exports and the Nonproliferation Regime." In Jed C. Snyder and Samuel F. Wells, eds., *Limiting Nuclear Proliferation* (pp. 183–212). Cambridge, MA: Ballinger for the Wilson Center.

Christian, Shirley. 1990. "Argentina and Brazil Renounce Atomic Weapons." *New York Times*, November 29, p. 1.

Cook, Don. 1968. "3 European Nations Pool Secrets." *International Herald Tribune*, December 3, p. 3.

Cook, Don. 1977. "Nuclear Policy Seen President's Biggest Hurdle." *Los Angeles Times*, May 6, p. 1.

Crossette, Barbara. 1995a. "Atom Arms Pact Runs into a Snag." *New York Times*, January 26, p. A1.

Crossette, Barbara. 1995b. "China Breaks Ranks with Other Nuclear Nations on Treaty." *New York Times*, April 19, p. A6.

Crossette, Barbara. 1995c. "Discord Is Rising Over Pact on Spread of Nuclear Arms." *New York Times*, April 17, p. A1.

Crossette, Barbara. 1995d. "Treaty Aimed at Halting Spread of Nuclear Weapons Extended." *New York Times*, May 12, p. A1.

De Gara, John P. 1970. "Nuclear Proliferation and Security." *International Conciliation* No. 578 (May).

"Deadlock Threatens Future of Nuclear Treaty." *New York Times*, September 16, 1990, p. A5: 1.

Dickson, David. 1982. "Europe's Fast Breeders Move to a Slow Track." *Science* 218 (December 10): 1094–1097.

Dufour, Jean-Paul. 1994. "L'Empire Nucléaire Eclaté." *Le Monde*, August 25, p. 1.

Epstein, William. 1975a. "The Proliferation of Nuclear Weapons." *Scientific American* 232(4) (April): 18–33.

Epstein, William. 1975b. *Retrospective on the NPT Review Conference: Proposals for the Future*. Stanley Foundation Occasional Paper No. 9.

Feiveson, Harold, and Theodore B. Taylor. 1977. "Alternative Strategies for International Control of Nuclear Power." In Ted Greenwood, Harold A. Feiveson, and Theodore B. Taylor, eds., *Nuclear Proliferation* (pp. 125–190). New York: McGraw-Hill for the Council on Foreign Relations.

Fox, Jeffrey L. 1984. "Nonproliferation Proposals Challenged." *Science* 226 (March 2): 915.

"French N-Sale Disappoints U.S. State Department." 1977. *Japan Times*, September 11.

Friedman, Thomas L. 1991a. "China Opposes Pressing North Korea on Reactor." *New York Times*, November 15, p. A7.

Friedman, Thomas L. 1991b. "U.S. Concerned Over Korean Atom Plan." *New York Times*, November 14, p. A6.

Friedman, Thomas L. 1993. "Beyond Start II: A New Level of Instability." *New York Times*, January 10, p. E1.

Garret, Banning N., and Bonnie S. Glaser. 1995–1996. "Chinese Perspectives on Nuclear Arms Control." *International Security* 20(3) (Winter): 50.

Gillette, Robert. 1975a. "Nuclear Exports: A U.S. Firm's Troublesome Flirtation with Brazil." *Science* 189 (July 25): 267–269.

Gillette, Robert. 1975b. "Nuclear Proliferation: India, Germany May Accelerate the Process." *Science* 188 (May 30): 911–914.

Gillette, Robert. 1975c. "Uranium Enrichment: With Help South Africa Is Progressing." *Science* 188 (June 13): 1090–1092.

Gilpin, Robert. 1962. *American Scientists and Nuclear Weapons Policy.* Princeton: Princeton University Press.

Goldschmidt, Bertrand. 1977. "A Historical Survey of Nonproliferation Policies." *International Security* 2(1) (Summer): 69–87.

Gordon, Michael R. 1990. "U.S. Concern Rises Over North Korea Atom Plant." *New York Times*, October 25.

Gordon, Michael R. 1993. "North Korea Rebuffs Nuclear Inspectors, Reviving U.S. Nervousness." *New York Times*, February 1, p. A6.

Gordon, Michael R., with Matthew L. Wald. 1994a. "Russian Controls on Bomb Material Are Leaky." *New York Times*, August 18, p. 1.

Gordon, Michael R., with Matthew L. Wald. 1994b. "Russia Treasures Plutonium But U.S. Wants to Destroy It." *New York Times*, August 19, p. 1.

Greenhouse, Steven. 1994. "Caution Urged on U.S.-North Korea Pact." *New York Times*, August 14, p. A8.

Greenwood, Ted. 1977. "Discouraging Proliferation in the Next Decade and Beyond." In Ted Greenwood, Harold A. Feiveson, and Theodore B. Taylor, eds., *Nuclear Proliferation* (pp. 25–124). New York: McGraw-Hill for the Council on Foreign Relations.

Greenwood, Ted, George W. Rathjens, and Jack Ruina. 1976. "Nuclear Power and Weapons Proliferation." *Adelphi Papers* No. 130 (Winter).

Hammond, Allen L. 1971. "Breeder Reactors: Power for the Future." *Science* 174 (November 19): 807–810.

Hammond, Allen L. 1972. "The Fast Breeder Reactor: Signs of a Critical Reaction." *Science* 176 (April 28): 391–393.

Hammond, Allen L. 1976. "Uranium: Will There Be a Shortage or an Embarrassment of Enrichment?" *Science* 192 (May 28): 866–867.

Hammond, Allen L. 1977. "Brazil's Nuclear Program: Carter's Nonproliferation Policy Backfires." *Science* 195 (February 18): 657–659.

Hanley, Charles J. 1995. "Nonaligned States Fail to Agree on Treaty's Fate." *Seattle Post-Intelligencer*, April 16, p. A3.

Hawkes, Nigel. 1977. "Science in Europe/Carter Nuclear Policy Finds Few Friends." *Science* 196 (June 3): 1067–1068.

Isnard, Jacques. 1993. "La Dissuasion se Cherche de Nouvelles Règles du Jeu." *Le Monde*, November 6, p. 12.

Jehl, Douglas. 1995. "U.S. in New Pledge on Atom Test Ban." *New York Times*, January 31, p. A1.

Kandell, Jonathan. 1977. "European Nations Sign Accords on Developing Breeder Reactors." *New York Times*, July 6.

Krass, Allan S. 1977. "Laser Enrichment of Uranium: The Proliferation Connection." *Science* 196 (May 13): 721–731.

Lawerence, Robert M., and Joel Larus. 1974. "A Historical Review of Nuclear Weapons Proliferation and the Development of the NPT." In ·· Lawrence and ·· Larus, eds., *Nuclear Proliferation: Phase II* (pp. 1–29). Lawrence: University Press of Kansas for the National Security Education Program of New York University.

Lellouche, Pierre. 1981. "Breaking the Rules Without Quite Stopping the Bomb: European Views." *International Organization* 35(1) (Winter): 46–47.

Lewis, Paul. 1991. "Atomic Agency Maps Plans to Go After Nuclear Cheats." *New York Times*, October 11.

Long, Clarence D. 1977. "Nuclear Proliferation: Can Congress Act in Tune." *International Security* 1(4) (Spring): 52–76.

Macilwain, Colin. 1994a. "The Plutonium Regime." *Nature* 367 (January 27): 301–302.

Macilwain, Colin. 1994b. "U.S. Panel Wants Global Body to Control Civil Plutonium Use." *Nature* 367 (January 27): 307.

Markham, James M. 1977a. "French N-Sale Disappoint U.S. State Department." *Japan Times*, September 11.

Markham, James M. 1977b. "U.S.-Pakistani Rift on Atom Fuel Grows." *New York Times*, May 8.

Marshall, Eliot. 1981. "Reagan's Plan for Nuclear Power." *Science* 214 (October 23): 419.

Metz, William D. 1975. "European Breeders (1): France Leads the Way." *Science* 190 (December 26): 1279–1281.

Metz, William D. 1977a. "Reprocessing Alternatives: The Options Multiply." *Science* 196 (April 15): 284–287.

Metz, William D. 1977b. "Reprocessing: How Necessary Is It for the Near Term?" *Science* 196 (April 1): 43–45.

Miller, Sarah. 1977. "Uranium Supply Breeds Doubts." *New York Times*, June 19, sec. 3, p. 1.

Milton R. Benjamin. 1978. "Carter's Nuclear Policy Wins Few Converts Aboard." *Washington Post*, December 6, p. A10.

Nolan, Janne E. 1991. "The Politics of Proliferation." *Issues in Science and Technology* 8(1) (Fall): 63–69.

Norman, Colin. 1991. "Iraq's Bomb Program: A Smoking Gun Emerges." *Science* 254 (November 1): 644–645.

Norman, Colin. 1992. "Inspectors Uncover New Data on Iraqis' Nuclear Program." *New York Times*, January 20.

Nuckolls, John H. 1995. "Post Cold War Nuclear Dangers: Proliferation and Terrorism." *Science* 267 (February 24): 1112–1114.

"Nuclear Reactors Seem to be Entering Worldwide Decline." 1989. *Japan Times*, March 4.

Nunn, Sam. 1995. "U.S. Investment in a Peaceful Russia." *Issues in Science and Technology* (Summer): 27–30.

Nye, Jr., Joseph S. 1977. "Nuclear Power Without Nuclear Proliferation: Avoiding False Alternatives." In The Stanley Foundation, *Eighteenth Strategy for Peace Conference Report*, October 13–16, pp. 61–69.

Nye, Joseph S. 1980. *The International Nonproliferation Regime.* Stanley Foundation Occasional Paper No. 23, July.

Nye, Joseph S. 1981. "Maintaining a Non-Proliferation Regime." *International Organization* 35(1) (Winter): 15–38.

Panofsky, Wolfgang K. H. 1994. "Safeguarding the Ingredients for Making Nuclear Weapons." *Issues in Science and Technology* (Spring): 67–73.

Pilat, Joseph F. 1992. "Iraq and the Future of Nuclear Nonproliferation: The Roles of Inspections and Treaties." *Science* 255 (March 6): 1224–1229.

Pollack, Andrew. 1994. "Seoul Offers Help on Nuclear Power to North Korea." *New York Times*, August 15, p. A1.

Pond, Elizabeth. 1980. "Nuclear 'Haves' and 'Have-Nots' Seek Joint Solutions in Geneva." *Christian Science Monitor*, September 3, p. 5.

Quester, George H. 1970. "The Nuclear Nonproliferation Treaty and the International Atomic Energy Agency." *International Organization* 24(2) (Spring): 163–182.

Riding, Alan. 1991. "Finally, France Will Sign Nuclear Pact. "*New York Times*, June 4, p. A4.

Roberts, Brad. 1993. "From Nonproliferation to Antiproliferation." *International Security* 18(1) (Summer): 139–173.

Rose, David J., and Richard K. Lester. 1978. "Nuclear Power, Nuclear Weapons, and International Stability." *Scientific American* 238(4) (April): 45–57.

Sanger, David E. 1991a. "Data Raise Fears of Nuclear Moves by North Koreans." *New York Times*, November 10, p. A1.

Sanger, David E. 1991b. "Japan's Plan to Import Plutonium Arouses Fear That Fuel Could Be Hijacked." *New York Times*, November 21.

Sanger, David E. 1991c. "North Korea to Allow International Inspection of Its Nuclear Sites." *New York Times*, June 9, p. A10.

Sanger, David E. 1992a. "Japan's Atom Fuel Shipment Is Worrying Asians." *New York Times*, November 9.

Sanger, David E. 1992b. "Japan's Nuclear Fiasco." *New York Times*, December 20, sec. 3, p. 1.

Sanger, David E. 1992c. "Japan Thinks Again About Its Plan to Build a Plutonium Stockpile." *New York Times*, August 3, 1992, p. A3.

Sanger, David E. 1992d. "North Korea A-Bomb: Fear Is Fading." *New York Times*, September 18, p. A4.

Sanger, David E. 1992e. "North Koreans Reveal Nuclear Sites to Agency." *New York Times*, May 7, p. A4.

Sanger, David E. 1992f. "North Korea to Part the Veil a Bit for Nuclear Inspectors." *New York Times*, May 4.

Sanger, David E. 1993a. "North Korea, Fighting Inspection, Renounces Nuclear Arms Treaty." *New York Times*, March 12, p. A1.

Sanger, David E. 1993b. "Reversing Its Earlier Stance, North Korea Bars Nuclear Inspectors." *New York Times*, February 9.

Sanger, David E. 1994a. "Japan, Bowing to Pressure, Defers Plutonium Projects." *New York Times*, February 22, p. A2.

Sanger, David E. 1994b. "North Korea Will Remove Nuclear Fuel." *New York Times*, April 22, p. A3.

Scheinman, Lawerence. 1969. "Nuclear Safeguards: The Peaceful Atom, and the IAEA." *International Conciliation* No. 572 (March).

Sciolino, Elaine. 1992a. "China to Build Nuclear Plant for Iran." September 11, p. A3.

Sciolino, Elaine. 1992b. "CIA Asserts Iran Is Making Progress on Nuclear Arms." *New York Times*, November 30, p. 1.

Sciolino, Elaine. 1992c. "U.S. Report Warns of Risk in Spread of Nuclear Skills." *New York Times*, January 1, p. A1.

Sciolino, Elaine. 1993. "U.S. Plan to Help Russia Destroy Nuclear Arsenal Is Bogged Down." *New York Times*, March 10, p. A4.

Sciolino, Elaine. 1995. "China Cancels a Sale to Iran, Pleasing U.S." *New York Times*, September 28, p. A1.

Shapley, Deborah. 1971. "Plutonium: Reactor Proliferation Threatens a Nuclear Black Market." *Science* 172 (April 9): 143–146.

Simpson, John, and Darryl Howlett. 1994. "The NPT Renewal Conference: Stumbling Toward 1995." *International Security* 19(1) (Summer): 41–71.

Spector, Leonard. 1990. *Nuclear Ambitions: The Spread of Nuclear Weapons, 1989–1990*. Boulder: Westview Press.

Spinrad, Bernard I. 1988. "U.S. Nuclear Power in the Next Twenty Years." *Science* 239 (February 12): 707.

Starr, Chauncey. 1984. "Uranium Power and Horizontal Proliferation of Nuclear Weapons." *Science* 224 (June 1): 952–957.

Sterngold, James. 1991. "North Korea to Allow Nuclear Inspections If U.S. Does." *New York Times*, November 27, p. A3.

Suzuki, Tatsujiro. 1991. "Japan's Nuclear Dilemma." *Technology Review* 94(7) (October): 42–49.

Swinbanks, David. 1994a. "Fast Reactor Delays Lead to Plutonium Cutbacks in Japan." *Nature* 369 (June 23): 596.

Swinbanks, David. 1994b. "Japan's Fast Reactor Heads for Slow Track." *Nature* 368 (April 14): 575.

Taubes, Gary. 1995a. "Cold War Rivals Find Common Ground." *Science* 268 (April 28): 488–491.

Taubes, Gary. 1995b. "The Defense Initiative of the 1990s." *Science* 267 (February 24): 1096–1100.

United States Department of State. 1946. *A Report on the International Control of Atomic Energy.* Washington, DC: U.S. Government Printing Office.

Van Cleve, William. 1974. "Nuclear Technology and Weapons." In Robert M. Lawrence and Joel Larus, eds., *Nuclear Proliferation: Phase II* (pp. 30–68). Lawrence: University Press of Kansas for the National Security Education Program of NYU.

Walsh, John. 1978. "Fuel Reprocessing Still the Focus of U.S. Nonproliferation Policy." *Science* 201 (August 25).

Walsh, John. 1981. "Reagan Outlines Nonproliferation Policy." *Science* 213 (July 31): 522–523.

Walsh, John. 1982. "Reagan Changes Course on Nonproliferation." *Science* 216 (June 25): 1388–1389.

Weisman, Steven R. 1991a. "North Korea Adds Barriers to A-Plant Inspection." *New York Times*, October 24, p. A7.

Weisman, Steven R. 1991b. "North Korea Digs in Its Heels on Nuclear Inspections." *New York Times*, October 27, p. A10.

Whitney, Craig R. 1994a. "Germans Seize Third Atom Sample, Smuggled by Plane from Russia." *New York Times*, August 14, p. A1.

Whitney, Craig R. 1994b. "Germans Suspect Russian Military in Plutonium Sale." *New York Times*, August 16, p. A1.

Wren, Christopher S. 1991. "Pretoria Accepts Atom Arms Ban and Agrees to Plant Inspections." *New York Times*, June 28, p. A1.

V

Regimes of Low Effectiveness

Introduction: Common Features of Ineffective Regimes

The configuration of circumstances that we expect will lead to low effectiveness is the inverse of that hypothesized for effective regimes. The combination of a poorly understood malignant problem and a system with low problem-solving capacity will most likely lead to low scores on both dimensions of effectiveness (both improvement and functional optimum). Neither of the three variables (type of problem, problem-solving capacity, or political context) is a necessary or sufficient condition for low effectiveness, but problems characterized by strong malignancy can be solved effectively only when at least one of the other two factors is favorable—that is, if attacked by a system with high problem-solving capacity or muted by selective incentives or links to other, benign issues (see table V.1).

Table V.1
Hypothesized configuration of scores for ineffective regimes

Independent Variable	Hypothesized Score
Type of problem	Predominantly malignant State of knowledge: poor
Problem-solving capacity	Low, as indicated by • Decision rules requiring unanimity or consensus • Weak IGO serving the regime • No epistemic community present • Distribution of power in favor of laggards or laggards and bystanders • Scant instrumental leadership provided by delegates or coalitions of delegates
Political context	Unfavorable, as indicated by • Linkages to other malign problems • No ulterior motives or selective incentives for cooperation

12

The Effectiveness of the Mediterranean Action Plan[1]

Jon Birger Skjærseth

Overall Performance: A Collaborative Success Without Much Substantial Behavioral Impact[1]

Between 1964 and 1974, the population of the coastal states surrounding the Mediterranean Sea increased by 50 million and thereby increased the pressure on the marine environment.[2] Despite strong indications of a "sick" sea, no reliable information concerning the extent of pollution was available. Thus, the Mediterranean states requested the United Nations Environmental Program (UNEP) to develop a program aimed at identifying the extent of pollution, its sources, and relevant measures to remedy these. In 1975, sixteen Mediterranean states plus the European Community[3] approved the Mediterranean Action Plan (MAP, or Med Plan). A year later, the Barcelona Framework Convention was signed. The main purpose of the Barcelona Convention is to combat marine pollution and to protect the marine environment.

In the first phase up to the mid-1980s, the parties to the Barcelona Convention developed and adopted various protocols including a protocol on land-based sources of marine pollution. Approximately 85 percent of all pollutants in the Mediterranean originate on land. An assessment component called the *Med Pol* was developed to conduct monitoring and research. In addition, a system for integrated planning (or economic development compatible with environmental sustainability), known as the Blue Plan and the Priority Actions Program, was developed. Against this backdrop, the Med Plan has apparently been a political collaborative success in that it resulted in a comprehensive plan for cleaning up the Mediterranean Sea.

One of the principal aims in the second phase was to translate the general legal texts into specific joint commitments. The protocol on land-based sources was to be covered by twenty-eight specific joint commitments by 1995. However, only a few vague commitments were actually adopted. Due to poor reporting on implementation and vague obligations, it is very hard to trace any impact of the joint commitments in terms of domestic implementation and actual behavioral change provided at the national level. Much of the formal implementation that has taken place, such as the adoption of legal and administrative measures, appears to be due to factors other than the Mediterranean Action Plan itself, such as other international bodies (including the European Community). Thus, measured against the backdrop of results in terms of behavioral change provided, the MAP does not appear to be very successful. Still, the Med Plan has probably been important because it has increased the general awareness and preparedness in the Mediterranean area by increasing knowledge and cooperation.

Evolution of the Problem: From Environment to Development

In 1972, the Stockholm Conference identified the Mediterranean as belonging to the category "particularly threatened bodies of water."[4] Despite strong visible indications of a polluted sea, such as tar deposits on beaches from tanker spills, no reliable information was available concerning the extent of pollution. Since the time Aristotle (384–322 B.C.) puzzled over the strange currents in the strait between the island of Euboea and the mainland north of Athens, our scientific understanding of the Mediterranean has increased significantly.[5] In 1977, UNEP issued a report on land-based pollutants, the first study to provide specific data. Even though this was a study based on secondary sources, experts have stated that it reduced the degree of uncertainty concerning the Mediterranean pollution situation from three orders of magnitude to one.[6] This Med X study, as well as subsequent studies, seemed to reject the Mediterranean collapse hypothesis. The main pollution problems were seen as local and consequently an important threat to the beaches and to tourism.

Despite currents being too weak to fully exchange the wastes, some exchange at least of persistent substances probably does occur, given the anticlockwise direction of the currents and the fact that the exchange rate between the Mediterranean and the Atlantic Ocean is approximately eighty years. This concern was expressed by some delegates from less developed countries (LDCs) that feared that developed countries (DCs) tried to exploit developing countries by making them pay for pollution caused by the former. While the exchange rate of pollution between the DCs and LDCs was highly uncertain, it was probably perceived as more certain that abatement measures would affect the international competitive situation. After land-based pollution was placed on the agenda, high costs were calculated for cleaning up the Mediterranean Sea. Identifying rivers as a major pollution source implied the need for action covering the entire territory of participating states. LDCs were concerned that pollution-control measures would divert resources from economic development.

Later studies rejected the export-import hypothesis and confirmed the suffering-from-own-pollution hypothesis. The industrialized countries in the northwestern basin, which includes Italy, France, and Spain, were singled out as the major polluters, and their coastal waters as most polluted. Thus, the general-knowledge situation related to the export-import dimension, and affectedness seems to have improved somewhat throughout the 1980s. While it became clear that cleaning up the Mediterranean would be expensive, the financial calculations also indicated the highest costs for industrialized countries in the northwestern basin. Construction of sewage plants for all Mediterranean coastal cities with a population exceeding 10,000 would, for instance, place approximately half of the total costs on Italy and Spain alone.[7] In addition, together with France and Greece, these countries would suffer the most from polluted beaches and a decline in tourism. Furthermore, common measures on annex I substances of the land-based protocol would probably also imply higher costs for industrialized countries due to higher production of persistent substances like mercury and cadmium.

Although the initial focus of MAP was on the control of marine pollution, this gradually shifted to socioeconomic trends and development planning. From 1976 to 1990 population growth has been substantial on the Mediterranean coast especially in the eastern and southern parts. According to one projection, population in the Mediterranean area can be expected to increase from 323 million in 1980 to 433 million in the year 2000 and 547 million by 2045[8] (UNEP 1989b). This general population expansion has led to increased urbanization, which in turn has meant greater pressure from household and industrial effluents. The Mediterranean is also the world's major tourist destination: 70 percent of all tourists choose Mediterranean beaches, approximately 80 million a year ("The Med" 1990). Moreover, tourism is still increasing, and figures are expected to soar to somewhere between 268 million and 409 million by the year 2000 (UNEP 1989b, 40). This will involve not only an increase in solid and liquid waste but also growing pressure on habitats. In addition to the rise of tourism there has also been a shift from traditional agriculture to modern farming with greater productivity and increased use of fertilizers, insecticides, and pesticides, all of which drain into the Mediterranean Sea.

A general picture of the situation can be drawn up. We have no evidence pointing toward a systematic asymmetrical relationship between the DCs and LDCs. Our scattered data indicate a weak correlation between affectedness and abatement costs, which implies high abatement costs for those particularly affected by their own pollution. In addition, due to the importance of the tourist sector, these countries will probably benefit the most from cleaning up their coasts. This picture does not indicate any clear incentive asymmetry between the DCs and the LDCs in relation to preferred strength of measures, generated from the problem itself. However, not all states had environment-related motives for participation in the MAP. First, the MAP would be the only formally organized political body covering the entire Mediterranean area. Compared to the North Sea area, for example, the Mediterranean area is politically explosive.[8] The biannual meetings of the parties and ad hoc conferences would be the only regional forums where representatives from countries in dispute could meet face to face. Second, the prospects of financial, technical, and educational aid from UNEP to strengthen the marine scientific capacity was probably an important motivation, especially for the LDCs. Thus the states probably had mixed motives for participating in the MAP.

Actor Preferences: From Incompatible to Compatible?

In the mid-1970s, LDCs *were* suspicious of DCs' motives and the Med Plan in general. France and the DCs wanted to control land-based sources of pollution and to develop integrated planning. In contrast, LDCs were most concerned with basic needs such as housing conditions and starvation. As the Algerian president Houari Boumedienne stated early in the 1970s: "If improving the environment means less bread for the Algerians, then I am against it" (Haas 1990). At the UNCHE Stockholm conference in 1972, the Egyptian dependence theoretician Samir Amin, urged "the consideration of the true problems facing developing countries, not the 'imported' ones from developed countries" (Haas 1990). These statements indicate that at least some of the LDCs perceived the "paying for north externalities/ hampering own development" hypothesis as correct.

LDC resistance to the control of industrial pollution seems to have been weakened along with their participation in producing scientific knowledge. Since all parties had signed the Barcelona Framework Convention, it was no longer possible to ignore the need to control land-based sources. This is emphasized by the common measures adopted to implement the LBS protocol. However, the main discrepancy

between DCs and LDCs came to the surface in the late 1970s when the parties negotiated the land-based protocol. The Europeans preferred emission controls, while
the south preferred environmental standards. This configuration of preferences indicates that it was a combination of economic factors and "self-affectedness" that
decided the actors' preferences and not asymmetrical import-export of pollution.
The LDCs argued that their coasts were cleaner and therefore had a higher assimilative capacity. Following the protocol provisions, the emission-control strategy
could lead to bans on emissions of several substances and possible negative economic effects for the LDCs. The DCs, which were most heavily affected by pollution, wanted more stringent measures. In addition, France and Italy were already
committed to following emission standards under the European Community. This
conflict was resolved by a compromise in 1980. Both emission standards and quality
standards were applied for the blacklist (annex I) and the greylist (annex II). Other
disputes arose from single states' interests in deposition of "their" wastes where
costs of stringent measures probably were perceived as higher than the benefits. The
common measures adopted to implement the land-based protocol are results of
political compromise. For example, France's interests can be read out of the common
measures adopted for control of radioactive pollution. The decision does not include
any targets or measures other than an invitation to follow recommendations by
competent international organizations (UNEP 1991, 21–22).

In summary: In the early 1970s, the actors faced a problem characterized by a
high overall uncertainty, high complexity, different incentives due to perceived asymmetry in relation to abatement costs, and affectedness for at least some of the parties.
So why did the parties sign the 1976 Barcelona Convention at all? The main reasons
for some of the actors were probably only slightly related to cleaning up the Mediterranean Sea. Since UNEP bore most of the financial burden until 1979, it was a
golden opportunity for the LDCs to receive training and equipment for monitoring
pollution. It was also an opportunity to establish diplomatic ties between traditionally hostile countries. Thus, the states probably had mixed motives, and
probably more important, signing conventions and protocols are inexpensive
accomplishments while implementing them domestically may be expensive. The
main change in the problem was probably the reduced economic and ecological
uncertainty, which led to a different and less asymmetrical perception of the
problem. As a result, some of the LDC suspicion about the DCs' exploiting motives
seems to have disappeared or at least been weakened.

Regime Implementation: From Convention to Protocols to . . . ?

The main purpose of the 1976 Barcelona Framework Convention may be summed up as an effort to prevent, abate, and combat pollution and to protect and enhance the marine environment (article 4). The field of application for the Barcelona Convention and its related protocols is the Mediterranean Sea beyond the internal waters of the littoral states, excluding the Black Sea. Until 1980, MAP was supported almost entirely by UNEP, which was accorded the main secretariat functions according to article 13 of the Convention. In recent years, UNEP has sought to transfer more responsibility to the involved parties (Skjærseth 1993). The Mediterranean Action Plan's management component includes ambitions of *integrated planning* (or economic development compatible with environmental sustainability), known as the Blue Plan (BP) and the Priority Actions Program (PAP). Med Pol, an assessment component, was designed to conduct *monitoring and research*. Med Pol was divided into Phase I (1976 to 1981), which was to ascertain the degree of pollution in the Mediterranean, and Phase II (1981 to 1995), which aimed at evaluating the effects of Med Plan measures and setting standards for the control of land-based sources. Another important goal of the Med Pol has been to promote scientific capacity, most notably among the LDCs. According to UNEP, Med Pol and the other MAP components have indeed succeeded in creating a collective awareness (UNEP 1995).

To fulfill the goal of the Barcelona Convention, a complex legal and institutional structure was set up, and two protocols were adopted in 1976. The first prohibits *dumping* of dangerous substances (the blacklist); less noxious substances are placed under precise controls (the greylist).[9] The second protocol deals with accidents or other *emergencies* resulting from discharges of oil or other harmful substances into the sea. In 1980, twelve states and the European Community signed the Protocol on Land-Based Sources (LBS) of Pollution, which also includes a blacklist and a greylist of substances. Two years later, the Mediterranean states approved a protocol aimed at creating a network of specially *protected areas* to safeguard natural resources. A protocol for the protection of the Mediterranean Sea against pollution resulting from *seabed activities* was adopted in 1994. At the Barcelona Conference of Plenipotentiaries in 1995, MAP members approved amendments to the Barcelona Convention and the Dumping Protocol. A new protocol concerning Specially Protected Areas and Biodiversity was also adopted. These amendments take into consideration the UNCED process in particular by introducing the concept of sustainable development. It is also expected that a new protocol on transboundary movements of hazardous waste as well as amendments to the protocol on land-based sources will soon be adopted.

After the Barcelona Convention entered into force in 1978, the parties also developed a complex institutional structure to coordinate Med Plan operations. In 1980, the Coordinating Unit for the entire Med Plan was established and transferred to its permanent headquarters in Athens in 1982. Further specialized units are situated around the Mediterranean Sea. The center for the Priority Action Program is located in former Yugoslavia, the Blue Plan center in France, the center dealing with specially protected areas in Tunis, and a Regional Oil Combating Center (ROCC) in Malta.[10] Italy established the latest regional activity center, on remote sensing, in 1993.

Achievements of the Barcelona Convention

The parties are obliged to report on their follow-up actions. Under article 20 of the Barcelona Convention, they report on measures taken for implementing the Convention and its related protocols—a general duty also spelt out in the various protocols. For example, according to article 7 of the Dumping Protocol, the organization (that is, the Coordinating Unit) is to receive records of dumping permits. The LBS protocol states in article 13 that the parties shall inform one another through the Coordinating Unit of measures taken and results achieved. In short, the level of collaboration was medium: coordination of action was formally based on explicit standards and a centralized appraisal of effectiveness.

In practice, however, the parties have shown great reluctance in following up the reporting procedures required. According to articles 4, 5, and 6 of the Dumping Protocol, dumping into the Mediterranean requires a permit from competent national authorities. Even though all parties have ratified this protocol, only three states have actually reported annually to the Coordinating Unit.[11] The situation is no more encouraging with regard to the LBS protocol. In 1989, a questionnaire on land-based sources of pollution was sent from the Coordinating Unit to the contracting parties. By the fall of 1992, only seven countries had returned complete or partial replies. Moreover, no replies were received from the contracting parties to a letter sent in July 1990 asking for information on the implementation of existing or new legislation related to the pollution measures adopted by the parties since 1985. Furthermore, only four countries sent annual reports on measures adopted to implement the Barcelona Convention and its related protocols, as required under article 20 of the Barcelona Convention (UNEP 1991). This situation had not improved significantly by 1993 (UNEP 1993).

The most important protocol to be adopted is the land-based protocol (LBS). This protocol was adopted in 1980 and entered into force in 1983. Six parties had still not ratified the protocol by 1989, but as of 1993 all parties except Lebanon and Syria had ratified it. The LBS Protocol states in article 5 that the contracting parties within the Protocol area are to eliminate pollution from land-based sources of substances listed in annex I. Toward this end, the parties shall elaborate and implement, jointly or individually, necessary programs and measures—including, in particular, *common emission standards* and standards of use. Article 6 stipulates that the parties shall restrict pollution of substances listed in annex II. Altogether, twenty-eight groups of substances are to be covered by common measures.

Following a ministerial meeting in 1985, the parties developed an implementation plan in 1987 that seeks to develop common measures for all annex I and II substances. According to the plan, twenty-eight measures should have been adopted by 1995. To date, however, the parties have adopted only thirteen common measures, which mainly cover annex I substances (UNEP 1995). The thirteen common measures actually adopted hitherto have been quite general with few fixed time limits related to specific reduction goals. Due to vague goals and only recent approval of the common measures, it is difficult to assess precisely how many of these thirteen common measures have actually been implemented. However, some scattered data exist. The parties adopted common measures on organohalogen compounds in 1989. By 1995, eleven countries had adopted legal provisions on these substances, five of them in accordance with E.C. directives (UNEP 1995). With regard to the common measures on radioactive substances adopted in 1991, the countries that intentionally release radionuclides into the environment follow principles and guidelines of the International Commission on Radiological Protection (ICRP) and the International Atomic Energy Agency (IAEA) (UNEP 1992). The parties also adopted measures on organophosphorus substances in 1991. At that time, the only countries with legislation on effluent standards or water quality were Italy and Yugoslavia, although some other countries had general legislation in this field (UNEP 1991). There is an increasing tendency among the Mediterranean countries to install waste-water treatment plants along the coastal zone.[12]

With regard to the Dumping Protocol, common guidelines are in preparation for dumping of sewage sludge and dredged materials. Lack of guidelines and poor reporting makes it extremely difficult to assess the actions taken by the parties. It should be noted that in some areas, hazardous residues like solvents, organic chemicals, acids, and alkalies are disposed of through open dumping (UNEP 1989b). At a general level it is somewhat easier to assess the emergency protocol. According to article 3 in the Dumping Protocol, the parties shall endeavor to maintain and promote their contingency plans and means for combating pollution of the seas by oil and other harmful substances. As of 1993, seven states still had no national contingency plans, which are a prerequisite for handling any form of intentional or accidental spill (UNEP 1993). Data from the Regional Oil Combating Center (ROCC) show broad variations from 1978 to 1987 in relation to accidents and operation discharges. Total reports show that eleven accidents and operating discharges were reported in 1978, twenty-two in 1981, and twelve in 1987 (UNEP 1989b). Two states had not ratified the Protocol on Specially Protected Areas as of 1993. The Protocol entered into force in 1986, and a regional activity center began operation in Tunis in 1985. The Protocol seeks to encourage the creation of marine parks to preserve regionally endangered species. Approximately 100 areas have been selected as specially protected areas, and fifty more are under preparation. In 1989, it was decided to create a network for 100 historic sites to deal with a corresponding number of coastal historic sites of common interest in the Mediterranean. In addition, action plans for the conservation of marine turtles, monk-seals, and cetaceans have been approved by the contracting parties. The present author does not have sufficient information to judge the extent to which these species and areas are protected in practice.

It is even more difficult to assess the impact of policy-coordinating activities from 1976 on the state of the marine environment. One major problem is the lack of reliable time-series data for input and water quality, although some limited data exist. This problem is emphasized by the general methodological problems of separating the effect of intended environmental measures from general socioeconomic and technological change, as well as natural environmental variation (UNEP 1989b). Eutrophication is a local rather than a regional problem in the Mediterranean Sea, frequent wherever the rate of input of domestic industrial wastewater exceeds that of the exchange with the open sea. We do not know whether this problem has increased or decreased over time. The general situation indicates an upward trend in the number of beaches of acceptable standard.[13] Public-health problems in connection with the consumption of raw shellfish still occur regularly, although no major epidemic has been reported for a number of years (UNEP 1989b). The situation regarding microbiological contamination of shellfish has improved. Mercury poisoning can still be described as the major hazard originating almost exclusively from contaminated seafood. Less clear is the situation regarding adverse health effects from consumption of seafood polluted by chemicals other than mercury. The 1989 assessment of the Mediterranean Sea states that "All confined or semi-confined Mediterranean localities adjacent to large urban centers appear to be the site of progressive build-up, as a result of continued uncontrolled anthropogenic release."[14]

From this fragmented picture of the state of the Mediterranean environment, it is evident that the data are inadequate and uncertainty is high. Some indications of improvement exist, as do some indications of degradation. The European Commission has stated: "It is now time to move from the pilot and preparatory phase to the action phase," which indicates that little has actually been done ("Talks Sink Plans" 1990). Statements like "Yet, since the Barcelona Convention was implemented in 1976, pollution has shown little sign of improvement" ("Mediterranean Holiday" 1990), are frequent, while others claim that "The Med Is Cleaner" (Haas and Zuckerman 1990).

Lack of substantial action within MAP on marine pollution is evident. A growing population and general socioeconomic progress have further increased the pressure on the Mediterranean Sea over the past two decades. On the other hand, there is little evidence that the Mediterranean Sea has degraded significantly overall. This would indicate that a number of steps have been taken mainly outside the framework of MAP. Both national vulnerability and international cooperation have been important for the action actually taken by the parties. Some particularly vulnerable countries have established their own national programs independent of MAP. Important here is the fact that heavily polluted beaches may result in decline in tourist revenues, which constitute an average of 6.5 percent of the region's GNP. Significant contributions have come from other conventions and institutions than those that are integrated parts of the Med Plan, especially the European Community. As we have seen, the European Community is the main reason that some of the parties have taken action on organohalogen compounds. Spain, France, Italy, and Greece are European Community members and thus obliged to follow the E.C. common environmental legislation on the marine environment. Given a very simplified assumption that the biggest coastal cities are those that pollute most, we can say that the European Community and the E.C. countries control most of the polluting point sources in the Mediterranean Sea: 70 percent of the 539 Mediterranean coastal cities with a population exceeding 10,000 are located in the E.C. countries. This represents 59 percent of the total population in the coastal settlements (UNEP 1989). Furthermore, the European Community, together with the World Bank, the European Investment Bank, and UNEP, has initiated projects aimed at cleaning up the Mediterranean Sea. These institutions will commit financial and practical resources in order to achieve: "for the year 2025 at the latest, an environment in the Mediterranean Basin compatible with sustainable development." The European Union and the two banks are to allocate approximately $1.5 billion toward achieving this goal ("Mediterranean Holiday" 1990).[15]

The Mediterranean Action Plan has probably had greatest impact outside the area of marine pollution control. Its scientific assessment component as well as the ambitions of integrated planning have stimulated a collective awareness of the Mediterranean as a common heritage. We should bear in mind that many of the parties are less developed countries lacking financial, technological, administrative, and scientific resources in the area of marine pollution control. Moreover, the Mediterranean area is politically explosive. The 1995 Barcelona Conference was marked by time-consuming bickering in the plenary sessions between Greece and Turkey over their historical territorial disputes in the Mediterranean. The biannual meetings of the parties and ad hoc conferences have been the only regional fora where representatives from disputing countries could meet face to face. It is reasonable to assume that a large share of nonenvironmental motivation for MAP participation has significantly increased the distance between symbol and substance.

An important collective step forward was taken in the Barcelona Resolution adopted in 1995. Under article 6, the parties agreed (UNEP 1995) "to the reduction by the year 2005 of discharges and emissions which could reach the marine environment, of substances which are toxic, persistent and liable to bioaccumulate, in particular organohalogens, to levels that are not harmful to man or nature, with a view to their gradual elimination."

Closer examination, however, shows that this goal is not easy to use as a benchmark. First, actual reductions can hardly be evaluated without a baseline year. Second, the specific substances covered by this goal were not identified, though it was decided to commence work on this in 1996.

Problem-Solving Capacity

The outcomes seem to reflect compromises between DC and LDC interests, and given that other factors are constant, a roughly equal distribution of capabilities between North and South can be expected. Considering possible *influence through pollution*, no actor seems to have had any significant objective capabilities and potential power over other actors due to low pollution exchange, and consequently minor interdependence exists between North and South. However, France was recognized as the major Mediterranean polluter, and French participation in the Med Plan was a necessary condition for success in cleaning up the Mediterranean. On the other hand, the somewhat false perception of the problem as a truly collective problem could be used by countries in the South to refuse to comply if the Med Plan did not satisfy them (Haas 1990). An increased and uncontrolled industrialization in the South could be seen as a major threat to pollution problems in the North. The North, represented by France, had a potential for *influence through the market*. France's role as a major trading partner for the Mediterranean countries, without being particularly dependent on other economies, gave France a significant leverage.[16] The asymmetrical relationship between North and South becomes even clearer if we look at the distribution of *scientific capabilities* in the mid-1970s (Haas 1990). It seems evident that France and Italy were the scientific leaders during the Barcelona negotiations and that the industrialized countries in the North had the scientific upper hand. France and Italy also sent large delegations to the Med Plan meetings, while the LDCs did not have the capacity to be represented by many negotiators and scientists. So it seems reasonable to conclude that we have strong indications that France was the structural leader and the DCs were more powerful than the LDCs.

The LDCs gained scientific capabilities through Med Pol participation and more relevant technology and training. There are also indications that a major change has occurred in the parties' political priorities concerning the Med Plan. At the seventh ordinary meeting of the contracting parties in Cairo 1991, France was represented by three delegates, Greece by four, and Spain by two. By comparison, Morocco and Malta sent four delegates, while Albania and Algeria sent three (UNEP 1991). Thus, it is reasonable to assume that LDC delegations have become relatively more competent negotiators with experience. Still, these changes were probably not sufficient to alter the overall asymmetrical distribution of capabilities between the DCs and the LDCs. Thus it seems as though the distribution of capabilities in the Med Plan cooperation has had little effect on the final outcomes.

The capacity to integrate actor interests and preferences is a question of the cooperating system's ability to design integrative solutions.[17] One way of doing this is through the building of scientific capacity. As we have already seen, the parties had little reliable information on pollution in the Mediterranean, its causes, and possible solutions at the time they approved the Med Plan in 1975. High *ecological* uncertainty is underlined by an inadequate marine scientific capacity throughout the region. In 1974, the total number of marine scientists around the Mediterranean Sea was less than 700.[18] In comparison, the United Kingdom, bordering on the North Sea, had 1,046 marine scientists in 1983, despite a considerable drop in marine scientific capacity from the end of the 1970s (Skjærseth 1992b). Another significant problem was the uneven distribution of scientific capacity and awareness concerning pollution. For instance, while Italy had thirty-one marine research centers in 1974, Syria had only one. The Med Pol program was established to reduce the ecological uncertainty by improving the scientific capacity principally among the LDCs.

Med Pol has been quite successful in reducing uncertainty. By 1995, UNEP had issued a total of ninety-two scientific and technical reports, covering a wide range of sources and effects of pollutants entering the Mediterranean.[19] Hence, at a general level, lack of knowledge can no longer explain lack of action. As early as in 1977, the Med X study concluded that over 80 percent of all municipal sewage entering the Mediterranean was untreated. Thus, the parties knew what they should do; effective sewage treatment technology has been available for decades. On the other hand, considerable uncertainty remains concerning the sources and effects of specific pollutants entering the Mediterranean Sea. Assessment publications issued by UNEP are full of conclusions emphasizing "lack of data" and "uncertainty." Med Pol's effort to address the asymmetrical distribution of scientific capacity has been less successful. Although the LDCs in particular have gained scientific experience through participating in Med Pol, this point should not be exaggerated. If we compare the scientific capabilities of 1974 with 1992, no major change in the asymmetrical distribution between the DCs and the LDCs has occurred.[20]

According to article 13 in the Barcelona Convention, which describes institutional arrangements, UNEP is given the main secretariat functions. Before a Trust Fund was created in 1979, the Med Plan was almost entirely supported by UNEP and to some extent by France. Between 1973 and 1975, 7.2 percent of UNEP's Environment Fund resources went to the Mediterranean region. After the Trust Fund was established, the parties pledged to contribute proportionally to their U.N. schedules. From the Med Plan's inception through 1986, the participating governments contributed $13.3 million, UNEP and other U.N. agencies $14.4 million, and the European Community $2.2 million (Haas 1990). Thus, the average share of the total budget that was divided between the participating states was 44 percent. In addition, the World Bank and the European Investment Bank launched the Environmental Program for the Mediterranean (EPM) in 1988. From 1980 to 1990, the World Bank granted 37 loans totaling $2.35 billion for projects directed at environmental protection in the region. The European Investment Bank has granted a total of 3 billion ECU since 1980 in the Mediterranean Basin alone (World Bank/European Investment Bank 1990). In this allocation of *external benefits*, we may probably find an important key to why the parties joined the cooperation at all and why it has been a political success. Few countries would have said no to this free lunch. In comparison, the North Sea conventions are based almost entirely on the participants' contributions.

The budget for 1992 shows that the parties' share of the total Trust Fund contributions will increase. The total budget for 1992 is $6,268,000 (UNEP 1991). Of this, the contracting parties are to contribute $3,850,000, which constitutes 61 percent of the total budget. This indicates that the contracting parties' share of the budget will increase compared to their average contributions up to 1986. At the same time, the lag in financial contributions mainly by the major contributors has increased. This was explicitly considered a "cause of serious embarrassment to the secretariat" at the sixth ordinary meeting in Athens in 1989 (UNEP 1989a). The status of contributions to the Trust Fund as of April 30, 1991, shows that unpaid pledges as of December 1990 were $1,376,132, which constitutes 39 percent of the total pledges for 1991, excluding the E.C. voluntary contribution. Unpaid pledges for 1991 and prior years exceed $4 million. UNEP Executive Director Mostafa Tolba stated in his opening speech to the Seventh Ordinary Meeting of the Contracting Parties in Cairo 1991 that "Delays in payment, especially by the major contributors, are very disturbing. They are persistently crippling the implementation of the program, causing unnecessary additional burdens on the administration of the program, and putting the UNEP staff members of the Coordinating Unit continuously on short-term contracts."[21] Furthermore, "If the Contracting Parties to the Barcelona Convention sincerely want to have a coordinated effort to help protect the Mediterranean environment, the sum required—less than 6.7 million U.S. dollars per year—is extremely low. It is the price of three battle tanks."[22]

Thus, while extensive external funding helps us to understand why the parties joined the cooperation at all and why it has been a political success, the low budget and the major contributors' unwillingness to pay their share helps explain the lack of collective problem solving, even though this conclusion disregards purely national projects.

Capacity to aggregate interests is largely determined by the decision rule, but the negotiating level also can be important in forging tradeoffs that reach beyond the authority of low-level officials (Underdal 1990). The fifth ordinary meeting of the contracting parties in 1985 was held at the ministerial level. The parties adopted ten targets to be achieved as a matter of priority during the second decade of the Mediterranean Action Plan, which, among other things, included concrete measures to achieve substantial reductions in industrial pollution and disposal of solid waste (UNEP 1995). After this meeting, the parties started to develop common measures to implement the LBS protocol. Against this backdrop, it is reasonable to assume that the 1985 ministerial meeting speeded up the cooperation process.

Since the Barcelona Convention does not establish a supranational regime, the basic decision rule is consensus. However, a system of tacit approval of annex amendments is established according to Article 17 in the Barcelona framework convention. The procedure requires a three-quarters majority vote for parties to adopt the change, and after some time the amendment becomes binding on all parties except those that actively object. Since many parties have been rather passive in the cooperation process, this procedure seems to constitute an effective tool for cutting through deadlock.

Not only was UNEP active in allocating money, but it has also been a driving force and active player in designing politically feasible solutions. A recent example is the opening speech of the Seventh Ordinary Meeting of the Parties by Mostafa Tolba in Cairo in 1991. In contrast to most diplomatic speeches, he pinpoints in detail the results achieved and, notably, the results *not* achieved in accordance with the commitments under the Barcelona Convention and its related protocols. UNEP, backed by other parts of the U.N. organization, has had a substantial influence on shaping the MAP program through its allocation of benefits, expecially in the early years of the cooperation. The establishment of the regional activity centers in different regions shows, for instance, that efficiency has been sacrificed on the altar of political feasibility. For example, the Regional Activity Center on Specially Protected Areas was established in Tunis in 1985. Activities related to this center have been hampered by problems related to the need to appoint a full-time director by Tunisian authorities, recruitment of Tunisian experts to the center, the legal status of the internationally recruited staff, and arrangements for more space for the center's premises. In addition, most of the LDCs received high net benefits from participating in the MAP (Haas 1990). Still, a main problem at both the scientific and the political levels has been a lack of active participation from all parties. More than half of the contracting parties do not actively take part in the cooperation but tacitly accept proposals put forth. The Coordinating Unit of the MAP is highly aware of this situation and is working to improve it. Thus, UNEP's skill lay, at least in part, in making the LDCs aware of their interests and designing a political program that would appeal to the interests of all the parties.

Conclusions: Principal Determinants of Success and Failure

The Mediterranean states have succeeded in developing a comprehensive plan for cleaning up the Mediterranean Sea. This plan has increased steadily in scope over time. Moreover, a slight increase in decision effectiveness can be witnessed from the mid-1980s measured in terms of joint commitments adopted to implement the

Table 12.1a
Summary scores: Chapter 12 (MEDPLAN)

| Regime | Component or Phase | Effectiveness | | Level of Collaboration |
		Improvement	Functional Optimum	
MEDPL	1976–1985	Very low	Low	Intermediate
MEDPL	1985–1995	Very low	Intermediate	Intermediate

protocol on land-based sources. The reasons for this collaborative success lie particularly in high problem-solving capacity provided in particular by UNEP. UNEP has provided legal, financial, scientific, and diplomatic assistance since the 1970s. Moreover, decision procedures and the 1985 summit were important in aggregating interests and preferences into joint commitments. In addition, we also have some indications that the problem itself—as a collective-action problem—became somewhat easier to handle over time due to improvement in knowledge. The scientific assessment component—the Med Pol—has been quite successful in reducing uncertainty with regard to main causes of marine pollution as well as transport patterns of contaminants. As a consequence, the asymmetry in perceived interests between LDCs and DCs in particular seems to have been weakened over time.

However, the Mediterranean states have succeeded in transforming the joint Med Plan commitments and ambitions into behavioral change at the domestic level *only to a very limited extent* (see table 12.1). The initiatives taken by the states are to a significant extent due to other factors than the MAP. Particularly important in this respect are E.C. directives and support from the international financial institutions. The reasons for this situation are partly related to the MAP itself and partly to ecological, economic, and political factors beyond the competence and scope of the MAP. With regard to the former, a major problem of MAP seems to have been too wide a scope at the expense of stringency of measures compared to resources actually available. Related to this are the large-scale plans launched by the international banks and the involvement of the European Community. Since MAP itself aims at including almost everything related to the environment and development, there is obviously a danger of overlap between the various initiatives. Its future effectiveness and efficiency will depend heavily on the parties' ability to coordinate their activities with other projects and to give priority to truly transnational problems among MAP activities.

Table 12.1b

Problem Type		Problem-Solving Capacity			
Malignancy	Uncertainty	Institutional Capacity	Power (basic)	Informal Leadership	Total
Strong	High	High	Balance	Strong	High
Moderate	Intermediate	High	Balance	Strong	High

With regard to factors outside the competence of the MAP, three factors stand out as particularly important. First, the parties had mixed motives for participating in the MAP. In general, it is reasonable to expect that the distance between joint obligations and actual implementation will increase when the share of nonenvironmental motivation increases. Second, the Mediterranean collapse hypothesis was refuted and the parties' incentives to implement stringent measures decreased accordingly. Third, many of the parties are LDCs with a low environment-related capacity at the outset. Since the parties' share of the budget has increased, delay in contributions to the budget has increased and caused problems for the implementation of the programme.

MAP is now entering Phase II, and a new action plan has been adopted (UNEP 1995). Strongly inspired by Agenda 21, it describes the pressing environmental needs that should be addressed. There are currently no priorities or timetables among the various items. Given that a collective awareness currently exists, a key to speeding up the effectiveness of MAP lies in making realistic priorities for the future. Regrettably, this would imply that other valuable projects would have to be sacrificed.

Notes

1. This chapter is largely based on Skjærseth (1993, 1996).

2. The total population of the eighteen nations bordering the Mediterranean was about 360 million in 1985 (World Bank 1990).

3. Spain, France, Italy, Yugoslavia, Libya, Israel, Greece, Cyprus, Malta, Algeria, Turkey, Tunisia, Syria, Morocco, Egypt, and Lebanon.

4. Conference on the Human Environment, Stockholm, June 5–6, 1972, Identification and Controls of Pollutants of Broad International Significance (Subject Area III, A/CONF.48/8 7, January 1972, p. 25). Other seas that were regarded as threatened included the Baltic, the Black, and the Caspian Seas.

5. According to (one) legend he finally threw himself into the strait to explain the pattern of the currents.

6. The Med X study produced the following major findings: industrial and municipal wastes exceeded oil pollution problems, 85 percent of all pollutants in the Mediterranean originate on land, 80 to 85 percent of land-based pollutants were emitted to the Mediterranean by rivers, and over 80 percent of all municipal sewage entering the Mediterranean was untreated (see Haas 1990, 101–103).

7. Total costs are estimated at $5.1 billion. Of this amount, Spain and Italy would have to invest $2.4 billion (UNEP 1989b).

8. Since 1948 four wars have been fought between Israel and its Arab neighbors. Added to this picture is the Greek-Turkish antipathy and conflict over Cyprus and the Moroccan-Algerian dispute with regard to Western Sahara. Furthermore, the civil war in former Yugoslavia does not stimulate active participation in international environmental cooperation. In addition, the Persian Gulf war, though not geographically involving the Mediterranean area, sharpened traditional Israeli-Arab conflicts and shaped new alliances across traditional enemy cleavages.

9. Blacklist substances include mercury, cadmium, DDT, PCBs, some plastics, used lubricating oils, and radioactive wastes. Greylist substances include lead, zinc, copper, arsenic, cobalt, silver, cyanide, fluorides, and pathogenic microorganisms.

10. The new name of ROCC is Regional Marine Emergency Response Center for the Mediterranean Sea (REMPEC).

11. This information is from 1992, when France, Italy, and Israel reported annually. This situation may have improved, but the author has not found any information on the number of licenses issued with regard to dumping of industrial waste, sewage sludge, or dredged materials.

12. There are no recent data to indicate to what extent the coastal population is currently served by treatment plants. A survey conducted in 1978 indicates that only 50 percent of the seventy-eight Mediterranean municipalities responding to the questionnaire possessed treatment facilities prior to 1978. This situation had improved somewhat by 1980. Recently, establishing sewage treatment plants has been adopted as a priority matter by all Mediterranean countries.

13. Some of the results are published by national authorities, which fear negative publicity.

14. This can be observed in the Bay of Algiers, the "lac de Tunis," the Bay of Abu-Kir near Alexandria, the Bay of Izmir in Turkey, the North Adriatic, and the coastal belt along the north coast of the West Mediterranean (UNEP 1989b, 74).

15. In addition, the World Bank in collaboration with the European Investment Bank launched the Environmental Program for the Mediterranean (EPM) in 1988.

16. In 1971, for instance, 31.7 percent of Algeria's trade was with France, whereas 1.7 percent of France's trade was with Algeria. This indicates high Algerian sensitivity to French environmental policy (Haas 1990, 168).

17. An integrative solution is a solution where no parties lose and at least one party wins.

18. Number of marine scientists with a B.A. or higher equivalent degree (Haas 1990, 85).

19. See MAP Technical Reports Series.

20. If we compare the scientific capabilities of 1974 with 1992, we find that no major change in the asymmetrical distribution between the DCs and the LDCs has occurred. Scientific

capability is operationalized as number of marine research center and number of marine scientists with a B.A. or higher equivalent degree. Data reliability has probably improved from 1974 to 1992. On the other hand, the data are based on the number of respondents to questionnaires and may therefore not be completely reliable. See Skjærseth (1992a).

21. Report of the Seventh Ordinary Meeting, annex III, p. 3.

22. For a complete version of the ten-point list, see UNEP (1985, table 82).

References

Conference on the Human Environment. 1972. Identification and Controls of Pollutants of Broad International Significance. Subject Area III, A/CONF. 48/8 (7 January). Stockholm, June 5–6.

Haas, Peter M. 1990. *Saving the Mediterranean. The Politics of International Environmental Cooperation.* New York: Columbia University Press.

Haas, Peter M., and Julie Zuckerman. 1990. "The Med Is Cleaner: It's Still Diseased But Not Terminally Ill." *Oceanus* 33, no. 1 (Spring), 40–43.

Julian, Maurice, and Paul R. Ryan. 1990. "The Med." *Oceanus* 33, no. 1 (Spring): 11.

Luke, Anthony. 1990. "Talks Sink Plans to Clean Up the Mediterranean." *New Scientist* (November 2): 22.

Skjærseth, J. B. 1992a. *The Mediterranean Action Plan: More Political Rhetorics Than Effective Problem-Solving?* Oslo: Fridtjof Nansen Institute.

Skjærseth, J. B. 1992b. "Towards the End of Dumping in the North Sea: An Example of Effective International Problem-Solving?" *Marine Policy* (March): 130–141.

Skjærseth, J. B. 1993. "The 'Effectiveness' of the Mediterranean Action Plan." *International Environmental Affairs* 5(4) (Fall): 313–335.

Skjærseth, J. B. 1996. "The Twentieth Anniversary of the Mediterranean Action Plan: Reason to Celebrate?" *Green Globe Yearbook* (pp. 47–53). Oxford: Oxford University Press.

Stansell, John. 1990. "A Mediterranean Holiday from Pollution." *New Scientist* (May 5): 28.

Underdal, A. 1990. "Negotiating Effective Solutions: The Art and Science of 'Political Engineering.'" Draft, Department of Political Science, University of Oslo.

United Nations Environment Program (UNEP). 1985. *State of the Marine Environment.* Athens.

United Nations Environment Program (UNEP). 1989a. *Report of the Sixth Ordinary Meeting of the Contracting Parties to the Convention for the Protection of the Mediterranean Sea Against Pollution and Its Related Protocols.* Athens.

United Nations Environment Program (UNEP). 1989b. *State of the Mediterranean Marine Environment.* Athens.

United Nations Environment Program (UNEP). 1990. *Common Measures Adopted by the Contracting Parties to the Convention for the Protection of the Mediterranean Sea Against Pollution.* MAP Technical Report Series *No. 38.* Athens.

United Nations Environment Program (UNEP). 1991. *MAP Technical Report Series No. 58.* Athens.

United Nations Environment Program (UNEP). 1992. *MAP Technical Report Series No. 62.* Athens.

United Nations Environment Program (UNEP). 1993. *MAP Technical Report Series No. 62.* Athens.

United Nations Environment Program (UNEP). 1995. *MAP Technical Report Series No. 92.* Athens.

World Bank/European Investment Bank. 1990. *The Environmental Program for the Mediterranean: Preserving a Shared Heritage and Managing a Common Resource.* Washington, DC.

World Bank/European Investment Bank. 1991. *Report of the Seventh Ordinary Meeting.* Athens.

World Bank/European Investment Bank. 1993. *Report of the Eight Ordinary Meeting.* Athens.

World Bank/European Investment Bank. 1995. *Report of the Ninth Ordinary Meeting.* Athens.

13

Oil Pollution from Ships at Sea: The Ability of Nations to Protect a Blue Planet[1]

Elaine M. Carlin

Introduction and Reasons for Success and Failure

With the advent of tapping the earth's oil reserves, in but a moment of geologic time humans became a new and significant source of oil in the marine environment. For a million years or more, oil has seeped naturally from the ocean floor, causing no apparent compromise to the magnificent explosion of life forms above. But as ships began to use oil as fuel and to transport oil, pollution began to appear.[2] As reported by Thor Heyerdahl during the Ra expeditions and by others, pollution was observed for long stretches in the open ocean and on most sandy beaches of the world. *The primary purpose of the oil-pollution regime considered here is to prevent one important source of this pollution, so-called operational pollution, which refers to the intentional discharge at sea of oil-water mixtures.* While tanker accidents are not the primary focus of this regime, they have been addressed by it, and more important for our purposes, major oil spills have been a critical factor in precipitating action.

The oil-pollution issue is highly political and difficult, with a long, rich history of collaborative efforts. Eventually a regime is created and implementation attempted—with mixed results—including significant success and significant failure. The character of the problem is intriguing because of its extreme dichotomies: aspects of the problem make it extremely simple to solve, others make it extremely difficult to solve. Likewise, the problem-solving capacity of the institution is intriguing and also characterized by dichotomies: it generates the strongest, most prolific kind of decision rule, and yet for much of its history this international body essentially has been an epiphenomenon (Mearsheimer 1995).

Only during moments of high problem-solving capacity has the regime been able to overcome the problem's politically malign nature. Unfortunately such moments are not easily reproducible, requiring in this case a hegemon willing and able to use coercive leadership to pursue environmental goals, exogenous shocks sufficient to compel the hegemon (and others) to act, and the prevalence of selective incentives for collaboration. Nation states chose not to create a regime with high problem-solving capacity—a choice that is not surprising given the high political and economic stakes. The evolution of the regime was more about raw power, coercion, hidden agendas, ulterior motives, and short-term economic interests than it was about learning, the building of concern, or entrepreneurial leadership.

The problem itself has many natural advantages for its resolution. Whether in the form of dramatic oil tanker accidents or beach tar sticking to beachcombers' feet, this issue is experienced viscerally, without the perception issues that plague so many global environmental problems. Moreover, the basic solution is simple and was articulated from the beginning: control the discharge of oil from ships, and the pollution from these ships will abate.[3] Despite these and other positive attributes, the pull of highly divergent interests and other negative forces has prevented this regime from achieving high effectiveness.

The oil-pollution regime produced explicit standards, but implementation remained exclusively with the nation state. The regime failed to achieve coordinated planning and centralized assessment of effectiveness. This rather low level of collaboration (scored at 2 on our scale of 6) remained the same over time, despite fundamental shifts in mission from regime formation to regime implementation, despite order-of-magnitude increases in the scope of the problem, and despite pronounced implementation and compliance failures.

Despite the rather low level of collaboration, the oil-pollution regime has produced some changes in human behavior in response to explicit standards requiring the shipping industry and the nation state to act. The industry was to control the concentration and location of its discharges, to employ specific on-board equipment, and to build new oil tankers to prescribed specifications. The nation state was to provide the necessary onshore reception facilities and to provide effective enforcement of the regulations. In short, the regime consisted of discharge standards, control technologies, onshore facilities, and enforcement provisions. These components are not mutually exclusive but work together to form a complete regulatory regime.

But the behavior of tanker operators has remained relatively unchanged in response to the discharge standards. In fact, one could not reasonably expect that behavior would have changed. Arising out of superficial efforts aimed at limiting the regime to a no-cost appearance of a solution, the rules initially required operators to take action that was impossible at the time and ultimately failed to provide workable compliance and enforcement mechanisms. In sharp contrast to the discharge standards, *requirements for control technologies have produced profound change*. New tankers are built according to construction and equipment standards set by the regime. Oil-price increases account for early adoption of the less expensive of the two major technologies, but it seems clear that employment of the more expensive technology has been in response to the international rules (Mitchell, McConnell, Roginko, and Barrett 1999).[4]

The behavior of governments in the regions of the world where onshore facilities are most critical has remained unchanged. Essential in oil-loading ports to receive tank-cleaning wastes before new oil is loaded, facilities are almost nonexistent in the Persian Gulf region. Capacity is also essential in several geographic areas designated by the regime as requiring special environmental protection. Yet special areas of the Baltic, Black, Mediterranean, and Red Seas and the Persian Gulf remain unimplemented twenty years after rule adoption (Mitchell, McConnell, Roginko, and Barrett 1999). Where onshore capacity has been provided, its development has been linked to nonregime factors,[5] although some analysts suggest that treaty requirements and ongoing discussions "appear to have helped motivate financially capable and environmentally concerned states to provide them" (Mitchell, McConnell, Roginko, and Barrett 1999). Domestic political forces, rather than the regime, have dictated behavioral change in the area of enforcement (Mitchell 1993). Most states conduct little if any enforcement, while a few states have strong programs (Mitchell 1993). The potential for truly effective enforcement by any coastal state remains limited by the failure of the regime to break away from traditional flag-state jurisdiction.

Environmental Impact

Has the behavioral change that has been accomplished been sufficient to solve the environmental problem of ship-generated oil pollution? Clearly the answer to this question is no. The goal of the regime was to prevent pollution of the sea by oil and to completely eliminate operational pollution; the regime's solution was to reduce the volume of wastes and retain waste on board for eventual discharge on shore. Today, ships continue to discharge substantial quantities of oil into the ocean, and tar from ships (and other sources) continues to pollute coastlines in many parts of the world (GESAMP 1993). The ecological effects of oil pollution continue to be of scientific concern. In fact, there is increasing evidence of mutagenic and carcinogenic effects of these compounds, and the ecological effects of chronic sources of contamination "may be much larger than anyone has yet estimated" (National Research Council 1985, cited in GESAMP 1993, 48). There is increased concern about longer-term effects on euphausids (krill) and other plankton that support the southern food webs, and many habitats are now known to be particularly vulnerable (GESAMP 1993). Recent study of natural oil seepage confirms "the fact of natural oil seepage in no way forgives oil pollution" ("Natural Oil Spills" 1998).

Ideally the regime would have set up a way to measure, over time, whether it was meeting its goal. Original wealthy parties certainly had the scientific and resource bases necessary to design and carry out such assessments. In the absence of such a program, our analysis is severely constrained by a lack of good data. However, the lack of onshore facilities alone makes significant noncompliance unavoidable, leading to this (overly simplified) question: how effective can a regime designed to eliminate wastes be if there is no place to put the waste? U.S. tanker crews report that illegal discharges are "the norm on the high seas" (Curtis 1985), and German port officials have reported that 50 percent of the tankers they inspect cannot explain where oil residues have gone.[6] Belgian and Dutch studies in 1987 found that most oil waste generated by tankers—85 and 65 percent, respectively—was still being disposed of at sea (Mitchell 1993). A 1993 report asserts that international rules have substantially reduced the amount of oil entering the sea but bases this conclusion on compliance estimates of 80 to 99 percent (GESAMP 1993).[7] If more realistic compliance assumptions are used, the estimated volume of oil discharged increases dramatically.[8] Enforcement data from those states most effectively applying treaty provisions confirm that as of 1994, violations are increasing in absolute and relative terms (Payoyo 1994). With this compliance picture in a region actively engaged in enforcement, "it is not difficult to imagine the magnitude of . . . [treaty] violations occurring in the rest of the world" (Payoyo 1994).

In summary, we rate this regime as having low effectiveness because it has failed to solve the problem, as evidenced by continuing environmental impairment. The regime produced a low level of collaboration and inadequate levels of behavioral change in three of the four major regime components. Other analysts have rated the effectiveness of this regime more highly than we do, in part as a result of taking into consideration the political difficulty of solving the issue (see Mitchell, McConnell, Roginko, and Barrett 1999).[9] It seems that if we were to consider political feasibility, we should also take other considerations into account that would also be mitigative, some in the other direction—for example, the fact that the solution was clear and understood in 1926. Depending on one's assumptions about a no-regime condition, the regime can be credited with producing relative improvement in behavior, particularly in its successful promotion of control technologies and to a much lesser extent in the moderately to severely unsuccessful implementation of the other three major regime components.[10] Moreover, we can deduce that relative environmental improvement is occurring as control technologies are implemented.

Type of Problem and Problem-Solving Capacity

Problem Character

With respect to intellectual complexity (that is, the amount of descriptive and theoretical uncertainty in the knowledge base), we find that in the oil-pollution case, both types of uncertainty are low. With the exception of an early period of the regime (when scientists disagreed over oil's impact, movement, and ability to break down naturally in the marine environment), understanding of the problem has been solid. As distinguished from the complexity of marine ecosystems, the distribution of oil and its immediate effects on wildlife are grossly evident at the macro level. Likewise, the cause-and-effect logic of a solution is simple: if ships refrain from discharging oil, the pollution from these ships will abate. Based on how tangible the pollution and its gross effects are and how simple the rationale for action is, this problem should be easy to solve. Despite its intellectual simplicity, the issue has been characterized by highly divergent actor interests: maritime nations host shipping; oil industries act to protect those industries or to promote legal regimes conducive to military or security interests; other maritime states with extensive pollution act to protect their coasts; and coastal states seek to protect their coasts from pollution or to use the issue to increase the extent of their jurisdiction over ocean space. As a result of this mix of interests, the oil-pollution problem became substantively linked with highly political, economic, security, and jurisdictional issues—issues of a similarly malign character.

The most formidable threat of serious costs to the industry came from unilateral controls, likely to be stringent and to cause competitive disadvantage to that segment of the industry most affected. To avoid this worst-case scenario, the shipping industry *supported* international collaboration while working to ensure that such collaboration produced substantively ineffective results. To the extent that this ineffective collaboration created a base on which to build meaningful agreements and ultimately accomplish their implementation, competition supported effective action, contradicting our assumption that competition, as a component of the malign problem, always makes collaboration more difficult.

Incongruity problems are particularly intractable when they are characterized by *asymmetry*. A problem is asymmetrical to the extent parties are, or perceive that they are, affected differently by the activity. The consequences of oil pollution are highly asymmetrical, with many states enjoying no benefits of the activity suffer pollution, while countries enjoying major economic benefits suffer little or no pollution. Geographically and temporally, pollution is often uncontrollable because oceans distribute oil from its point of release and in a pattern that varies according to oceanic and atmospheric conditions (UNEP 1982). A problem is also considered asymmetrical if the structure of the system of activities being regulated is asymmetrical. Nation states concerned about oil pollution vary greatly in their level of involvement in shipping. According to both measures of asymmetry, ship-generated oil pollution is a strongly asymmetrical issue.

In summary, the issue of ship-generated oil pollution is intellectually simple but politically malign based on its incongruity. Because it is infected with strong competition, significant externalities, and high asymmetry, it is highly malign. These characteristics do not change over time. Together, our findings on regime effectiveness and type of problem support our expectation that in this malign case we would find low effectiveness. However, we discover that one characteristic of the malign issue—competition—supported international collaboration. Below we explore the merits of our second hypothesis, which links high effectiveness to the presence of selective incentives for collaboration, linkages to more benign issues, and high problem-solving capacity.

Problem-Solving Capacity
In the oil-pollution case, the institutional setting and power relationships are the important elements of problem-solving capacity. Coercive leadership (a component of the power variable) rather than instrumental leadership produced results. This finding is consistent with our hypothesis that in the malign problem, instrumental leadership is more difficult, and coercive leadership more likely (if there is a hegemon). We have also identified several intervening variables that make operative, act as an agent of, or otherwise influence the independent variables. In this case, the pattern of coalition formation and negotiating strategies helps to make power and the institutional setting operative.

Institutional Setting

Of particular importance when actor preferences diverge, as in the malign problem, is an institution's ability to aggregate those preferences into collective decisions. Chief among the important determinants of this capacity are the decision rules and procedures. The oil-pollution regime enjoyed the important advantage of decision making by majority rule. We postulate that majority, as opposed to consensus, voting rules will lead to more effective regimes, including more ambitious regulations. However, to the extent ambitious rules are accomplished at the cost of sacrificing significant actor interests, compliance will suffer. For agreements to enter into force, the regime required ratification by a certain percentage of the major players in the field, giving an effective veto to a small number of maritime states with large shipping industries. Eventually tacit acceptance procedures and fixed effective dates for certain treaty provisions were introduced to remove obstacles to ratification.

The mere existence of a formal institution over an extended period of time, an institution functioning as an actor itself, and the development and operation of a transnational network of experts are additional components of the variable institutional setting that we believe will, *ceteris paribus*, facilitate cooperation and enhance the effectiveness of international regimes. A United Nations agency provided an enduring arena, but nation states did not give the agency the autonomy necessary to act independently. And, as expected, given the low level of intellectual complexity of the problem, we do not find a transnational community of experts operating in this case. In summary, the oil-pollution regime's institutional setting was characterized by strong decision rules, a long-standing arena for decision-making, and a weak organization without actor status.

Distribution of Power

In the oil-pollution arena, states with large shipping and oil interests are powerful in the "basic game"; states with market or political power are capable of playing in the "policy or decision game." The distribution of power was basically bipolar: two coalitions of actors formed around two issue-specific sets of interests. But power in this case can also be thought of as unipolar in the sense that a uniquely powerful actor used its power coercively to control outcomes and that for much of the regime's history, the shipping nations' coalition controlled decision making and the entry into force of agreements.

When formal institutional power is lacking, nation states can transform their power into coercive leadership in a number of ways: by providing selective incentives (for example, threatening to take action it would not actually contemplate except for the purpose of influencing behavior), by taking unilateral measures that set a pace others find in their interests to adopt, or by causing a restructuring of domestic interests within other countries (for example, by using market power to change the structure of opportunities available to others). In the oil-pollution case, coercive leadership was pivotal to the adoption of potentially effective regulations.

The Preregime Phase[11]

In the early 1920s, chronic oil pollution from the operation of nontankers off North America's eastern shores led to public demands for national action and the passage of domestic legislation. The United States called the first international conference on oil pollution in 1926 and, together with the United Kingdom and Canada, proposed limiting operational discharges oceanwide and requiring the installation of on-board technology (oily-water separators) to retain wastes for disposal in reception facilities on shore. America and Britain were major shipping nations at the time, but they promoted strong action in response to environmental-interest pressures at home.[12]

Most of the thirteen participating states, however, either did not suffer pollution or were not compelled to address it and opposed the cost of the ship and shore technologies. At the same time, these nations were in favor of some sort of international program for the advantages it would bring. International rules would prevent an uneven playing field and a patchwork of diverse regulations and were likely to be less stringent (meaning less interference with shipping) than national rules. Unilateral laws were also to be avoided because they might contain claims of extended jurisdiction, which had implications for maritime governments well beyond the oil-pollution issue. Spain and Portugal had already banned the discharge of oil in waters extending six miles from their coasts (Pritchard 1987). A no-cost international agreement held the further advantage of quieting domestic environmental pressure without unduly affecting shipping (Pritchard 1987). These benefits served as selective incentives toward the establishment of an international regime. Here one can observe how the competitive character of the problem provided an impetus for international action. At the same time, the costs associated with an effective solution served as disincentives to a meaningful program. The impact of these disincentives is easily identified in the results of the 1926 negotiations. Rather than oceanwide restrictions, discharges would be allowed oceanwide, except for 50-mile coastal zones in which discharges would also be allowed but limited in concentration. Wastes would have to be retained on board for discharge outside the zone or onshore, but neither on-board equipment nor onshore facilities were required. The United States made an initial foray into coercive leadership (in this issue area) by threatening to ban from its ports any ship violating the draft agreement. In the meantime, pollution had apparently improved, and the Americans no longer believed oil to be persistent in the marine environment. The 1926 agreement was never formally adopted, nor was a similar agreement developed in the 1930s.

In summary, in the absence of a formal regime, a technically appropriate solution was devised (albeit this was a straightforward task), and nation state actors formed coalitions based on domestic environmental pressure or the lack thereof, economic incentives based on shipping interests, and the larger jurisdictional issue. Selective incentives to avoid unilateral regulation and jurisdictional claims influenced actors to support some sort of international collaboration.

The Process of Regime Creation

By the early 1950s, worsening pollution, including the killing of thousands of sea birds in the United Kingdom's coastal waters, fueled public-interest-group pressure on the British government. Concerned that similar pressure in other states might result in a patchwork of unilateral legislation certain to interfere with shipping, the United Kingdom convened a global conference in 1954. The potential for significant costs and coastal states' interests in expanding their jurisdiction, brought the vast majority of players to the table: thirty-two states representing 95 percent of the world's shipping tonnage. Despite extensive lobbying by the British prior to the Conference, the United Kingdom's proposal of oceanwide restrictions was rejected as it had been in 1926, and a compromise package developed and debated: 50-mile coastal zones (and some larger regional zones) in which discharges under 100 parts per million (ppm) in concentration would be permitted and a 100 ppm limit anywhere for tankers going to ports with reception facilities.

Supporters of oceanwide restrictions were an eclectic coalition of West European, Commonwealth, and Communist bloc states, including two European maritime states suffering pollution from heavy traffic in the English Channel and the North Sea. Most of the states lent support under diplomatic pressure because they lacked large shipping and oil industries or because of other environmental concerns (the Soviet Union was concerned over oil's effects on fishes, fish roe and spawn, and fish migratory routes).

The United States believed that voluntary adherence to its coastal zones was effective in controlling pollution; the status of the problem was measured parochially—

on its own shores. Most of the delegates objected strongly to both the original ocean-wide restrictions and the compromise package, signaling their unwillingness to incur the expense of onshore facilities or to impose costs on their shipping and oil industries. Basic game power equaled decision game power as the major shipping states first defeated the oceanwide limits, then the limits for tankers using ports with reception facilities (a stepping stone to oceanwide limits), and then the larger regional zones. In the end, only the 50-mile zones and two reduced-in-size regional zones were adopted. There were basically no restrictions for nontankers, as they were allowed to discharge in the zones if using ports without onshore reception facilities.

Self-reporting was adopted as the basic enforcement mechanism. Ship operators were required to report on themselves in the event of an illegal discharge, recording the offense in a log book that could be read by a port officer provided the ship was not delayed. If a violation was detected, the port state was required to turn the matter over to the flag state for prosecution. Given these constraints and with the fox guarding the hen house, it is not surprising that noncompliance was pervasive.[13]

The only costs imposed by the final agreement were those resulting from the need for tankers to spend extra time outside prohibition zones to discharge oily ballast water and tank cleanings. The International Convention for the Prevention of the Sea by Oil was adopted in 1954 and entered into force in 1958, after ratification by the requisite ten states, five possessing at least 500,000 tons of ship tonnage. An initial regime was in place. A formal arena for consideration of shipping matters generally was also agreed on in 1958 with the creation of a United Nations agency: the International Maritime Organization (IMO) (originally the Intergovernmental Consultative Maritime Organization or IMCO).

The Conference on the Law of the Sea in Geneva in 1958 and 1960, worsening pollution, and a 1954 Convention recommendation led IMO to convene a global conference in 1962. Maritime oil movements had doubled since 1954, and there was now a general recognition that there had been no real improvement in pollution for any nation since 1954, and in most areas—particularly the Mediterranean—pollution had become much worse. Thirty-eight states participated, and sixteen of these had ratified the 1954 Treaty. The United Kingdom proposed oceanwide restrictions of 100 ppm for *new* vessels over 20,000 gross registered tons (grt) of capacity (mainly tankers) but did not propose the on-board equipment that would be necessary for ships to comply. Britain also explicitly rejected requirements for onshore facilities and proposed to exempt ships from discharge restrictions when going to ports without facilities (Mitchell 1994). The United Kingdom was operating as both a pusher and a laggard—pushing for international rules but opposing effective ones.

Japan, the Netherlands, Norway, and the United States (with large shipping and oil industries) opposed the proposal, but Mediterranean, Commonwealth, and Communist bloc states were supportive and the amendment passed. For *existing* ships, zones would be extended from 50 to 100 miles if states requested the change. Many nations extended zones, principally around the polluted countries of Western Europe and North America. The 100-mile zone left the central Mediterranean (and most of ocean space) without restrictions for all existing vessels, allowing discharges to continue largely unabated. While the continued fouling of the Mediterranean made life easier for tanker operators in the short run, its environmental—and therefore political—impacts in countries whose votes the maritime bloc might need was certain to eventually make life more difficult. Again, no control equipment was required.

In the area of enforcement, the United Kingdom introduced a number of important proposals. Together with France (concerned about the declining state of the Mediterranean), the United Kingdom pushed for port state inspection powers in opposition to many of its maritime colleagues (which feared such powers would be used to discriminate against the competition). Liberia and the Soviet Union in particular attacked this notion. Proposals for a reversed burden of proof, for sanctions capable of deterrence, and for the IMO to be informed of flag-state enforcement decisions met with strong opposition: states were relentless in their protection of the "thick wall" of exclusive flag-state sovereignty surrounding their ships (M'Gonigle and Zacher 1979). The debate over enforcement provisions is an excellent example of how power controlled outcomes in the development of this regime. It was not a lack of know-how (high-quality proposals were offered) but an ability of those not interested in a solution to use their power that stymied progress. In fact, as early as the 1930s, solutions to the enforcement problem were recognized and advanced.[14]

The 1962 amendments entered into force in 1967. After the 1962 Conference, the oil industry began to appreciate the economic implications of the oceanwide discharge restrictions for new ships. To ward off equipment (control-technology) requirements, the industry proposed an alternative shipboard procedure called load on top (LOT), which involved letting water settle out from oil, decanting it, and then adding oil on top (rather than washing out tanks producing oil-laden wastewater). The procedure depended on human judgment and frequently produced discharges greatly exceeding the concentration limits. But if used properly, it could be very effective at pollution prevention.

Exogenous shocks began to influence the evolution of the regime with the grounding of the *Torrey Canyon* oil tanker in 1967. The worst ecological disaster on record led to important oil-spill intervention, liability, and compensation treaties, but with regard to controlling operational pollution, industry power continued to prevail at the bargaining table. Amendments to the 1954 Convention could now be handled in committee, where the industry, with British and French support, was quite successful in ratcheting back the 1962 regulations: the oceanwide restriction for new vessels was replaced with a rule allowing discharges up to a specified number of liters of oil per mile (removing the need for on-board equipment), and proposed requirements for onshore facilities were blocked. Industry tried to have the zones scrapped, but 50-mile zones were retained in which ships could not leave a visible sheen on the water. Woven together with these changes was the adoption of the LOT system. States opposing the maritime bloc were led by the United States and managed to secure a requirement that total discharges be limited to 1/15,000 of cargo capacity. Empty tanks in ports would thereby constitute a violation (but inspection of tanks did not become legal until 1983 and would have to be conducted in oil-loading port states, which lacked any incentive to do so).

The *Torrey Canyon* spill quickened the pace of negotiations on enforcement within IMO, but the fifteen-year-old compliance system was not amended. France reacted strongly to the spill's damage to its Normandy coast, making proposals that "went right to the heart" (M'Gonigle and Zacher 1979) of the enforcement problem (rigid port-state and coastal-state inspection). These proposals and all proposals to improve the prosecution of violations were rejected. The 1969 amendments entered into force in 1978.

In summary, the emerging regime was unable to jump the hurdle of the malign problem. The first exogenous shock did not lead to immediate progress, but it did bring the United States back to a pusher role, after a half century hiatus. Negotiations and results continued to be about power and the avoidance of costs. LOT's promise would go unrealized because an essential ingredient was missing: ship owners and operators had to care about pollution or be otherwise compelled to use the system and to use it properly. "To most individual shipowners engaged in this massive and highly competitive business, the state of the marine environment—and their small contribution to it—were just not important" (M'Gonigle and Zacher 1979).

Process of Regime Implementation and Change

For the first two decades of the regime, the combination of the decision rules and the distribution of power gave effective control to a small number of maritime states. Basic game power equaled decision game power. But in the early 1970s, the balance of power and the positive correlation between basic and decision game power changed. Maritime nations with large shipping industries were no longer able always to control the vote because many more nation-state players had entered the arena and because some maritime states were coerced or had other incentives to defect from the traditional maritime bloc. An exogenous event—the entry of developing countries into the international arena—led these nations, with little or no shipping power, to cast pivotal votes to enact regulations strongly opposed by those states with basic game capabilities. As a major ocean trader with many shipping nations dependent on the use of its ports, as a major oil importer, and as the result of its overarching political and economic strength, the United States had powerful decision game strength. This was true in spite of no unique or even particular shipping capability. In fact, the lack of major shipping interests and the economic buffering provided by flag-of-convenience arrangements gave the United States the freedom to pursue environmental initiatives.

With a backdrop of increasing environmental consciousness the world over, worsening oil pollution, attention to vessel-source oil pollution at the highest levels of the U.S. government, and a very deficient international regulatory system, President Nixon directed that U.S. initiatives be developed. If the international community failed to adopt more acceptable controls, the United States would proceed unilaterally (as ordered in U.S. legislation). The United States lobbied intensively for its proposals in Europe and at IMO, prior to their formal consideration at a 1973 IMO conference on oil pollution and other polluting substances. The oil industry was an active player in negotiations, and once it read clearly the U.S. handwriting on the wall, decided to support one of the major American proposals—segregated ballast tanks (SBTs)[15]—bringing with it a number of maritime states. These countries (including Britain, the Netherlands, Belgium, and Italy) responded to the selective incentives associated with support of their industries. The industry responded to its own set of selective incentives—to avoid even more expensive requirements and U.S. unilateral action and to avert blame over the failure of LOT.[16]

Developing states were in the majority at the 1973 Conference and, although illogical for most of this group, supported the SBT proposal.[17] Egypt was influential, supporting the United States in response to selective incentives inherent in SBT;

the new technology would reduce the necessary size (and therefore cost) of reception facilities. Only a few coastal Arab and African developing states had an environmental interest, experiencing a rapid increase in oil pollution after the closure of the Suez Canal in 1967.[18] Eastern European countries also strongly supported the United States, as the result of environmental (their location on semienclosed seas), the lack of economic (only small players in the global oil transportation industry), and political (détente) interests. The U.S. proposal was opposed to the end by those countries (Norway, Denmark, Sweden, Germany, and Greece) that were home to large independent shipowning interests and by France and Japan, which were anticipating construction of ultralarge crude-oil carriers. The SBT requirement for new tankers over 70,000 dead weight tons (dwt) passed by a vote of thirty to seven.

The U.S. delegation was also firmly committed to "double-bottom" ship construction for large tankers—controversial because of its cost and risks.[19] In the absence of persuasive technical arguments, the United States lost it bid for this requirement. A third major U.S. proposal would extend the regulations to refined or "white" oils. American scientists claimed toxic, carcinogenic, and bioaccumulative properties of white oils posed serious threats to marine life and human health. European states, particularly the United Kingdom, claimed evidence of such effects did not exist. The United States had the support of its traditional environmental allies (the Eastern European bloc, Canada, and Australia) and a number of developing countries, but it needed some crossover support. In a narrow vote of twenty-three to nineteen, the proposal passed with the help of two conservative maritime states, Greece and Italy.[20] The vote on white oils is a vivid example of how ambitious regulations are made possible by the majority voting rules and how they would be impossible without them.

Participants in 1973 easily adopted the 1969 discharge standards with one important addition: ecologically sensitive "special areas" could now be created in which no discharge greater than 15 ppm would be allowed. While costly to implement, opposition was minimal because implementation would take place only after bordering states provided onshore facilities and because the rules protected against unilateral designations.[21] Difficult for the Europeans (who had been promoting LOT over SBT) to oppose and necessary for ship operators to be able to comply with discharge rules, the ship and shore technologies associated with LOT were finally adopted. States supported the shore-technology (reception-facility) requirement for all the wrong reasons: because they read the rule as legally nonbinding (creating only an obligation to urge local authorities and industry to build facilities), because they read it as legally binding on other parties such as industry, or because there was little likelihood their governments would ratify the treaty.

In 1973, filling the void of effective enforcement provisions was a tall order but a task that offered plenty of room for accomplishment. Unfortunately, the 1973 IMO Conference produced only "an adaptation of the long-ineffective traditional regime" (M'Gonigle and Zacher 1979). Despite twenty years of indifferent behavior by flag states, exclusive flag-state enforcement remained, with no transfer of authority to coastal states, which as the victims of noncompliance, had the proper incentives to investigate and prosecute violators. The outcome of the negotiations over enforcement turned on votes by developing countries: these states voted with the environmentalist states on measures supportive of coastal-states rights (and obligations if they were not too costly or could be avoided). But any threat to their larger jurisdictional or economic interests moved them to the maritime bloc. A proposal for port-state authority was flatly rejected. Offered in lieu of the more ambitious coastal-state jurisdiction, which would allow prosecution of noncompliant vessels in the coastal zone (potentially up to 200 miles), this authority would allow a port state to prosecute vessels *in its port* for violations committed anywhere. Afraid of discriminatory treatment by governments that were competitors in the field, European tanker owners, Liberia, and developing maritime states opposed the proposal, as did Soviet bloc and maritime states that did not want to concede any bargaining chips, which might be more effectively applied at the United Nations Conference on the Law of the Sea (UNCLOS III). Many developing coastal states, aiming for coastal-state jurisdiction at UNCLOS III, believed port-state authority via the oil-pollution regime might compromise this broader goal.

Japan, the United States, and the oil industry supported port-state authority in response to selective incentives to reduce the need and demand for coastal-state jurisdiction, which they opposed because of its interference with navigation (vessels could be delayed for environmental or political reasons).[22] Only a general right of inspection for port authorities to enforce discharge standards was provided: if a violation was observed, the vessel could be inspected to confirm the violation.[23] European maritime states, in response to domestic pressure, were willing to accede to this right and to modest increases in prosecution to control pollution. For the new equipment and construction requirements, maritime states would agree to inspection only by the flag states, which were obligated to issue certificates of compliance that port states were required to accept absent "clear grounds" for believing a certificate to be invalid. Coastal states were obligated to prevent a ship from sailing if it presented an unreasonable pollution threat, to report coastal-zone and high-seas violations to flag states, and to initiate proceedings or report to flag states for violations in territorial seas and inland waters. A U.S. proposal to use a visible sheen around a vessel as evidence, unless disproved, was withdrawn in the face of opposition. The Convention for the Prevention of Pollution from Ships (MARPOL) was approved—a watershed agreement given the issue's history. MARPOL did not enter into force until 1983 because of strong resistance to the equipment and reception-facility requirements.

In summary, the 1973 negotiations were all about power, including the coercive leadership of a uniquely powerful actor. When present, coercion produced the first potentially effective regulations; when absent, as in the negotiations on enforcement, no significant progress was possible. Coercion was effective only up to a point: the hegemon was unable to force its entire agenda. In addition to coercion, selective incentives and exogenous events brought necessary support from developing countries and crossover maritime states.

Two different chains of events, beginning with exogenous shocks, led to the next major development in international oil-pollution regulation. In 1973, the OPEC oil price increase led to decreased demand for oil and a major slump in the tanker market (overbuilt during a boom in the late 1960s). An environmental NGO offered an idea that could help alleviate both the laid-up tonnage problem and the continuing oil pollution—extend the SBT requirements to *existing* tankers. States with independent tanker operators (bearing the brunt of the slump) embraced the idea because of powerful selective incentives: SBT reduced cargo capacity, which would allow them to get their laid-up fleets back in the water. The proposal was considered in committee at IMO in December of 1976. Norway, Greece, and Sweden (independent tanker owners), a few Mediterranean countries (suffering significant pollution), and a few oil-exporting states (viewing SBT as an alternative to reception facilities) were opposed by maritime states whose fleets were largely owned by the oil industry, developing maritime nations, most oil importers (the slump meant lower transport costs), and surprisingly Canada (the price of oil might increase even further with the cost of SBT and the reduction in ships' cargo capacity). The United States (conducting its own study) was silent. The proposal was defeated.

"No sooner had the gavel fallen to end the MEPC [Environment Committee] session than nearby in the 'City' the bell was ringing in Lloyd's. Another shipping disaster" (M'Gonigle and Zacher 1979). The second chain of dominos began to fall. The *Argo Merchant* had gone aground off Cape Cod (on the northeastern U.S. shore). Two days later, on December 17, the *Torrey Canyon*'s sister ship, the *Sansinema*, exploded in Los Angeles harbor, killing nine people. Three more accidents occurred on December 24, 27, and 29. A total of fifteen ship casualties between December 15 and March 27, 1977, many in U.S. waters, spurred another American offensive for better environmental protection. The United States again threatened to proceed unilaterally unless international negotiations produced stronger standards. The IMO Council, "clearly shaken by yet another threat of U.S. unilateralism," agreed to convene another international conference ((M'Gonigle and Zacher 1979). Retrofitting existing ships with SBT, double-bottom ship construction, and inert-gas systems (to prevent explosions) were all insisted on by the United States, which visited a dozen countries to argue the merits of its proposals prior to the 1978 Conference. A total of fifty-eight states participated—twenty-two developed Western states, four Eastern European bloc states, and thirty-three developing countries.

In response to the oil-price rise, the industry had perfected a system of washing tanks with crude oil instead of water. This technique saved oil and significantly reduced the volume of residual oil that would end up mixed with sea water and discharged. A group of countries led by the United Kingdom, in collaboration with the oil industry, proposed the new system (called, crude oil washing [COW]) as an alternative to retrofitting vessels with SBT.[24] Most states greatly preferred COW, but they also had been influenced by the U.S. visit or at least appreciated the need for a compromise to prevent a U.S. walkout (leading to unilateral action), and—as gravely believed by IMO's secretary-general—to preserve IMO. Four days of intense, private discussions produced a compromise proposal that was adopted. The most important provisions were for new tankers: SBT "protectively located" (the alternative to double bottoms), COW, and inert-gas systems were all required for a wide universe of new vessels, down to 20,000 dwt. A fixed effective date was set, regardless of when entry into force occurred. For existing tankers, retrofitting was basically rejected. The 1978 Conference did not focus on enforcement. The final agreement was adopted as a protocol to MARPOL. MARPOL 73/78 entered into force in 1983, after being ratified by fifteen states, which together possessed at least half of the world's merchant tonnage.[25]

In summary, at the end of two chains of events that began with coincidental, exogenous shocks, coercion forced more stringent international rules (and in record time), even though most states had been strongly reluctant to consider any changes to the 1973 Convention. Selective incentives brought the support of crossover maritime and oil-exporting states. The compromise nature of the final result is testament to the continued substantial influence of the maritime community.

Regime Implementation: Evaluating Performance Over the Life Cycle of the MARPOL Regime[26]

With extensive regulations in place, many nations and IMOs attempted to turn the agency's attention toward compliance (Mitchell 1994). We expect a malign problem to require higher levels (and perhaps more complex arrangements) of cooperation, including more attention to procedures for monitoring and enforcing compliance. But nation states did not increase the level of collaboration despite their vision of full implementation: the IMO Assembly pledged in 1973 that intentional discharges of oil would be eliminated by the year 1980 (Sebek 1980). With no authority to act independently and a low level of collaboration, IMO approached implementation challenges with its hands effectively tied behind its back. In addition to these constraints, ambiguous and unworkable regulations, ambitious regulations that had sacrificed actor interests, a lack of concern, and a lack of incentives also made implementation problematic.

The institution had done little to promote concern for the problem, and as a result, implementation depended on the motivation of individual governments, which often lacked concern. The coercive leadership that produced forceful selective incentives in regime formation was absent; as a result the problem-solving capacity of the regime was low. Without high problem-solving capacity, linkages to benign issues, or selective incentives, we believe high effectiveness will not occur. This hypothesis is again supported by the evidence: for the three regime components lacking these attributes, implementation has been inadequate. For the one regime component that has been effectively implemented, the nature of the regulation itself provides strong selective incentives for compliance.

The fact of great transparency explains the highly successful implementation of the on-board control technologies for new vessels rather than some prescribed enforcement mechanism. While enforcement mechanisms fit neatly into existing inspection and certification regimens, and port states were allowed—even obligated—to detain and bar noncompliant ships, ultimate prosecuting power remained with the flag state. It is primarily the built-in selective incentives for compliance that compel ship owners to act: the ship builder, classifier, and insurer would all need to collude with the owner in an illegal act to build a noncompliant vessel. Moreover, building a vessel that could be unattractive for sale poses an enormous financial risk. A moment of high problem-solving capacity during regime formation allowed the passage of an ambitious, high-quality regulation; powerful selective incentives for its implementation overcame the fact that significant actor interests were sacrificed.

Ambitious prescriptions for governments have met a far different fate. Because of its high cost, the requirement for reception facilities is an ambitious one that sacrifices significant actor interests, and therefore we would expect that compliance may be compromised. Without coercive leadership (the United States was not threatening to ban vessels from its ports from countries that did not provide these facilities, for example) and without selective incentives for compliance, governments have failed to provide onshore facilities where they are most needed, and where they have been provided, nonregime factors (the price of oil and domestic political pressures) have driven their development (Mitchell 1994). Governments have failed to address noncompliance, explicitly rejecting sanctions and other compliance incentives (Mitchell, McConnell, Roginko, and Barrett 1999). The regime has relied on self-reporting, despite the dismal track record of this approach generally and its consistent failure to identify ports without adequate facilities, specifically.

Here we can observe how the regime's low level of collaboration directly influenced the organization's ability to promote compliance. A centralized appraisal function would have allowed IMO to construct a database of facilities, promoting industry's efficient use of available capacity and rewarding those ports where ships were serviced without delay, as required under the regulations. In the absence of IMO action, the industry itself eventually developed a database, offering anonymity to tanker operators who provided information on ports found lacking (Mitchell, McConnell, Roginko, and Barrett 1999).

The overall impact of the regime on implementation of the discharge standards and on enforcement rights and obligations has been minimal. Rising oil prices (Mitchell 1994) and domestic political pressure (Mitchell 1994), respectfully, have been found to be the major causative factors underlying what implementation has occurred. A lack of coercion or other incentives, workable enforcement mechanisms, and (for discharge standards) reception facilities helps to explain why governments fail to enforce and ship operators continue to discharge oil at sea. For the relatively few states interested in enforcement, the regime has been found to be helpful in reducing certain barriers and, by adopting equipment-based rules, in making the detection of violations easier (Mitchell 1994). In the absence of effective enforcement mechanisms, without higher levels of collaboration at IMO (such as integrated implementation and centralized appraisal), and in response to yet another exogenous shock, European coastal states developed their own compliance mechanism.[27] In contrast to the regime, this mechanism "has increased the attention and resources states dedicate to enforcement."[28]

The continued basic reliance of the regime on flag-state enforcement means that traditional obstacles to direct action by the coastal state remain, akin to other historical but outdated impediments to safety and pollution prevention.[29] It would be politically naive to suggest that actors believed flag-state enforcement would result in effective levels of behavioral change; rather, it protects and preserves an existing order, rewarding industry by preventing costly disruptions and maritime governments by preserving traditional freedom-of-the-seas privileges. The jurisdictional issue prevented the coercive hegemon from pushing for effective enforcement in the same way she pushed for effective regulation. For the United States, freedom of navigation for military and commercial vessels is paramount: "any legitimization of unilateral extensions of jurisdiction—despite its potential for dramatic impact on marine pollution control—is to be avoided (M'Gonigle and Zacher 1979).

Conclusions

What changed to allow nation states to achieve some success? The problem's character remained the same, but problem-solving capacity changed (see table 13.1). At one point in time, despite a weak organization and a low level of collaboration, the

Table 13.1a
Summary scores: Chapter 13 (OILPOL)

| Regime | Component or Phase | Effectiveness | | Level of Collaboration |
		Improvement	Functional Optimum	
OILPOL	1: 1920–1954	NA	NA	NA
OILPOL	1: 1954–1969	Very low	Low	Intermediate
OILPOL	1: 1969–1996	Very low	Low	Intermediate
OILPOL	2: 1969–1978	Significant	Low	Intermediate
OILPOL	2: 1978–1996	Significant	Low	Intermediate

Note: This table shows overall problem-solving capacity increasing despite no visible change in any of the indicators. This paradoxical result is due to rounding error in the scores. These small increases are disguised in our crude trichotomies, but they are sufficient to tip the scale.

system had increased problem-solving capacity due to the combination of its majority decision rule and coercive leadership by an actor uniquely powerful in the decision game. Exogenous events set the coercive leader into action. Exogenous events, coercive leadership, and selective incentives provided two new and essential sets of supportive actors—the developing countries and crossover maritime states. When coercive leadership was lacking, despite the prevalence of selective incentives and exogenous events, problem-solving capacity was insufficient to overcome the resistance of maritime and industrial interests and produce effective measures protective of the environment. In these instances, the malign character of the problem predominated and was predictive of outcomes.

The combination of increased problem-solving capacity, exogenous events, and selective incentives produced a change in the balance of power. Basic game power had to be decoupled from decision game power to achieve potentially effective regulations. Basic game players lacked incentives to protect the environment and had strong disincentives to do so to avoid costs in a fiercely competitive industry. While the competitive nature of the basic game worked against strong regulations, it fostered international as opposed to unilateral action. With an issue this malign, majority decision rule was absolutely essential to achieving some success. Even during the moments of increased problem-solving capacity, there was always a maritime (basic game) bloc opposing the majority and the evolution of the regime. The hypothesis that the more complex and the more malign a problem, the more intractable it will

Table 13.1b

Problem Type		Problem-Solving Capacity			
Malignancy	Uncertainty	Institutional Capacity	Power (basic)	Informal Leadership	Total
Strong	Low	—	—	—	—
Strong	Low	Intermediate	Laggard	Weak	Low
Strong	Low	Intermediate	Laggard	Weak	Low
Strong	Low	Intermediate	Laggard	Weak	Intermediate
Strong	Low	Intermediate	Laggard	Weak	Intermediate

be is supported based on the issue's malignancy. The intellectual simplicity of the problem was overshadowed by its highly malign character; the result was a regime of low effectiveness.

We also find support for our assertion that a regime dealing with truly malign problems will achieve high effectiveness only if selective incentives for cooperative behavior, linkages to more benign issues, or systems with high problem-solving capacity are present. Both selective incentives and high problem-solving capacity were necessary to produce some effective outcomes in regime formation. In regime implementation, powerful selective incentives were sufficient to achieve compliance with a well-designed regulation: the regulation itself was the offspring of high problem-solving capacity and selective incentives. We also postulated that majority, as opposed to consensus, voting rules will lead to more effective regimes, including more ambitious regulations. However, to the extent ambitious rules are accomplished at the cost of sacrificing significant actor interests, compliance will suffer. The majority rule was critical to what effectiveness was achieved, and compliance has been low for ambitious rules that sacrificed actor interests. The one regime component that has been successfully implemented is an important exception; the nature of the regulation itself provided powerful selective incentives for compliance. IMO provided an enduring arena supporting discussion and negotiation, but the benefits of the arena were insufficient to offset the fact that nation states chose not to create a regime with high problem-solving capacity. By failing to provide IMO actor status and by restricting the level of collaboration, governments all but ensured the problem would not be solved. These are not surprising choices on the part of the interested players: there was simply too much at stake, given the malign nature of the problem, to give IMO much independence, authority, or power.

The challenge of collaboration on malign problems comes sharply into focus in considering this case. The coincidence of exogenous events, the coincidence of a hegemon able and willing to push environmental interests, and the coincidence of an issue for which unilateral action was anathema (providing powerful selective incentives for collaboration) were necessary catalysts for successful outcomes. A hegemon can control results, but only to a point and only under circumstances that are likely to be largely uncontrollable. For the malign problem, selective incentives may be necessary, as they were in this case, at every step of regime evolution. The level of collaboration is an important determinant of a regime's potential, especially in the areas of implementation and enforcement. To move from precipitating exogenous events to successful outcomes, chains of dominos had to fall. One set of dominos, for example, began with major tanker accidents and worsening operational pollution, which caused enough public and interest-group concern to generate sufficient domestic political pressure to compel a uniquely powerful actor to use coercive leadership to obtain results and to move some maritime states over to the environmental bloc. Another set of dominos began with the OPEC oil-price increase, which led to a slump in the tanker market, which in turn produced selective incentives for some maritime bloc states to cross over and support the environmental coalition, and also led to the development of a technology that provided a compromise necessary to render the coercive leader's demands more palatable.

While the power and coincidence that tell this story cannot be constructed, we can improve the prospects for collaborative success if we can find a way to reduce the degree of political malignancy of problems or increase some aspect of problem-solving capacity. A majority decision rule to enhance regime formation and a higher level of collaboration to support implementation and compliance are obvious choices. Selective incentives *can* be designed and constructed, and it is in this area that creative persons, unafraid of experimentation, may make real progress in promoting the ability of international regimes to protect a blue planet.

Notes

1. I am grateful to Rear Admiral Sidney Wallace (USCG-Ret.), former chair of the Marine Environment Protection Committee of the International Maritime Organization, for his gracious assistance; his reflections and comments were most helpful. In addition, I am grateful to Ronald Mitchell, who read and commented on an earlier draft of this chapter, and to Moira McConnell, who generously let me read and reference with profit a prepublication draft of her chapter entitled International Vessel-Source Oil Pollution. See Mitchell, McConnell, Roginko, and Barrett (1999).

2. These wastes are produced from the cleaning of cargo tanks with sea water and the use of cargo tanks for ballast (stability) by filling empty tanks with water during a return voyage (and then discharging this water at sea before arriving in a loading port). In addition, ships release oil from their bilges and from the (largely past) practice of using fuel tanks for ballast. These oily-water wastes can be retained on board for separation and decanting of the water and for disposal of the oil residues on shore. But ship operators find it easier and cheaper to discharge at sea.

3. However, it is important to note that most pollutants entering the ocean follow two principal pathways: river runoff (44 percent) and tropospheric transport (33 percent). Pollution from at-sea activities like shipping and oil production accounts for approximately 13 percent of the total problem (GESAMP 1990).

4. Behavioral change by government authorities and ship-classification societies, which issue certificates documenting compliance with these rules, has also occurred.

5. These factors include the rising price of oil during the 1970s (making private-sector reclamation attractive) and domestic political forces.

6. Second International Conference on the Protection of the North Sea, *Quality Status of the North Sea: A Report by the Scientific and Technical Working Group* (London: Her Majesty's Stationary Office, 1987), p. 14, as cited in Mitchell, McConnell, Roginko, and Barrett (1999, 24).

7. The conclusion is based on a 1990 report (IMO 1990).

8. Noncompliant ships are assumed in the report to discharge from 1/1,000 to 1/2,000 of their load, compared to from 1/25,000 to 1/100,000 for compliant ships.

9. There are a number of additional reasons that our effectiveness rating is lower than some other rankings: (1) we use a more demanding measure; (2) other analysts have rated this regime relative to other international efforts; (3) other analysts appear to weight the regime components differently than we do (we consider all four major components as essential to an effective regime); (4) we expect an effective regime to overcome or mitigate hurdles key to the true success of the regime despite their apparent intractability in this and other arenas (for example, flag-state jurisdictional issues); and (5) some analysts give credit to the regime for the Paris MOU (discussed below) and its sister agreements, while we view the MOU as clear evidence of the failure of the regime to provide enforcement and compliance mechanisms—failure of such significance that countries had to design and implement their own strategies outside the regime to obtain results.

10. In the absence of a regime, it is likely that unilateral regulations would have been both more timely and more stringent but would have applied to a smaller universe of vessels. The tradeoffs between the stringency and timing of regulation, and the size of the universe, appear impossible to compute meaningfully. If one assumes the no-regime condition would have produced less effective results than the regime, then one can give credit to the regime for relative improvement.

11. The story of collaborative efforts used for the analysis and presented here is based (except where otherwise indicated) on the acclaimed work of M'Gonigle and Zacher (1979). Their exhaustive recounting of the complexity of interests and political dynamics involved in six decades of collaborative effort to resolve the problem of oil pollution by vessels provides our raw data.

12. The term *environmental* is used here to include those interested in preventing pollution, including bird-protection societies, antipollution groups, resorts and tourists, local authorities responsible for cleanup, and insurance companies paying claims for port fires.

13. For a discussion of noncompliance with the discharge standards and a variety of sources of documentation of noncompliance, see Mitchell, McConnell, Roginko, and Barrett (1999).

14. France had proposed coastal-state jurisdiction for the prosecution of treaty violations, and a proposal requiring the imposition of fines severe enough to deter violations was accepted in a draft agreement (Pritchard 1987, 57, as cited in Mitchell 1994, 198).

15. Ships would be constructed with separate (segregated) tanks for cargo and for sea water used for ballast (to stabilize the ship). As a result, ballast tanks would never become contaminated with oil and therefore never produce oily wastes from cleaning. Cargo tanks, however, would still require cleaning, and this process would continue to produce wastes.

16. Secret surveys by major oil companies had revealed that LOT was not used at all or used very poorly by two-thirds of tankers using the companies' oil-loading terminals in the Middle East.

17. Preoccupied with jurisdictional and enforcement issues (below), without the type of fleets for which these rules would apply, and assuming they could avoid costs by not ratifying the agreement, these states did not consider the indirect costs of higher-priced shipping.

18. Developing states were supported and often led by nonmaritime developed countries such as Canada (which had begun an aggressive diplomatic campaign for coastal states' rights), Australia, and New Zealand—states that had worked to persuade developing countries to attend the 1973 conference (Mitchell 1994).

19. For states with major investments in tankers, "the prospect of spending vast sums of money for a structural change which would increase the likelihood of the loss of a vessel was an abhorrent specter" (M'Gonigle and Zacher 1979, 119).

20. These two nations were alarmed by the chronic poor health of the Mediterranean, very disturbed by the U.S. scientific claims, and worried that Arab nations might begin to refine more of their oil (resulting in more refined oil being transported and discharged in the Mediterranean).

21. A number of special areas were designated in the enclosed seas surrounding Europe and the Middle East.

22. The United States took other measures to forestall coastal-state jurisdiction. In proposing the creation of an environmental committee within IMO in 1973, it hoped to promote alternatives to coastal-state authority by focusing on improving flag-state enforcement.

23. The value of this right depended on the development of monitoring devices that would produce records acceptable in court and the ability and willingness of states to undertake meaningful surveillance and inspection programs.

24. COW addressed a different problem: SBT eliminated wastes from ballast tanks; COW reduced wastes from cargo tanks.

25. MARPOL was amended in 1984 to improve equipment regulations and in 1987 and 1990 to designate additional special areas. Major amendments to MARPOL regulations on tanker design were adopted in 1992. The United States had proposed these changes following its passage of the Oil Pollution Act of 1990 (OPA), a direct response to yet another major tanker spill (the *Exxon Valdez*) in Prince William Sound, Alaska.

26. Although regime formation continued with amendments to MARPOL in the 1980s and 1990s, we now assess implementation of the regime that was in place with the adoption of MARPOL 73/78. In this way, we have allowed the regime time to achieve implementation of its important provisions before we assess its effectiveness.

27. Just one month after the 1978 conference, another tanker accident, the *Amoco Cadiz*, occurred off the coast of France. At the urging of the French government, the European Community adopted a model enforcement tool of rare quality in its potential for effectiveness. Fourteen European states agreed to inspect 25 percent of ships in their ports and report violations to a central database, which could then be consulted before a ship's arrival to better target and address problem vessels. The 1982 Paris Memorandum of Understanding on Port-State Control (Paris MOU) covered inspections for certificate or equipment violations of MARPOL and compliance with a total of six IMO Conventions.

28. Secretariat of the Memorandum of Understanding on Port State Control, *Annual Report 1990* (The Hague: Netherlands Government Printing Office, 1990), as cited in Mitchell (1994, 231).

29. The marine insurer's reluctance to recognize investments in safety and limits to shipowner liability in light of the risk-reducing widespread use of insurance and third-party coverages are two examples (King 1995; Gauci 1995).

References

Curtis, J. B. 1985. "Vessel-Source Oil Pollution and MARPOL 73/78: An International Success Story?" *Environmental Law* 15(4) (Summer): 679–710.

Gauci, G. 1995. "Limitation of Liability in Maritime Law: An Anachronism?" *Marine Policy* 19(1): 65–74.

Group of Experts on the Scientific Aspects of Marine Pollution (GESAMP). 1990. *The State of the Marine Environment*. Reports and Studies No. 39. London.

Group of Experts on the Scientific Aspects of Marine Pollution (GESAMP). 1993. "Impact of Oil and Related Chemicals and Wastes on the Marine Environment." *GESAMP Reports and Studies* 50. London: IMO.

International Maritime Organization (IMO). 1990. *Petroleum in the Marine Environment*. MEPC 30/INF.13. London: International Maritime Organization.

King, J. 1995. "An Inquiry into the Causes of Shipwrecks: Its Implications for the Prevention of Pollution." *Marine Policy* 9(6): 473–474.

Mearsheimer, J. J. 1995. "The False Promise of International Institutions." *International Security* 19: 5–49.

M'Gonigle, R. M., and M. W. Zacher. 1979. *Pollution, Politics, and International Law. Tankers at Sea*. Berkeley: University of California Press.

Mitchell, R. 1993. "Intentional Oil Pollution of the Oceans." In P. M. Haas, R. O. Keohane, and M. A. Levy, eds., *Institutions for the Earth*. Cambridge, MA: MIT Press.

Mitchell, R. B. 1994. *Intentional Oil Pollution at Sea*. Cambridge, MA: MIT Press.

Mitchell, R., M. L. McConnell, A. Roginko, and A. Barrett. 1999. "International Vessel-Source Oil Pollution." In O. R. Young, ed., *The Effectiveness of International Regimes: Causal Connections and Behavioral Mechanisms* (pp. 35–37). Cambridge, MA: MIT Press.

National Research Council (NRC). 1985. *Oil in the Sea: Inputs, Fates, and Effects*. Washington, DC: National Academy Press.

"Natural Oil Spills." 1998. *Scientific American* (November): 57–61.

Payoyo, P. B. 1994. "Implementation of International Conventions Through Port State Control." *Marine Policy* 8(5): 379–392.

Pritchard, S. Z. 1987. *Oil Pollution Control*. London: Croom Helm.

Sebek, V. 1980. "Pollution Law: The Way Forward." *Marine Policy* 4(3): 158.

United Nations Environment Program (UNEP). 1982. *Pollution and the Marine Environment in the Indian Ocean*. UNEP Regional Seas Reports and Studies 13.

14

International Trade in Endangered Species: The CITES Regime

Maaria Curlier and Steinar Andresen

Assessing Performance: A Low-Effectiveness Regime?

This chapter looks at the regime governing international trade in endangered and threatened species—based on the Convention on International Trade in Endangered Species of Wild Flora and Fauna (CITES). If we were to assess effectiveness on the basis of *output*, it would deserve a high score, as a number of far-reaching decisions to protect endangered species through restricting trade have been passed. This is not the perspective adopted in this study, however. Assessed in terms of its success in improving target-group behavior and achieving functionally good solutions, the picture is different. By these criteria, the effectiveness of the CITES regime is difficult to assess due to insufficient knowledge about the nature and magnitude of the problem. The problem is partly that the number of species covered by CITES is very high but also that trade in endangered species constitutes a very small part of the problem of species extinction. If assessed exclusively in relation to limitation of trade, effectiveness would have been somewhat higher. If considered also in relation to preventing species extinction, as we have done, effectiveness is rather low. Advice from scientists has tended to be inconclusive, and the listing of species has often been politicized. Still, over time understanding of the complexity of the problem has gradually improved, and overall effectiveness has probably increased somewhat through the adoption of a more sophisticated and flexible approach to regulation.

At this stage the question about effectiveness can be answered only in relation to a few high-visibility species, such as the elephant, the tiger, the rhinoceros, and some others. For these species we know that populations have decreased dramatically over the last few decades, despite the prohibition on trade. It is hard to determine whether depletion would have gone even faster without the regime, but based on existing knowledge we have to conclude that the impact of the regime has been modest. Over the last decade, however, the numbers of a few species have increased. The most notable example is the crocodilians. Some observers also see the emerging limited trade in elephant products as a step in the right direction, but this trade is more controversial.

In short, it is very hard to assess the overall effectiveness of the regime as knowledge is insufficient about most of the species listed. In relation to a few selected species where we have more knowledge, there are indications of a slow increase in effectiveness, although effectiveness overall is rather low.

Background, Process of Regime Creation, and the Institutional Framework

Humankind's concern for nature and its species is not a recent concern. At the turn of this century, there were already books outlining endangered and extinct species (Scott, Burton, and Fitter 1987). The modern version of these books is the International Union for the Conservation of Nature (IUCN), now the World Conservation Union (WCU) Red Data Book series, "defined as a register of threatened wildlife that includes definitions of degrees of threat" (Scott, Burton, and Fitter 1987, 1). Many nations also had legislation or regulations protecting species; however, in the 1950s, as a result of increasing wildlife trade, it was perceived that uncontrolled exploitation was leading to significant reductions of wildlife populations (Huxley n.d.). Regional treaties attempted to place some control on this trade. For example, article 9 of the Western Hemisphere Convention requires export permits before any importation or exportation is allowed for a protected species (U.N. Treaty Series 1953). Thus, what was about to become the central theme of CITES—controlling trade through system permits—was not a new one.

During the 1960s, IUCN repeatedly urged importing governments to support the conservation efforts of exporting countries by restricting the importation of rare species that were protected by exporting countries (IUCN 1960, 1964). The IUCN Legislation and Administration Committee subsequently contacted 125 countries regarding implementing legislation to control and restrict imports, and together with other organizations and government departments was successful in pushing through the Animals (Restriction of Importation) Act in United Kingdom. This act, passed into law in July 1964, had three principal objectives: to help preserve animals in danger of extinction by controlling their importation, to set an example to other countries, and to support protective legislation in the countries of origin by the removal of a market for illegally caught or smuggled animals (Scott, Burton, and Fitter 1987, 251).[1]

The U.K. act illustrated that such control and regulation is feasible and that other countries could follow this lead. One country already had: the United States had in place the Lacey Act, which prohibited the importation of species that had been illegally exported (Burhenne 1968). However, to the extent that illegal trade was a problem, it could not be solved by a few key states acting unilaterally. International cooperation was needed to control the trade in all endangered species.

The 1972 Stockholm Conference on the Human Environment was a turning point for the development of the CITES regime. The Stockholm Conference resulted in the creation of the United Nations Environment Program (UNEP), which ultimately came to oversee the CITES regime. Members of the Stockholm Conference also called for a conference to be held as soon as possible "to prepare and adopt a convention on export, import, and transit of certain species of wild animals and plants" (U.N. General Assembly 1972, 52). The result was the Washington Conference, where the IUCN draft convention was discussed by eighty-eight countries and ultimately signed by twenty-one on March 3, 1973.[2] Thirty-five more signed the Convention by December 31, 1974 (Wijnstekers 1994).[3]

The Convention eventually went through six drafts before the final text was agreed on. It is interesting to note that the term *trade* does not appear in the convention until the fifth draft (although reference was made in earlier drafts to the elements *import*, *export*, and *transit*). The word *trade* may have been included to recognize the special status of the European Economic Community and of federal countries, where jurisdiction may be available only if trade is mentioned (Mitchell n.d., n.p.). The main players in the negotiations were some Western European countries, the United States, Canada, and Australia—all northern, rich, consumer countries. With the exception of the IUCN, nongovernmental organizations (NGOs) do not seem to have played a major role in the negotiations. The Convention was nevertheless drafted to allow for easy access by such organizations.

The issue does not seem to have generated great controversies or confrontations. Trade in endangered animals was initially perceived to be a largely benign problem. There were no significant counterforces to the dominant coalition, so establishing the regime was a fairly easy task. The terms of the Convention basically reflect the concerns and preferences of these northern states. The needs, concerns, and especially the institutional capabilities of the exporting countries—mainly less well-developed southern states—seem to have been largely neglected. Thus, with the benefit of hindsight, we can say that the real nature of the problem was poorly understood. Paradoxically, this may in fact have *facilitated* the formation of the regime.

The scientific evidence provided by IUCN served as the basis for the negotiations, but this information was insufficient for the purpose. There was a general recognition that trade could affect the state of endangered species, but the exact nature of this link was not known. This uncertainty seems to have been one main reason for adopting a precautionary approach to regulation.

The Convention provides a legal structure to control international trade in species of flora and fauna through a system of permits and quotas established by the Conference of the Parties and based on listings of species in various appendices. The structure makes the producer and consumer countries responsible for the control of the trade, provides for the monitoring of trade, and provides a mechanism by which threatened and endangered species can be internationally recognized.

The two main provisions of the Convention are articles II(4) and VIII(1). Article II(4) requires that parties prohibit trade of species listed in appendix I or II, except as authorized by the Convention. This provision is further elaborated in article VIII(1), which requires parties to enforce the Convention by prohibiting trade not otherwise permitted and by providing measures to penalize violations and confiscate and return specimens. In addition, parties are required to establish national scientific and management authorities. Scientific authorities are mainly responsible for monitoring species to see whether trade in a certain species would be detrimental to its survival and to advise the management authorities that have the responsibility for issuing permits.

Species are listed under one of three appendices, depending on the level of threat of extinction. Article II of the Convention specifies that appendix I shall consist of species "threatened with extinction which are or may be affected by trade" (Wijnstekers 1994, 406). Trade in appendix I species is strictly regulated, is permitted only in exceptional circumstances, and requires both exporting and importing permits. Appendix II includes (1) "all species which although not necessarily now threatened with extinction may become so unless trade in specimens of such species is subject to strict regulation in order to avoid utilization incompatible with their survival" (ibid.) and (2) "look-alike species" that may not be threatened themselves but that resemble a threatened species so closely that distinctions between the two species are difficult. The species listed in this appendix are required to have only exporting permits. Appendix III contains "all species which any Party identifies as being subject to regulation within its jurisdiction for the purpose of preventing or restricting exploitation, and as needing the cooperation of other Parties in the control of trade" (ibid.). These species also must have export permits from the exporting country or certificates of origin from other countries before they can be traded internationally.

The main decision-making body of CITES is the Conference of the Parties (COP), held approximately every two years. Initially the executive director of UNEP asked IUCN to perform the functions of the Secretariat for an initial period of one year.

Assessing the Effectiveness of CITES: A Difficult Task

As we have pointed out above, assessing CITES in terms of the effectiveness criteria specified in chapter 1 is a real challenge. Expert opinion is uncertain and inconclusive. It is difficult to measure the effect of international trade on the state of endangered species, and it is virtually impossible to examine the effect on *all* species—more than 30,000—listed in the CITES appendices. A recent evaluation of the regime looked at a sample of twelve species and concluded that CITES been effective for only two species (the Hawksbill turtle and the Nile crocodile) and had been moderately effective for another four. For the remaining six, the impact cannot be determined or may have been marginal (OECD 1997).

Any assessment of effectiveness must build on an assumption about the causal relationship between trade and species extinction. The three principal causes of extinction are habitat loss, species introductions, and exploitation. Exploitation can be further defined as subsistence use, domestic trade, and international trade (Burgess 1994). It is difficult to measure the extent of these uses, and especially domestic trade, since there are very few records. It is, however, estimated that most exploitation occurs for domestic purposes. Thus, international trade can be blamed for only a very small percentage of the problem.

Even so, it may be a significant factor for a few key species. Every year millions of birds, orchids, cacti, and numerous other species of flora and fauna are traded (at an estimated value of $20 billion to $50 billion, with one-quarter thought to be illegal around the world (Burgess 1994; WWF 1993). While the numbers traded are large, in general the trade is composed of a few key species. Only "six species make up 75 percent of reptile leather trade and less than 25 species make up 85 percent of tropical fish trade" (Burgess 1994, 128). Moreover, trade can drive a species into *commercial extinction* before it reaches *biological extinction*. Commercial extinction occurs when trade is not controlled and a species is exploited until the population has declined to the point where further exploitation is not profitable (Reid and Miller 1993). If the population is still large enough to ensure the continuation of the species, this level may be reached well before the species is in danger of biological extinction. However, if the population is relatively small and the species has a high commercial value and is easy to find, then trade could induce biological extinction.

Considering the relatively weak relationship between trade and species extinction, not too much can be expected from CITES in terms of alleviating the problem. CITES is best known for protection (or the attempts at protection) of certain high-visibility species, such as the elephant, the tiger, the rhinoceros, and some others. In one sense, the listing of a species can in itself be considered a failure since the objective of CITES and other organizations, such as IUCN and the World Wildlife Fund (WWF), is to prevent species from becoming endangered in the first place. CITES and other institutions cannot, though, be blamed for what happened before they came into existence. The key question therefore pertains to the *effect* of the listings. On the basis of what can be observed, we have to conclude that results are not impressive. Populations of several species have declined substantially despite a CITES ban on trade. Over half of Africa's wild elephants have been lost since 1979. The populations of tigers have also declined dramatically—with an estimated range of 5,300 to 7,400 tigers worldwide (two subspecies have become extinct since the 1940s) (Jackson 1994).

On the bright side, we may note that the record regarding crocodilians is much better. In this case another approach has been chosen in that limited trade is now permitted, in line with scientific advice. This species is not threatened with extinction, although trade has been increasing. In a similar vein, limited trade of ivory from some African countries has recently been permitted by CITES because although the elephant in general is threatened, not all subspecies are in danger. Many observers see this as a step in the direction of a more sustainable and science-based approach, but others disagree. The same kind of controversy can be found in the case of whale products. Incomplete information has been a serious problem, leading to politicization of listings. Political premises seem to have been decisive in some cases where conclusive scientific evidence *was* available. In fact, it is believed that many of the species listed in the appendices, particularly in the early listings, were included mainly for political reasons (Wijnstekers 1994). More recently, however, there are indications that the parties rely more systematically on scientific advice, although the pattern is by no means uniform. Even though the will to use inputs from science may have increased, it is probably fair to say that the extension of the appendices has outpaced the growth in knowledge.

In sum, then, we have to conclude that the scattered evidence available for a few species that have attracted much public attention indicates that the performance of CITES has not been impressive at all. There is some good news, though: CITES appears to be moving slowly in the right direction.

Regime Implementation

Phase 1 (Early 1970s to Mid-1980s): Institution Building

The Conference of the Parties (COP) reviews the implementation of the Convention and adopts provisions designed to meet the objectives of the regime. Reports by various committees and working groups assist the parties in their deliberations. The listing of species in the appendices necessitated the development of criteria to determine which species were in need of protection. These criteria require a species to meet, first, certain biological criteria. If these biological criteria have been met, then the species must meet also a set of trade-related criteria. Before the criteria for the listing of species in the appendices were adopted, over 1,000 species were listed, on the suggestion from states or international organizations (Wijnstekers 1994). At its first meeting in Berne 1976, however, the COP adopted criteria for the inclusion of species in appendix I and appendix II, for the transfer of species from one appendix to the other, and for the deletion of species from the appendixes: "It must be recognized, and it was, that under the Berne criteria, the transfer of species from one appendix to the other and or the deletion of a species from these appendices, was almost impossible" (Berney 1997, 101). Gradually, the parties came to realize this, but not until the early 1990s were the criteria revised to allow for easier transfer between the listings. Thus, during this first phase of its history CITES grew through a continous adding of new species to the appendices—often without much knowledge or scientific justification.

A characteristic feature of CITES is its open-access structure enabling non-governmental organizations (NGOs) and intergovernmental organizations (IGOs) to participate in its work. CITES is unique in the sense that it relies heavily on NGOs as well as international organizations in its performance of various regime functions, from the building of a knowledge base to controlling trade and helping oversee whether the parties are in compliance or not. A brief account of some of these actors and their role in relation to CITES is given here.[4] Some of the most important actors are Trade Records Analysis of Flora and Fauna in Commerce (TRAFFIC), IUCN, World Conservation Monitoring Center (WCMC), The International Criminal Police Organization (INTERPOL), and WWF. The close relation between CITES and the International Whaling Commission (IWC) regarding trade in cetaceans will also be touched on.

Trade Records Analysis of Flora and Fauna in Commerce (TRAFFIC) was established in 1976 by the IUCN "pour assurer la surveillance du commerce de la faune et de la flore et pour appuyer la mise en œuvre des dispositions CITES" (WWF 1993). WWF provides for the majority of the funding for this network, but the IUCN, various NGOs, private organizations, and countries also contribute. It consists of a network that spans five continents, linking government agencies, academics, and local conservation organizations. TRAFFIC has brought to light trade that local authorities were not aware of.

The *Species Survival Commission (SSC) Wildlife Trade Program* is a component of the IUCN. It consists of a volunteer network of thousands of scientists, researchers, government officials, and leaders of conservation initiatives from 169 countries (Haywood 1996). As the number of species listed under CITES and the number of parties have grown, so too has the importance of information concerning which species should be listed. "With its network of scientists expert in their various fields and operating within an established, functional framework, the SSC is ideally equipped to mobilize its membership for the purpose of responding to CITES listing questions" (Brautigam 1987, 14). This program was established in 1986 to coordinate the IUCN's contribution to CITES. Its main goal is "to promote conservation and sound management of plant and animal species in international trade" (Haywood 1996, 2). It does so by providing assessments of the status of species subject to international trade, by maintaining links with the scientific community regarding sustainable management policies, and by identifying measures that will give internationally traded species the level of protection that they need to survive. It is the main focus of the IUCN's support to CITES.

The *World Conservation Monitoring Center (WCMC)* is a nonprofit nongovernmental organization based in Cambridge, United Kingdom. It provides information and advice services to U.N. organizations, treaty secretariats, NGOs, industry, business, scientists, and the media. It has provided services to the CITES Secretariat since 1980. These services include the computerization of trade statistics for listed species (the information spans from 1975 to the present). Each year, over 200,000 trade records are submitted to the organization. In addition to the listed species, WCMC also provides information on other species of conservation concern, protected areas, and habitats.

Another organization with which CITES has developed close ties is *INTERPOL*. Although most infractions of CITES are due to the lack of national enforcing legislation, a big issue is that most enforcement and customs officers are not properly trained to cope with CITES issues. At the Eighth COP, the secretariat reported that INTERPOL was financing a bulletin of major infractions (CITES Secretariat 1992). In addition, the Secretariat is also working closely with the Customs Cooperation Council to develop a Customs Training Package and a brochure on customs and wildlife issues.

Finally, the *World Wide Fund for Nature (WWF)* is another organization with which the Secretariat has developed close ties. WWF works with CITES not only through TRAFFIC but also as a source of information and assistance. The organization plays an important role in the education and awareness of the global public about the status of species around the world. The organization funds numerous projects and is an advocate for endangered and threatened species.

Based on the characteristics and functions of these actors and the relation between them, one or more of these networks probably meets Peter Haas's criteria for epistemic communities (Haas 1992, 3). Considering the high number of actors and many and intertwining networks, some streamlining might be called for.

Marine species fall under a different category in CITES, as they do not necessarily belong to any one country. To import such species requires a certificate of introduction from the sea from the importing state. The state must determine that the import will not be detrimental to the survival of the species, that it will be suitably transported and housed, and, for appendix I species, that it is not to be used for primarily commercial purposes (Wijnstekers 1994). Following the 1982 IWC moratorium, the Seychelles at the CITES COP in 1983 proposed that all species covered by the IWC management regime should be added to appendix I. The CITES Secretariat recommended that the proposal be rejected because it met neither the Berne criteria nor the provisions of the Convention. A number of countries, including the United States, pointed out that the biological criteria for such a listing were not met, but the proposal was nonetheless adopted. This issue has, however, resurfaced in the 1990s.

As illustrated by the whaling issue, the Secretariat is more activist and takes a stand on controversial issues. Whaling is not the only issue where the Secretariat has spoken out against the majority of the parties. The Secretariat seems to consider its role to be that of a caretaker of the Convention and its rules and procedures. When these rules are violated, the Secretariat speaks out. Generally, however, the Secretariat is the clearing-house for the regime, whose functions include arranging meetings of the parties, calling COPs every two years and circulating any proposals before the meeting, undertaking studies to improve the implementation of the Convention, and publishing and distributing current editions of the appendices and information to help in the identification of species (the Identification Manual).[5] As previously stated, the executive director of UNEP asked the IUCN to perform the functions of the Secretariat for an initial period of one year. This contract was subsequently renewed regularly until October 1984 (Wijnstekers 1994). In 1984 the parties requested the executive director of UNEP to take more direct control of the secretarial functions. From their first COP in Bern, the parties have acknowledged that the effectiveness of the Convention depends in large part on the effectiveness of the Secretariat. Over the years, the parties requested continued strengthening of the Secretariat from UNEP. However, in 1984 "the arrangements [with the IUCN] were no longer appropriate in the situation where the Parties were themselves providing the funds for the conduct of business and where the Convention represented a substantial proportion of the governments of the world, not all of which were members of IUCN" (CITES Secretariat 1989, 224). To alleviate difficulties with sharing of space and costs with the IUCN, the parties wanted the Secretariat moved to another location that would not be too far from the IUCN as they still wanted to retain technical and professional association with the organization. This was accomplished through a memorandum of understanding between UNEP and the IUCN "setting out the framework for the legal, scientific, and technical cooperation between the IUCN and CITES Secretariats" (Wheeler 1985, 25).

The financial conditions of the regime are also established during this phase. UNEP had agreed to provide regular funding for the Secretariat until 1983, after which time the parties would be solely responsible for its funding, as well as that of the COP meetings.

Summing up this initial phase of the history of CITES, we can say that it was characterized mainly by *institution building*. The Secretariat was created, as were the rules and procedures governing the regime, including the Berne Criteria for the listing of species. The financial mechanisms were put in place. The decision to rely on the IUCN to serve as Secretariat seems to have been based primarily on financial considerations, although politics seems to have played a role, too. The dominant parties may well have seen the IUCN as an actor with basically the same diagnosis of the problems and the same ideas about solutions. However, gradually the Secretariat came out as a stronger and more independent actor than most other secretariats serving international environmental regimes.

The nature of the problem was generally perceived to be benign in this phase. The few controversies that did occur pertained to the institutional set up of the regime (for example, regarding the role of IUCN), and they were settled rather peacefully. The export (developing) countries in general were still mainly voiceless, and the regime was basically run from the North. The level of uncertainty was still high, and scientific knowledge continued to be insufficient for informed listing, but this was not seen as an obstacle to adding new species to the various appendices. In a few instances disputed science led to political confrontations, but this was to become much more frequent in the next stage.

Phase II (Mid-1980s to the Present): More Conflict and New Perceptions

Starting from the mid-1980s, CITES is gradually changing. Conflicts are increasing as the developing countries begin to raise their voices. The monopoly of some of the key developed countries and international organizations with regard to problem diagnosis and policy development is thereby gradually eroded. Developing export states realized that their interests were not necessarily well served by severe restrictions on trade. Moreover, it became evident that the regime required sizable investments, especially in institutional capacity. Most developing nations simply could not find the financial resources required. Finally, new alliances are built as some northern countries also realize that their interests are threatened by the very restrictive approach generally adopted by CITES. In the words of the CITES Secretariat, "The problems facing CITES begin to be identified and discussed openly" (CITES Secretariat 1992).

The crux of the traditional CITES approach is well captured by the IUCN-SSC Crocodile Working Group, stating that the main assumption of CITES is "that trade is bad and incompatible with conservation." (Ross n.d., 10). The survival of many species and many communities in developing countries, however, depends on the high economic value of certain species, especially for sustainable utilization programs. Therefore, an increasing number of developing countries started to argue for limited trade, if it could be proven beyond doubt that this would not be harmful for the species in question. While key northern countries have worked within CITES to reduce or prohibit trade, their demand for these products has caused the problem of species extinction and illegal trade. That is, most of the *consumer countries* are in the North (Western Europe, North America, and Japan), while most of the *producer countries* are in the South. This indicated that the problem structure, at least in relation to some key species, was less benign than originally perceived. From the mid-1980s CITES was no longer the "consensus club" it used to be. It still took some time, however, before the main developing countries got more support for their views.

As Arturo Martinez, vice president of Argentina, pointed out at the fifth COP in Buenos Aires, poor economic conditions in the South hampered implementation and enforcement and encouraged smuggling of banned species to feed northern consumer demand (CITES Secretariat 1985). The South gradually realized that the North had imposed its will (for example, by placing species on appendix I) without adequate consideration of the effects in the South, particularly on local populations. At COP 6 in Ottawa, it was reported that trade in some species is needed to be able to protect them: "CITES can help take money out of poacher's pockets, and put it into the pockets of the Third World's rural poor" (CITES Secretariat 1987). The delegate from Saint Lucia was concerned over the watering down of proposals to favor developed countries and "expressed hope that, in future meetings of the Conference of the Parties, less developed countries would also be treated leniently and that documents would be 'watered down,' which . . . had been the case for developed countries at the present meeting" (CITES Secretariat 1987, Plenary 6.7). Thus, according to some developing countries the prohibitive approach of the CITES majority in fact contributes to increase the problem of species extinction and illegal trade as the demand is strong and there is a lot of money involved for poachers as well as smugglers. A report by IUCN concluded that "[r]egarding the African elephant, the picture emerging, is one of continuing, and possibly increasing, illegal off-take, coupled with firm evidence of the existence of ivory markets" (Dublin, Milliken, and Barnes 1995, 84)—four years after a CITES ban on trade was put in place. Moreover, overall bans penalize countries with sound management programs and especially those who could use the funds generated from legal trade for further conservation initiatives. For example, while poaching continues and elephant populations decline in certain countries, in others, such as Zimbabwe, they are flourishing to the point where they have become a problem ('t Sas-Rolfes n.d.).

Another problem reducing the effectiveness of CITES was the absence of compliance and reporting. By 1993, fewer than 15 percent of the parties had adequate legislation to implement CITES (De Klemm 1993). To assist parties into compliance, the Secretariat worked with the Environmental Law Center at the IUCN on a project providing guidelines for legislation as well as the reasons why such legislation is crucial to the effectiveness of CITES (culminating in the book, *Guidelines for Legislation to Implement CITES*, De Klemm 1993).

In addition, parties are required to turn in annual reports of trade records. In 1984, 81 percent of the parties submitted reports. In 1989, 71 percent did so (CITES Secretariat 1992). Not only is the submission of reports as such a problem; equally severe are the difficulties encountered in getting reports in on time and in the required format. In 1984, 65 percent of the parties submitted their reports on time. By 1989 and 1990, fewer than 30 percent of the parties were doing so (this subsequently increased to 50 percent in 1992 (OECD 1997). In addition, for 1988 and 1989, only six parties fully complied with the required format, illustrating that this is not only a developing-country problem. Most of the others either omitted data or did not generally follow the guidelines. Not only is this an implementation problem, but it also undermines CITES's ability to assess the conservation status of species and the value to their trade (OECD 1997). It also illustrates that the level of cooperation has continued to be rather low.

Although the networks previously mentioned contribute substantially in providing assistance as well as in monitoring, it is not enough to make the regime function effectively. On the other hand, without the support of the nongovernmental actors, CITES would probably not be able to function at all. Financial contributions by member states (which are based on the United Nations sliding scale) have been small and irregular. In addition, "[a]part from one or two projects, CITES has not benefited from multilateral financial and technical assistance, despite calls to international aid agencies in general by the COPs to support Convention-related activities" (OECD 1997, 40). The United States is currently in the midst of a lively debate about its Endangered Species Act and its contributions to international organizations (such as the United Nations) in particular. The critical questions raised by skeptics have no doubt been one important factor behind the recent cut in U.S. financial support of CITES. The United States is also the largest market for endangered species and their products, and in 1994 it imposed unilateral trade sanctions for the first time in response to another country's violation of an international environmental agreement.[6] Australia in turn bans all live native wildlife exports and requires an import permit for species on all the appendices to be used commercially. So does the European Union (OECD 1997). In sum, these unilateral measures may contribute to undermine the collective-action capacity of CITES.

CITES stood forth as a regime with high ability to adopt ambitious goals but weak will and ability to reach these goals. In addition, the legitimacy of the regime was weakened as an increasing number of developing countries realized that the regime had been run from the North—and in part contrary to their inxterests. These problems spurred a gradual change of policy within CITES toward a more flexible and maybe also sustainable approach, more in line with the interests of the southern states.

One of the most important changes related to the change of the mechanisms for delisting species. These problems were seen to be caused by the unclear meaning of *threatened with extinction* (Wijnstekers 1994). As these issues were recognized by the parties, it became one of the main topics of the Fort Lauderdale COP 9 in 1994, where the main criteria for listing species were revised. It was stated that (Wijnstekers 1994, 28)

· The appendices to the Convention now include a very large number of species, many of which may not be threatened by commercial trade;
· Certain species may not be appropriately listed in the appendices; and
· The mechanisms approved by the Conference of the Parties to delete from the appendices or to transfer between appendices inappropriately listed species were inadequate.

An additional responsibility was added—the consultation with range states and intergovernmental bodies that have a function related to a species, such as the IWC. Thus, after two decades a procedure was finally established for downlisting species from appendix I to appendix II.

This paved the way for an interesting development within CITES, first visible in relation to the crocodilians—the first case where the traditional CITES approach was modified. The success of the new sustainable-use strategy has been due to their high commercial value and the fact that these animals are not attractive to humans, in contrast to the panda, the cetaceans, or the elephants. The current legal trade of crocodile and alligator skins is estimated at 360,000 skins per year. What is more important, this *legal* trade has virtually wiped out illegal trade as a result of constant supply and improved enforcement.

Due to the high visibility of the elephant, many countries and green NGOs put up a much harder fight before limited trade in ivory products was allowed. Regarding the African elephant, many of the range states (the Southern African states) felt that they had in place adequate controls to protect the species and thus were not in favor of an appendix I listing. They were particularly concerned that there would not be an adequate mechanism to delist the species in the future. The delegate from Zimbabwe stated that "without criteria, money, and procedures, the elephant would simply be put in Appendix I and asked to protect itself" (CITES Secretariat 1989,

98). In a similar vein, "Our basic understanding of cause and effect between developments in end-use countries and the reality for elephants in Africa's range states became distorted by the lag in time and the inadequacy of reliable information" (Dublin, Milliken, and Barnes 1995, 8). Despite repeated proposals from some African states to allow regulated trade in certain populations and a favorable TRAFFIC report, these were rejected at every new COP. The elephant populations remained on appendix I until COP 10, when it was eventually downlisted.

Thus, three meetings after CITES decided to transfer the African elephant from appendix II to appendix I (CITES 1989), against the protests of most African range states, the elephant populations of Botswana, Namibia, and Zimbabwe were returned to appendix II (CITES 1997). This was, however, coupled with extremely strict conditions, and the trade in raw ivory was limited to the export to Japan of government stockpiles of ivory on an experimental basis (Berney 2000). The four countries concerned took strict measures to meet the conditions imposed on them, and the CITES Secretariat authorized the experimental trade even though a number of states as well as NGOs were opposed to any trade in elephant products. It appears that the system has functioned well, but opposition is still strong, and this battle is probably not over yet.

The same goes for the issue of export of products from cetaceans. As it became clear that the IWC Scientific Committee unanimously agreed that certain whale species could be harvested without danger of depletion, provided the carefully elaborated management procedures were followed, the question of trade of such products resurfaced in CITES in the 1990s (Andresen 2000). Both Japan and Norway proposed downlisting of various stocks of minke whale species—first at the 1994 CITES meeting, then at COP 10 in Harare—and then at COP 11 in Nairobi, Kenya, in April 2000. Support for these proposals have been increasing strongly: the majority favored a downlisting of the Northeast Atlantic minke whale in 1997, but the required two-thirds majority was not achieved. In this case, the main conflict is not between the North and the South but within the northern block, just as in the IWC. It is interesting for the future development of CITES that more developing countries support the position of the whaling nations, while the latter in turn support the position of the developing-country range states in their demand for sustainable-use policies. The general North-South divide is therefore muted by crosscutting cleavages.

IWC is not the only other international organization with significance for the performance and future of CITES. Two others are the world trade regime General Agreement to Tariffs and Trade (GATT) and the World Trade Organization (WTO) and the 1992 Convention on Biodiversity (CBD). While governments lobby for environmental agreements, they are at the same time lobbying for the removal of *trade barriers*—a policy that could undermine environmental agreements. For example, intraregional trade is technically international as it passes from one country to another, but it is still different from trade outside the region. The Convention was amended to allow for regional economic integration organizations (such as the European Union) to become Party to the Convention. The free flow of goods within the European Union is a matter of some concern to CITES, particularly since some E.U. member countries do not have adequate legislation to implement CITES regulations (Wijnstekers 1994).

In addition, conservation seems to be moving from a species-by-species approach, to one encompassing entire habitats and ecosystems. This could facilitate cooperation between CITES and the CBD as the two conventions complement each other. However, such cooperation might imply that CITES would be subordinated to the CBD since the latter convention is broader, is politically more important, and has a higher international profile (OECD 1997). So far, however, the CBD does not appear to have had much practical impact on the operation of CITES, but this may change in the future.

Summing up, we find that conflicts have increased over time and that new coalitions have been formed. The major conflict is still the North-South division, and the major problem is lack of reporting, implementation, and compliance. More than 80 percent of the parties do not have adequate legislation and enforcement systems. To a large extent the regime has relied on various epistemic communities and other supportive networks to perform its functions. Although the parties, in particular the developed states, have been keen on adopting ambitious goals, they have been less inclined to equip the regime and the developing countries with sufficient means

Table 14.1a
Summary scores: Chapter 14 (CITES)

| Regime | Component or Phase | Effectiveness | | Level of Collaboration |
		Improvement	Functional Optimum	
CITES	1900–1960	NA	NA	Low
CITES	1960–1973	Very low	Low	Intermediate
CITES	1973–1996	Low	Intermediate	Intermediate

to secure implementation and compliance. In this perspective, the reduced support from the United States may be serious for the future of the regime. An important question is whether the tendency toward unilateral legislation will be strengthened. Such a development may significantly affect the effectiveness of the CITES regime.[7]

The most important change in this phase is that it has brought to light the stakes of the southern countries and demonstrated their ability to affect outcomes. The South seems to have become more unified. Although there are discrepancies between the various range states with respect to protection policies, they at least agree that they should have more control over what is said and done in their own territories. Politics, more than science, for long tended to be the driver of the regime, at least with regard to some high-visibility species. With the new political energy mobilized by the South and with new alliances with some northern countries, a sustainable-use approach has come more to the forefront, but counterforces are strong and future conflicts are likely.

Conclusions: Determinants of Success and Failure

How can we explain the successes and failures of the regime (see table 14.1)? One of the most serious flaws has been the basic approach of responding to threats by completely banning trade in particular species. This recipe seems to have done more harm than good, as was the case for the African elephant populations. Although CITES can call public attention to particular species, this publicity and the funds it generates are—according to the IUCN-SSC Crocodile Specialist group—often directed to species with the *highest commercial value, rather than those that are most endangered* (Ross n.d.) (emphasis added): "It is a perpetual problem to find

Table 14.1b

Problem Type		Problem-Solving Capacity			
Malignancy	Uncertainty	Institutional Capacity	Power (basic)	Informal Leadership	Total
Benign	High	—	Pushers	Weak	—
Benign	High	High	Pushers	Intermediate	High
Mixed	Intermediate	—	Balance	Intermediate	—

funding for species on appendix I, while species on appendix II, especially those threatened with a trade ban, absorb huge resources" (Ross n.d., 2). In addition, most of the species on appendix I are not subject to monitoring and management requirements. The development of these policies was driven by some of the rich developed countries that were concerned with domestic agendas heavily influenced by strong green NGOs. For a long time, the South by and large accepted this perception of the problem.

One of the most serious weaknesses of the regime is that it generates significant costs that have been, and continue to be, to a large extent unforeseen: "Many listings or resolutions passed by the Parties have implications in management or financial terms which are not fully understood at the time, or if understood, are not fully revealed until after the action" (Ross n.d., 8). These costs lead to poor implementation of the Convention and subsequent resolutions. A number of international organizations and epistemic communities have worked hard to address these and other problems. Their work has been successful, given their roles and resources, but by no means sufficient to rectify these problems. The rich northern countries have been more eager to adopt far-reaching and visible policies than to provide the financial resources necessary to implement them.

The trend toward increasing effectiveness can be attributed in large part to the more flexible approach adopted in recent years. The regime is designed to allow, if the parties so wish, a flexible use of the various appendices, split listings, ranching, and quotas. For a long time the dominant coalition was not willing to use the scope of flexibility built into these instruments, but this changed when the developing countries mobilized resources, in concert with some of the developed countries. The crocodilians case illustrates that a downlisting from appendix I to II *can* work. The IUCN-SSC Crocodile Specialist group also believes that the crocodilians' success would not have been possible without the flexibility permitted under the reservation system (Ross n.d.). Unfortunately, only a few species have so far benefited from such flexibility, but the downlisting of some African elephant populations indicates that a more flexible approach can be expected in the future.

Has CITES been an effective response to the problem of species loss caused or exacerbated by international trade? The short answer is, only to a modest degree. The structure of the regime is complex, and in actual operation it has to a large extent lost sight of the original aim—which was to *control*, not necessarily *ban*, trade that threatened endangered species. If the regime continues to evolve in the direction of greater flexibility and more systematic reliance on research-based knowledge, its overall effectiveness may well continue to improve.

One final observation: In a comparative perspective, CITES is an interesting case in that it is among the few regimes studied in this book where we see some increase in effectiveness despite increasing problem malignancy. *Learning* seems to be one clue to understanding this development. As the level of conflict increased, the parties realized that new approaches were necessary to deal more effectively with the problem and also to prevent the regime from total collapse. A moderate amount of stress can help improve regime performance.

Notes

See William T. Hornaday's *Our Vanishing Wildlife* (New York Zoological Society, New York, 1913), J. E. Harting's, *British Animals Extinct Within Historic Times* (1880), W. H. Hudson's *Lost British Birds* (1884), as cited in Scott, Burton and Fitter (1987).

1. The act does not cover birds since it was thought that they were already controlled by the Protection of Birds Act 1954.

2. These twenty-one countries were Argentina, Belgium, Brazil, Costa Rica, Cyprus, Denmark, France, the Federal Republic of Germany, Guatemala, Iran, Italy, Luxembourg, Mauritius, Panama, Philippines, Socialist Republic of Vietnam, South Africa, Thailand, the United Kingdom, the United States, and Venezuela.

3. Ger van Vliet, CITES Secretariat, personal communication, May 2, 1996.

4. Greenpeace became an observer to CITES in 1981 but does not appear to have had as much influence as the network described in the text.

5. As modified from article XII, as reprinted in Wijnstekers (1994).

6. This action was against Taiwan for continued trade in rhinoceros horn and tiger parts (Jenkins 1996). This raises the question of whether this is a new trend in U.S. international relations or whether there were some other incentive for the U.S. action (that is, a non-CITES issue masked as a CITES issue).

7. Some people feel that the unilateral actions taken by certain nations, particularly in the realm of stronger domestic legislation, could be one direction of evolution for the treaty since the disparities between nations (rich and poor, strong or weak enforcement capacity, and so on) are so numerous. Thus CITES would evolve to looking not at the global status of a species but rather its status in individual nations (Martin n.d.). Others feel that unilateral action, especially with stronger legislation, harms conservation efforts in range states. Ross points to the *Caiman crocodilus yacare* as a case where stricter U.S. legislation threatens conservation programs in the range states despite expert opinions on the importance of trade as a conservation tool.

References

Andresen, Steinar. 2000. "Utilization or Preservation? The Battle over the Management of Whales." Paper presented at the Second (IWMC) Symposium, Chengdu, China, November 22–26, 1999.

Berney, Jaques. 1997. "CITES and International Trade in Whale Products." In G. Peturs-dottir, ed., *Whaling in the North Atlantic* (pp. 99–113). Reykjarik: Fisheries Research Institute, University of Iceland.

Berney, Jaques. 2000. "Outcome of the Re-opening of a Limited Trade in Ivory." Paper presented at the Second (IWMC) Symposium, Chengdu, China, November 22–26.

Brautigam, Amie. 1987. "The New Trade Group's Job." *Species (Newsletter of the SSC)* 8 (February).

Burgess, Joanne. 1994. "The Environmental Effects of Trade in Endangered Species." In *The Environmental Effects of Trade*. Paris: OECD.

Burhenne, Wolfgang. 1968. "The Draft Convention on the Import, Export, and Transit of Certain Species." *Biological Conservation* 1(1).

CITES Secretariat. 1978. *Proceedings of the Special Working Session of the Conference of the Parties, Geeneva, Switzerland 1977*. Morges, Switz.: IUCN.

CITES Secretariat. 1985. *Proceedings of the Fourth Meeting of the Conference of the Parties*. Buenos Aires.

CITES Secretariat. 1987. *Proceedings of the Sixth Meeting of the Conference of the Parties*. Ottawa.

CITES Secretariat. 1989. "Strategic Plan for the Secretariat of the CITES." In *Proceedings of the Seventh Meeting of the Conference of the Parties*. Lausanne.

CITES Secretariat. 1992. "Sixteenth Annual Report of the Secretariat (1 Jan.–31 Dec.)." In *Proceedings of the Eighth Meeting of the Conference of the Parties*. Kyoto.

CITES Secretariat. 1997. *Proceedings of the 10th Meeting of the Conference of the Parties*. Harare.

De Klemm, Cyrille. 1993. *Guidelines for Legislation to Implement CITES*. Gland, Switz.: IUCN.

Dublin, H. T., T. Milliken, and R. F. W. Barnes. 1995. *Four Years After the CITES Ban: Illegal Killing of Elephants, Ivory Trade and Stockpiles*. Cambridge: IUCN.

Fauna and Flora International. 1997. *A Short History of Fauna and Flora International*. ⟨http://www.wcmw.org.uk/ffi/history.htm⟩.

Haas, Peter M. 1992. "Introduction: Epistemic Communities and International Policy Coordination." *International Organization* 46: 1–35.

Harting, J. E. 1880. *British Animals Extinct Within Historic Times*. London: Trubner and Co.

Haywood, Many. 1996. *IUCN/SSC Wildlife Trade Programme*. Cambridge: IUCN/SSC.

Hind, Sharon A. 1989. "In the Market of Extinction: International Trade in Endangered Species." Master's thesis, Sonoma State University.

Hornaday, William R. 1913. *Our Vanishing Wildlife*. New York: New York Zoological Society.

Hudson, W. H. 1884. *Lost British Birds*.

Huxley, Chris. n.d. *CITES: The Vision*. ⟨http://www.wildnetafrica.co.za/cites/info/vfs_essay_005.html⟩.

International Union for the Conservation of Nature (IUCN). 1960. *Seventh General Assembly: Proceedings, Warsaw June 1960*. Brussels: IUCN.

International Union for the Conservation of Nature (IUCN). 1964. *Eighth General Assembly: Proceedings, Nairobi September 1963*. IUCN Publications New Series, Supplementary Papers No. 1. Morge, Switz.: IUCN.

International Union for the Conservation of Nature (IUCN). 1970. *Tenth General Assembly: Proceedings, Vigyan Bhavan, New Delhi, 24 Nov.–1 Dec. 1969*. IUCN Publications New Series, Supplementary Papers No. 27. Morge, Switz.: IUCN.

International Union for the Conservation of Nature (IUCN). 1980. "Species Thought to Be Extinct (Conference Resolution 2.21)." *Proceedings of the Second Meeting of the Conference of the Parties, San José, Costa Rica, March 1979*. (IUCN: Gland).

International Union for the Conservation of Nature (IUCN). 1986. *Proceedings of the Latin American Conference on the Conservation of Renewable Resources*. Argentina 1986, IUCN Publication New Series No. 13. Gland, Switz.: IUCN.

Jackson, Peter. 1994. "The Tiger: On the Brink of Extinction." *CITES/C&M International Magazine* 1(2) (October–December).

Jenkins, Leesteffy. 1996. "Using Trade Measures to Protect Biodiversity." In William J. Snape III, ed., *Biodiversity and the Law*. Washington, DC: Island Press.

Kemf, Elizabeth, and Peter Jackson. 1994. *Rhinos in the Wild*. WWF Status Report. Gland, Switz.: WWF.

Martin, Rowan B. n.d. *When CITES Works—and When It Doesn't*. ⟨http://www.wildnetafrica.co.za./cites/info/vfs_essay_009.html⟩.

Mence, Tony. 1981. "IUCN: How It Began, How It Is Growing Up." Gland, Switz.: IUCN.

Miles, Edward L. et al., 1994. "Science, Technology, and International Collaboration: Conditions for Effective Global and Regional Action Concerning the Problem of Global Climate Change." Project proposal 1991, revised 1994.

Mitchell, Heather. n.d. *History of CITES*.

Organization for Economic Co-operation and Development (OECD). 1997. *Experience with the Use of Trade Measures in the Convention on International Trade in Endangered Species of Wild Fauna and Flora (CITES)*. OECD/GD(97)106.

Reid, Walter, and Kenton R. Miller. 1993. *Keeping Options Alive: The Scientific Basis for Conserving Biodiversity*. New York: World Resources Institute, 1989, reprinted 1993.

Ross, James P., ed. n.d. *The Seventeen Lessons for CITES from the Crocodilian Conservation Worldwide: A Report from a Workshop held in Santa Fe, Argentina*. IUCN/SSC Crocodile Specialist Group. Available at ⟨http://www.wildnetafrica.co.za/cites/info/iss_005_croclessons.html⟩.

Scott, Sir Peter, John A. Burton, and Richard Fitter. 1987. "Red Data Books: The Historical Background." In Richard and Maisie Fitter, eds., *The Road to Extinction: Problems of Categorizing the Status of Taxa Threatened with Extinction*. Gland, Switz.: IUCN/UNEP.

United Kingdom Secretary of State for Foreign Affairs. 1951–1952 and 1955–1956. *Accounts and Papers*. Vol. 12, *State Papers General and International Treaties: Austria–France*, Vol. 30, 1951–1952; Vol. 41, 1955–1956.

't Sas-Rolfes, Michael. n.d. *Does CITES Work? Four Case Studies.* http://www.wildnetafrica.co.za/cites/info/vfs_essay_003.html.

Tolba, Mostafa K. 1992. "Counting the Cost." In *Proceedings of the Eighth Meeting of the Conference of Parties, Kyoto, Japan.*

UICN-PNUE-WWF. 1991. *Sauver la Planète: Stratégie pour l'Avenir de la Vie.* Gland, Switz.: UICN-PNUE-WWF.

United Nations Environment Program (UNEP). 1995. *Global Biodiversity Assessment.* Edited by V. H. Heywood. New York: Cambridge University.

United Nations Environment Program (UNEP). 1990. *UNEP Profile* Nairobi: Information and Public Affairs Branch, UNEP.

United Nations General Assembly. 1972. *Report of the United Nations Conference on the Human Environment held at Stockholm 5–16 June 1972, A/CONF.48/14.*

United Nations Treaty Series. 1953. No. 485.

Wheeler, Joseph. 1985. "Speech at the Fifth Meeting of the Parties." In *Proceedings of the Fifth Meeting of the Conference of the Parties, Buenos Aires, Argentina.*

Wijnstekers, Willem. 1994. *The Evolution of CITES: A Reference to the Convention on International Trade in Endangered Species of Wild Fauna and Flora* (4th ed.). Châtelaine-Geneva, Switz.: CITES Secretariat.

World Resources Institute. 1994. *World Resources 1994–95.* New York: Oxford University Press.

World Wildlife Fund (WWF) International. 1993. *Portefeuilles de Projets: TRAFFIC.* Gland, Switz.: WWF.

World Wildlife Fund. *CITES Fact Sheet.* Gland, Switz.: WWF.

World Wildlife Fund. *What We Do.* http://www.wwf.org.

15

The International Whaling Commission (IWC): More Failure Than Success?

Steinar Andresen

Assessing Performance: More Failure Than Success?

In this chapter, the fifty-year history of the International Whaling Commission (IWC) is divided into three distinct phases with varying degrees of effectiveness. It does not make much sense to aggregate these and assign an average score for the performance of IWC.[1]

During its first phase (late 1940 to the early 1960s), the IWC was a failure both in economic and ecological terms. By the end of this period, the whaling industry was virtually wiped out, and so were most of the large Antarctic whales. In the middle period of its history (the early 1960s to the late 1970s), the score is markedly higher, not least in biological terms. A number of measures were gradually introduced to secure a better management system for the remaining whales.

While the evaluation of these first two phases is a fairly straightforward exercise, the performance of the IWC in its third, more recent period is difficult to measure because of the controversy and polarization surrounding the issue. The majority of IWC nations ("whale lovers") tend to attribute very high scores because they essentially protect whales from any kind of exploitation. The minority of IWC members (the "whale killers"), on the other hand, claim that the present protectionist regime is ineffective from both biological and economic perspectives.[2] In fact, the score depends on the measurement device applied. Even when the same criterion is used, scores vary in relation to various components of the whaling regime. In relation to the collective optimum criterion, as a point of departure, effectiveness is low. The low score is particularly relevant for the last five years or so, when the scientific message has become much more robust as well as consensual; cautious management represents no threat to certain whale stocks. The majority of the IWC members, however, are not ready to accept this message and continue to oppose *commercial* whaling. The only catch that is presently regulated by the IWC is the aboriginal catch. In the 1990s, this has been conducted very much in line with scientific advice, rendering a fairly high score regarding this type of catch. Finally, the scientific catch is not under the authority of the IWC but is left to the individual nation to decide, even though the IWC frequently voices its opinion on the matter. Although the majority of IWC members oppose catches for scientific purposes, it is presently done in a sustainable manner.

What about the *behavioral impact* of the regime? Initially, the IWC had no or only marginal impact on the behavior of its members. However, this has increased steadily over time, and over the last fifteen years the IWC has been quite effective in this regard. The IWC desired that commercial whaling should cease, and so it did. This is not commonplace within international regimes. Still, it should be kept in mind that whaling was in rapid decline at the time when the moratorium on commercial whaling was adopted in 1982. Nevertheless, it is highly unlikely that whaling would have been as minimal as it is today in the absence of the IWC. To further complicate matters, although there is no doubt that the regime has had behavioral impact, the *direction* of that impact may be relevant. Is a regime effective where a majority imposes its will on a very reluctant minority and threatens economic sanctions if the majority will is not complied with? Probably most analysts would answer yes to this question if members shared consensus over the goal of the regime. If laggards were slow to phase out their CFC emissions as agreed, a regime contributing to change this would be seen as highly effective. In the whaling regime, however, there is no consensus over the true purpose of the organization. It is fairly straightforward to assign a score if we take the *official purpose* of the Convention as our point of departure. The stated goal of the International Convention for the Regulation of Whaling (ICRW), otherwise known as the Whaling Convention, is "to conserve whales in order to secure an orderly development of the whaling industry" (Preamble to the ICRW).

As the IWC is against commercial whaling even if it can be done on a sustainable basis for some species, clearly this is no "orderly development of the whaling industry." But does it make sense to use a goal that was adopted more than fifty years ago, under completely different circumstances, as a valid measurement tool when it is obvious that "orderly development of the whaling industry" is no longer the *real goal* of the IWC majority? This would clearly be a violation of the spirit of the present whaling regime, as perceived by most members.

To wind up on the achievements of the regime, on the substantive side the IWC has to some extent contributed to stopping the depletion of the largest marine mammals in the world. Although other factors are more important for this development, this is in itself no small achievement. However, if we set out to try to measure what has been achieved more systematically, we are confronted with severe methodological challenges. Not only are we unable to aggregate the whole history of the IWC into a mean score, but we are also unable to aggregate a common score for the last phase as scores vary significantly within the different indicators. Under such circumstances, the most credible and intellectual approach may be to say either that the score is undetermined due to the difficulty of making value-free judgments or that the score varies depending on the indicator used.

Type of Problem, the Process of Creation, and the Institutional Frameworks

While most international resource-management and environmental treaties are of relatively recent date, the IWC is the oldest global regime within this realm, having celebrated its fiftieth anniversary at the IWC in Oman, 1998. Compared to the lengthy history of whaling, however, the IWC is a creation of a very recent date. In Norway, there is written documentation of whaling taking place as early as the late ninth century (Stenseth et al. 1993). At some time or another most important seagoing nations have a history as whaling nations, including Spain, the Netherlands, France, Germany, the United States, Australia, New Zealand, Norway, the United Kingdom, Japan, Russia, and Canada. However, it was not until the technological progress of this century that it became a major industry. Whaling moved into Antarctic waters, and by the end of the 1930s the Antarctic seas were producing some 85 percent of the world catch. Although some eight to ten countries undertook pelagic whaling in the 1930s, the industry was completely dominated by Norway and the United Kingdom, which between them accounted for more than 95 percent of the catch (Tønnesen and Johnsen 1982). By the 1930s the whaling industry constituted a significant segment of these countries' national economies and accounted for one-sixth of the world sea catch by weight (Holt 1985, 193).

By then various actors had long worked for some kind of regulation of the whaling industry. The first attempt at regulating the whale catch was introduced in 1896 by Norway, which required vessels to register to obtain licenses. The United Kingdom and Canada soon followed suit. Just after the turn of the century, scientists started to express concern that this "most important source of marine wealth would mathematically be exhausted within a short time" (Birnie 1985, 106). The newly established International Council for the Exploration of the Sea (ICES) also continually drew attention to the need for regulation of whaling to conserve stocks.

ICES cooperated with the League of Nations toward this end. Whaling nations as well as whaling companies also gradually came to the conclusion that international cooperation and regulation was needed. The first Convention for the Regulation of Whaling was signed in 1931. The Convention built on Norwegian legislation but emerged as a less comprehensive legislation than that of Norway. The Convention did not come into force until 1935 due to reluctance on part of the United Kingdom to ratify it.

Although scientists were concerned about the state of important whale stocks, the economic dimension was all that mattered for the whaling companies and the whaling nations. Whales were valued and measured according to the amount of whale oil they yielded. The peak in this respect was reached by 1931 when 40,000 large whales yielded 600,000 tons of whale oil. The huge expansion of the whaling industry, however, produced a supply of whale oil that exceeded what the market could absorb, and in the 1931 to 1932 season, the whale-oil market collapsed, and prices fell drastically. The main players realized that agreements between the whaling companies on production restrictions were needed, and private company agreements were introduced. This had a short-term positive effect on catch as well as on prices. However, the emerging whaling nations, Germany and Japan, did not participate in any of these agreements. Efforts were made to include them in the 1937 whaling convention, but with no success. This convention is said to have been an improvement compared to the 1931 Convention, but the overall effects of these and other measures were limited, and generally "all sizes and all species were fair game" prior to World War II (McHugh 1974, 322). In this rather unsuccessful process, Norway stood forth as the dominant player. Norwegian whaling legislation was highly advanced, and Norway was also the key player among the whaling nations in pushing for company agreements as well as international agreements.[3]

Since the two countries that had demonstrated the strongest resistance to international regulation of whaling during the 1930s, Japan and Germany, were no longer active players in the whaling business, "the period following World War II marked a tremendous opportunity for whale conservation" (Scharff 1977, 351). The political discussions leading up to the present International Whaling Convention started at an international conference in 1944. The initiative was again taken by Norway. At a corresponding conference in the United States in 1945, a quota of 16,000 blue whale units was agreed on.[4] In 1946 the new Convention was adopted. The initiative and leadership role had now gradually been taken over by the United States, and the Convention built on a U.S. draft formulated by the biologist Brian Kellogg (Røssum 1984). While the traditional Antarctic whaling nations stressed the need for food and fat in the postwar era, the United States also underlined the need for conservation of whales. This *balance* of interests can also be found in the Preamble to the 1946 Convention. Fourteen nations, including all major pelagic whaling nations, signed and later ratified it, and the ICRW came into force in 1948.

The purpose of the ICRW is to provide for the *proper conservation* of whale stocks and thus enable the *orderly development* of the whaling industry. The Whaling Convention consists of two parts—the Convention, which is short and rather general, and the Schedule (articles I and V), which is an integral part of the Convention and an instrument to ensure the flexibility of the Convention. In article V(1) the provisions of the Schedule are spelled out, such as fixing protected and unprotected species, open and closed seasons and waters, size limits, and maximum catch of whales for the season. A three-fourths majority is required to amend the Schedule, while the procedural rules need only a simple majority to be amended. The amendments become effective for the contracting governments unless they lodge an objection to the proposed amendments within ninety days. Article III provides for the establishment of an International Whaling Commission, operative in 1948, "to be composed of one member from each Contracting Government." Each member casts one vote and may be accompanied by one or more experts and advisers. The main function of the IWC is to implement the ICRW. The Commission has established three permanent committees—Finance and Administration, Scientific, and Technical. In the two latter, the member nations may have any number of representatives. The Scientific Committee (SciCom) reviews catch data and research programs and makes scientific recommendations based on that review. All states may become members of the IWC, irrespective of their whaling interests. Initially, the IWC had no independent secretarial facilities, but in 1975, a Secretariat for IWC was set up in Cambridge, United Kingdom. These main institutional structures have been remarkably stable over time. Attempts have been made to change the Convention, but in essence this has not amounted to much. Nevertheless, the IWC has been turned into a very different regime due, among other things to changes of the Schedule and the Rules of Procedures.

Assessing the Effectiveness of the IWC

Phase 1 (1946 to Early 1960s): No Conservation and No Orderly Development

If performance is to be judged by the end of this period, it is bound to be low, since by the mid-1960s whaling in Antarctic waters—the area within the realm of IWC regulation—was about to become history. Catches were down to a fraction of what they had been; 4,500 BWU in 1965 compared to approximately 16,000 BWU the two preceding decades and a mere 10 to 15 percent of the catch immediately prior to World War II (Tønnesen and Johnsen 1982). Moreover, the whaling nations were no longer able to fulfil the strongly reduced quotas (McHugh 1974). Thus, the IWC was unsuccessful in managing the whale stocks both as regards "conservation" and "orderly development."

Scientific uncertainty was very high at the conception of the ICRW. The quota, set at 16,000 BWU, was little more than a guesstimate arrived at by three scientists (Tønnesen and Johnsen 1982, 157). According to one of them, "The two others were pleased that I suggested this figure instead of 15,000 or 20,000 BWU. It looked more reliable." Thus, at least until the first half of the 1950s, the regulators were following the scientific advice very closely. In fact, initially there was a perfect match between scientific advice and regulations adopted: the IWC adopted the scientific advice and set the quota at 16,000 BWU. Moreover, based on the catch reports, the level of actual catch followed the quotas very closely (Birnie 1985).[5] Thus, during the first decade of the IWC, unfortunate regulations were enacted, mostly due to inadequate knowledge. Later on, however, when more regular procedures for scientific advice were adopted, the majority of scientists on the Scientific Committee declared that the quotas were too high, but they found it difficult to quantify the necessary reductions (Schweder 1994). This concern did not result in reduced quotas. Beginning in the early 1960s specific quota reductions were given, but again the IWC failed to follow the advice of the scientists (Andresen 1989, 104). It later was suggested that even a quota of 3,000 BWU might have been too high (McHugh 1974, 309).

What about the behavioral impact of the regime? Obviously, nobody knows what would have happened to the Antarctic whales in the absence of international regulations. Still, some speculations are warranted. On the one hand, the quota set by the IWC and the actual catch were quite modest compared to catches in the period immediately prior to World War II, when some 30,000 BWU were taken—indicating that depletion would have been even faster with no international regulation. On the other hand, it was a common assumption that the resources were already seriously depleted as a result of intensive catches in the prewar period (Tønnesen and Johnsen 1982). It is therefore an open question whether the whalers would have been able to catch more than the stipulated quota had they been allowed to. In addition, the adoption of the BWU as a regulative tool, based on the whaling companies' prewar oil production agreements, as well as the fact that there were no national quotas and no limitation of number of factory ships and land stations, intensified competition and laid the ground for the "whaling Olympics" in the 1950s and the first part of the 1960s (Scarff 1977). The result was that the whaling companies built up excessive capacity with increasing pressure on the total quota and falling profits. It may have been that without regulations the whaling companies would have been even more "efficient" in depleting the large whales in the early 1950s. But at least by the beginning of the 1960s the situation would have been no worse in the absence of any regulation at all as quotas could not be fulfilled. Thus, overall, the IWC stands forth in this period as a low-effectiveness regime along both indicators.

Phase II (Mid-1960s to Late 1970s): Effectiveness on the Rise
While the first phase of the history of the IWC is a failure, by the latter half of the 1960s, its effectiveness was clearly on the increase. A number of concrete measures were taken to protect the remaining Antarctic whales as well as whales in other parts of the oceans in this period.

As important Antarctic stocks were depleted, the whaling industry gradually shifted focus from the Antarctic to the North Pacific and the North Atlantic. While approximately 75 percent of all whales taken in the 1950s came from Antarctic waters, this had declined to 62 percent in 1966–1967 (Hoel 1985, 54). However, the IWC responded to this development by including new species and areas that had previously been unregulated, and by 1976 all major whale stocks had their own quotas. In this period, the BWU as a management unit was finally abolished, and separate quotas were set for different species. Moreover, partial moratoria were introduced as commercial whaling was prohibited for the blue, gray, bowhead, and right whales. From the mid-1960s there was also a stronger inclination to follow scientific advice, and in 1967, "twenty-one years after it had been established, the IWC had agreed on a quota that was below scientific estimates of the sustainable yield of the stock" (Scarff 1977, 366). Another important decision made in the mid-1970s was the adoption of a new management procedure (NMP). Based on advice from the SciCom, the stocks were divided into three categories: initial management stocks, sustained management stocks, and protection stocks, depending on their "health." The different stocks were to be managed in such a manner that they could be classified as sustained management stocks.

It is said that the adoption of the NMP "marked the strongest and most specific commitment to conservation that the IWC had ever undertaken" (Scarff 1977, 366). It also has been argued, and rightfully so, that there were serious difficulties in applying the NMP: "Unfortunately, although the procedure looked very attractive in principle, the Scientific Committee found that full implementation was difficult" (Gambell 1993). There was no doubt, however, that the adoption of the NMP was a manifestation of a strong *inclination* on the part of the IWC to give science a much more prominent place in the decision-making system.[6] This is not to say that scientific advice was always adhered to or that the scientific message was always consensual or clear (Andresen 2000).

Overall, these and other measures not only increased the weight attributed to conservation; they also provided for a more orderly development of the remaining whaling industry. The introduction of national quotas reduced competition, and from the end of the 1960s there remained only two Antarctic whaling nations (the USSR and Japan). The problem of pirate whaling was also reduced, as many of the states conducting whaling outside the realm of the IWC eventually became IWC members.

There was considerable improvement in the achievements of the IWC in this period compared to the status quo. It is hardly conceivable that the Antarctic whaling nations or the smaller coastal whaling nations would have been able to accomplish by themselves what was accomplished by the IWC in this period both in terms of conservation and orderly development. Experience from the early and unregulated days of whaling as well as from open-access regimes for fisheries supports such an assumption. Thus, in contrast to the previous period, now the IWC had a real impact on the behavior of its members, and there was a much better balance between conservation and utilization.[7]

Phase III (Late-1970s to the Present): Increased or Decreased Effectiveness?

In 1982 a proposal for the cessation of commercial whaling (zero quotas, not a moratorium as such)—from 1985–1996 for pelagic whaling and from 1986 for coastal whaling—was approved by the required three-fourths majority. To assess the moratorium, we first have to look at the reasons it was imposed. Was this a necessary move to conserve the whale stocks? That is, what was the role of science in adopting the so-called moratorium? This question is difficult to answer due to polarization and blurring of the lines between science and policy in the Commission as well as within the SciCom at this time. On the one hand, there is no doubt that there was scientific disagreement as well as uncertainty, not the least demonstrated when the NMP was to be implemented (Gambell 1993). Nevertheless, it is quite clear that science was *not* a driving force behind the adoption of the moratorium. Emphasis should be placed on a statement by the United Nations Food

and Agriculture Organization (FAO)—presumably not a part of the controversy—
at the IWC special meeting in 1982: "the conservation of commercial whaling
can also be threatened by management measures that are too restrictive. The
most extreme example is a moratorium on all commercial whaling. . . . There seems
to be no scientific justification for a global moratorium. . . . A justification . . . can
be put forward on aesthetic or moral grounds, but these seem outside the terms
of reference of the Commission. The best, if not the only way to determine the
sustainable yield of the whale stocks is carefully monitored harvesting" (Birnie
1983, 64).

As mentioned, selective moratoria on threatened whale stocks had already been
introduced. This being said, others have reached the opposite conclusion on this
point: the so-called moratorium was indeed scientifically justified. For example, a
key person in the controversy over management of whales, Sidney Holt, maintained
that a moratorium would enable the IWC at least to get down to real science; social
relations of whales could be studied only if there was no whaling (Birnie 1985).
Although we conclude that science was not the driving force behind the moratorium, when large national research programs were launched to assess the status of
different stocks, it became clear at least in some cases, the status of the stocks had
been strongly overestimated by some whaling nations.[8] Thus, during the 1980s there
were *real* scientific uncertainties over the status of many stocks. The assessments
carried out were part of the so-called comprehensive (scientific) assessments to be
conducted within the framework of the IWC SciCom. When the moratorium was
adopted, it was made clear that it was a temporary measure pending a comprehensive stock assessment to be completed no later than 1990: "This was defined by
the scientists as an in-depth evaluation of the status and trends of all whale stocks
in light of management objectives and procedures" (Gambell 1993, 101). As a direct
result of this initiative, research started to revise and improve the 1975 management procedure, and in 1991 the SciCom (with a minority in dissent) recommended
that the Commission adopt a new procedure—the Revised Management Procedure.
After some adjustments, the Commission in 1994 adopted this procedure for the
management of whales. Through this process it became clear that certain whale
species could be harvested commercially without any danger to the status of the
stocks, provided that very conservative management procedures were followed:
"The procedure is very conservative compared with anything that had gone before,
and also by comparison with management regimes for other wildlife or fisheries
resources" (Gambell 1993, 101).

Although the Commission has adopted the new procedure in *principle*, the majority of the IWC members have not been willing to implement it in practice so that quotas could be set for limited commercial catch of the most abundant species: "The Commission accepted the scientific engine that will calculate catch quotas, but added numerous features that need to be in place before the engine is used" (Cherfas 1992, 9). What may seem as a deliberate strategy of prolongation in stalling whaling in 1993 caused the U.K. chair of the IWC SciCom to resign. In a letter to the IWC Secretary (May 26, 1993) referring to the NMP, he wrote that "the future of this unique piece of work, for which the Commission has been waiting for many years, was left in the air." Another illustration of the lack of concern for the role of science and scientific advice is the adoption of the Southern Ocean Sanctuary in 1994 (Burke 1996). The Convention states that management shall be based on scientific advice and that its goal is to conserve whales to secure the orderly development of the whaling industry; therefore, its recent performance must be judged as ineffective in relation to these criteria. If judged by the de facto goal of the majority, however, effectiveness is high.

The behavioral impact of the IWC is considerable in this period. There can be no doubt that whaling would have been conducted on a considerably larger scale had it not been for IWC decisions. However, contrary to common perceptions, catches were reduced by some 60 percent during the decade *before* the adoption of zero quotas. In 1975 some 40,000 whales were caught. Five years later the figure was down to 15,000 (Andresen 1998). This was also due to increasingly strict regulations and selective moratoria. This development continued in this period as the catch *regulated by the IWC* went down to zero.[9] The moratorium has been very effectively implemented, and the whaling industry has almost disappeared. As late as 1986, twelve whaling nations were still hunting, but in 1988 Japan became the last nation to quit commercial whaling. During the period from 1989 to 1992 there was no commercial whaling whatsoever, until it was reopened by Norway in 1993, still the only "traditional" whaling nation.[10] Japan has kept some of its whaling fleet intact, but only for catching whales for scientific purposes (some 500 animals in 1999). Apart from this, *whaling nation* today is synonymous with *aboriginal whaling nation*—the United States, Denmark (through Greenland and the Faroes), Russia, St. Vincent, and the Grenadines.[11] Thus, over the last decade the function of the IWC has not been the management of whales through stipulating quotas for commercial whaling as it used to be. Issues now being decided relate to issues like aboriginal rights to land whales, small cetaceans, humane killing, and evaluation of proposed plans for scientific catches.

In short, the IWC in this most recent period scores very differently on our indicators of effectiveness. It has been quite effective in inducing behavioral change among the previous whaling members but ineffective if judged in relation to the scientific basis (*collective optimum*). In relation to the official goal, the score is low. In relation to how the majority of member countries perceive the real goal of the IWC, the score is high. However, when explaining this development, we show that it is not necessarily the IWC as such that has been the most important driving force in this game.

The Process of Regime Implementation

Phase 1: Difficult Problem and Weak Institutions

What was the nature of the problem facing the IWC nations at its inauguration? The major challenge for the IWC was to reduce the competitive element in a way that secured an "orderly development of the industry" whaling and avoided pressure on the total quota. The Antarctic whaling nations were the dominant players in this period. They decided the direction and the pace of the IWC, and they were neither able nor willing to address these problems due to the prominence of their short-term economic interests (Tønnesen and Johnsen 1982). As for the other IWC members, although they had no material interests in Antarctic whaling, they were generally engaged in some kind of whaling and were anxious that their interests should not suffer due to a stronger IWC (Birnie 1985). However, if we look at the positions of the various players on a conservation and harvesting dimension, the non-Antarctic nations were more conservation-oriented overall, but this did not have any impact on decisions reached (Røssum 1984).

The most important regulatory issues in this period—the question of national quotas and the limitations of factory ships and land stations—were dealt with outside the realm of the IWC. This was not necessarily just because the main Antarctic nations wanted a more exclusive approach but that the ICRW explicitly *prohibited* the IWC from involvement in this issue (article V(2)(c)). The IWC could only set a total quota, and already by 1947 and 1948 the catch capacity of the whaling industry was sufficient to take the total quota of 16,000 BWU (Tønnesen and Johnsen 1982). Thus, instead of contributing to reduce a malign problem these institutional features contributed to make the problem even more difficult. This problem was aggravated by the significant differences among the whaling nations as to the size and structure of their whaling industries. Japan and the Soviet Union were upcoming whaling nations with new equipment and were steadily expanding their capacity. Some were small, like South Africa and the Netherlands, but eager to stay in operation, while others, such as Great Britain and Norway, were still big but with old ships and equipment. Thus the whaling nations had different interests and consequently variously affected by the regulations adopted. Obviously, Japan and the USSR as the stronger parties saw their interests best served by a total quota. The other Antarctic nations gradually came to favor national quotas, but if this could not be achieved, they wanted the total quotas to be as large as possible. In fact, the most reluctant player or chief laggard in this period was the Netherlands (Schweder 1994).[12] When the whaling nations finally reached an agreement on national quotas, this did not result in a reduced total quota but rather the contrary. Moreover, another reason for agreement being finally reached after many years of difficult negotiations in the early 1960s was that some of the Antarctic whaling nations (the United Kingdom, South Africa, and the Netherlands) no longer found it profitable to continue their operations. Japan and the USSR bought the quotas from the others that left the Antarctic scene. Thus, the national quota agreements were a typical example of too little, too late (Tønnesen and Johnsen 1982).

These conflicting national interests constitute the main explanation for the poor performance of the IWC in this period. It has also been noted that institutional deficiencies aggravated the problem. What about other institutional provisions? The de facto veto right of the IWC members through the objection procedure made the "law of the least ambitious program" (Underdal 1980) work unfailingly in the greater part of this period—that is, no decision went beyond the interests of the least enthusiastic (Antarctic whaling) member. The Netherlands and Norway also left the IWC for a short time around 1960 due to differences with other whaling nations (Birnie 1985). As the independent institutional basis in the larger part of this initial phase in the history of the IWC was so weak, the IWC as such had no means of checking the power and interests of the whaling nations. The IWC did not have any independent secretariat in this period; it operated from the U.K. Ministry of Agriculture and Fisheries, using part-time staff only. The important game was played outside the IWC over the question of quotas. Also, meetings of the Scientific Committee were held only infrequently, and a number of member countries did not participate (Allen 1980). In the Scientific Sub-Committee, where most of the basic work was being done, the mean attendance per meeting (1953 to 1959) was only seven scientists (Schweder 1994). In sum, the IWC was a very weak regime, entirely in the hands of the main whaling nations without the will or ability to halt continued depletion.

Phase 2: Stronger Institutions Deal with an Easier Problem

In the early 1960s the first signs of a somewhat stronger role for the IWC can be observed in the process of setting quotas. First, only the whaling *industry* was involved in the negotiations outside the IWC, then the *states* were brought in, and finally the IWC was given an independent role of its own and no longer simply reflected the interests of the major whaling nations. The whaling companies as well as the whaling states seemed to realize that the IWC might be a necessary instrument for reducing conflicts among them (Tønnesen and Johnsen 1982).

Gradually, the commission also realized that the IWC was in need of improved scientific advice (Andresen 2001). In 1961, a committee of three independent scientists (later extended to four) was drawn from countries not engaged in pelagic whaling in Antarctic waters. The group was appointed by the chair of the Commission and vice chair of the Scientific Committee. This small group of scientists, working from 1961 to 1964, was able to do what the SciCom had failed to do hitherto—provide specific and detailed advice. Why were they able to do what the SciCom had apparently not been able to do? The infusion of politics into the SciCom, which prevented this body from reaching consensus (Andresen 1993), did not exist in the new small group of very able scientists.[13] When this ad hoc group ended its work in 1964, another outside organ was brought in: the FAO carried out stock assessments on behalf of the Commission until 1969 (Andresen 1989, 104). This improvement in procedures and substance in knowledge production no doubt strengthened the authority and legitimacy of the IWC. The scientists also gradually exerted stronger and more independent pressure on the Antarctic whaling nations. Due to frustration over the unwillingness of these countries to accept reduced quotas, the chair of the SciCom threatened to resign. Similarly, the FAO representative of the Committee of Three made it clear that the FAO would not continue to cooperate with the IWC if scientific findings were used merely for a more efficient depletion of whale stocks (Røssum 1984). The more stringent advice was probably also one reason for the reduction of quotas gradually taking place in the 1960s, although at least of equal importance was surely the fact that the whaling nations were no longer able to fulfill their quotas (McHugh 1974).

Generally, the institutional problem-solving capacity of the IWC continued to increase in the 1970s. In light of the strongly increased activity, especially on the scientific side, the question of strengthening the administrative capacity was raised in 1973, and two years later the IWC aquired its own Secretariat with full-time staff. According to Birnie (1985, 465), "backed by the services of the improved Secretariat, both the meetings and the reports on them became longer and more detailed as the IWC expanded its role and tasks." A major problem in implementing the new and more ambitious New Management Procedure (NMP) was that all necessary data had to be provided by the whaling nations, which were often "weak and confused" (Birnie 1985, 472). In the latter half of the 1970s, however, procedures were gradually changed to allow for a stronger independent element among the scientists (Andresen 1993). In short, the IWC as an organization seemed to have grown up, and problem-solving capacity had increased considerably.

Does this mean that a stronger and more independent IWC brought about a change in the behavior of the whaling nations? To answer this question, we need to look at the development in the nature of the whaling problem. Catches of whales in Antarctic waters continued to decrease throughout the 1970s, and the competitive element was strongly reduced as only the Soviet Union and Japan remained on the Antarctic scene where their national quotas were taken. Economic stakes were no longer so high for these two nations, and the other whaling nations were mostly involved in limited whaling, often along their coasts. By 1980, whaling constituted less than 0.2 percent of the global marine harvest in terms of economic value (Allen 1980, 17). Thus, the system of activities was far more benign than it had been in the first period of the history of the IWC. We would maintain that the *main* reason behind this development was *resource depletion* brought about by the whaling members themselves. Large-scale whaling was simply not economically interesting for most previous Antarctic whaling nations because too few whales were left. This may seem to lead to the rather depressing conclusion that "it only becomes easy to get regulations once we have depleted a resource beyond economic viability."[14] To some extent this is true, a depressing but maybe not really surprising lesson. Pressure on resources also was reduced when other products were substituted for whale oil and whale meat.

Yet even if the problem is easier to handle from an economic point of view, this does not necessarily imply that *politically* the problems are reduced correspondingly. As whaling in other ocean areas was brought under IWC regulations, the number of whaling nations affected by IWC regulations increased (Hoel 1985). An illustration of this development is that after Norway quit its Antarctic operations in 1967, it was not considered a whaling nation by IWC standards until the Norwegian catch of minke (or piked) whales in the Northeast Atlantic was included in IWC regulations in 1976. On the other hand, countries that quit commercial whaling in this period, such as the United States and New Zealand, came out strongly against commercial whaling. A stronger antiwhaling block pushed for stricter regulations, and Japan and the Soviet Union tended to become more isolated. Thus, the weakening of the whaling side and the strengthening of the nonwhaling block, under the leadership of the United States explain why regulations gradually became more stringent. Previously, voting had not been used very actively "as most members had some relation to whaling" (Hoel 1985, 112). As this was gradually changing, voting was more frequently used, illustrating that the IWC was no longer the intimate whaling club it had previously been.

Still, the most important political development regarding whaling in the 1970s did not take place within the IWC: the whale was about to be adopted as possibly the most important symbol for the environmental movement (Andresen 1989). During the U.N. Conference on the Human Environment in Stockholm in 1972, a ten-year moratorium on all commercial whaling was called for. Thus, by the early 1970s whaling was no longer the concern of only the limited group of IWC members. The gradually changing perception of the whale strengthened the position of the nonwhaling nations. However, it was not until the next period that this new antiwhaling landslide found its way into the IWC with full force.

Thus, on the one hand, the basic problem had become much easier because competition had almost been eliminated and the remaining economic stakes had become much smaller. On the other hand, the political game had become more conflict-oriented due to the gradual strengthening of the nonwhaling faction. The Antarctic whaling nations in particular fought hard to avoid stricter regulations, but they were on the losing side, not the least due to the role of the United States and increased external pressure. However, they might have put on an even tougher fight if the material stakes had been higher. The stronger and more independent role of the IWC as an institution, not the least on the scientific side, was also instrumental to the interests of the nonwhaling nations. Thus, it seemed the issue of whaling was about to be solved in the sense that whaling was being carried out on a comparatively modest level along the lines of more legitimate scientific advice. However, the strong emotional element of conservation that had started to grow concerning whaling was soon to enter the IWC with full force, in essence creating a completely new setting for the management of whales.

Phase 3: New Perceptions and Power Coalition

In the most recent history of the IWC, the dynamics of the battle over whaling have changed completely. The problem facing the IWC is gradually becoming more benign as the catch is continuing to decline and the number of whaling nations is reduced. In material terms, the whaling issue is getting marginalized; it means less and less for fewer actors. In practice there is no competition and no externalities.[15]

On the other hand, the IWC has certainly been facing a most difficult political problem, as a shift of *values* in the perception of whales has occurred.[16] This shift of values does not apply only to the nonwhaling nations; for the previous whaling nations, the question of whaling became a question both of self-determination and of sustainability. The challenge for the IWC is no longer one of regulating a marine living resource according to the best scientific criteria but of tackling a complex political and environmental problem in an atmosphere where "saving the whale is for millions of people a crucial test of their political ability to halt environmental destruction" (Holt 1985, 12). The large coalition of NGOs and nonwhaling nations—with backing from the public, especially in countries such as the United States, the United Kingdom, and Germany—has succeeded in changing the agenda and the very thinking about whales and whaling. It has been turned into a moral question, where saving the whale has become associated with the morally good approach, whereas catching them is morally inferior or bad. A strong antiwhaling position has become a rather convenient way of acquiring a green image, as no (material) costs were involved for the nonwhaling nations. This may be particularly pronounced for some of the major industrialized countries like the United States, the United Kingdom, and France, all vehemently opposed to commercial whaling but otherwise not known for green environmental policies. The fact that public opinion in these countries also comes out strongly against whaling adds momentum to their strong stance in the IWC (Freeman and Kellert 1992).

In this new atmosphere, the issue-specific power of the whaling nations (the fact that they were catching whale)—which was usually a significant power basis in international negotiations—lost most of its relevance. Traditional whaling countries such as Iceland, Norway, and Japan argued that management was both justifiable and reasonable, provided it took place within the limit of sustainable yield of the relevant species, but they were not able to sell this idea in the polarized atmosphere in the 1980s.

Nor did it help them much that they had the most knowledge and had done the most research on the issue of whaling, as science lost all or much of its relevance as a decision premise in the 1980s (Andresen 2000). The traditional commercial whaling nations opposed to the aboriginal whaling nations, like the United States, were not only strongly outnumbered; they were also outvoted. In contrast to most international regimes, this was not only an option in principle but became a very common practice in this period (Andresen 2001).

On the global scene, not the least within various U.N. forums, it is commonplace that ambitious commitments of various kinds are adopted by the majority but disregarded by the minority. As the IWC is no more supranational than the United Nations, there is no more reason that the "harassed" minority should comply with the wishes of the majority. Why did the whaling countries abide by the moratorium at all? They could simply lodge an objection to the so-called moratorium and continue to whale, or they could leave the IWC and seek to build an alternative to the IWC. Both options had previously been used on several occasions.

Although the roles of public opinion, the media, the force of the moral argument, as well as the new power base in the IWC should not be underestimated, it is well known that another major reason behind this development is the policeman role of the United States within the IWC (Aron 1988). The United States has put legal, political, and economic instruments to use to stop whaling. Of particular importance is the Pelly Amendment, requiring that the U.S. Secretary of Commerce certify any country found to be diminishing the effectiveness of an international conservation agreement to which the United States is a party.[17] In the 1980s there were also legal provisions to reduce by up to 50 percent fisheries allocations in the U.S. economic zone for countries that continued to whale commercially.[18]

Was it the threats and/or implementation of these U.S. measures that brought an end to commercial whaling? The answer is yes, to a large extent, but there are additional important elements to consider. First, there is a need to distinguish between the whaling states. At the time of the adoption of the moratorium, as mentioned, there were still twelve whaling nations.[19] They were not a homogenous category. First, there were the large Antarctic whaling nations—the USSR and Japan—catching the larger share of the 11,000 animals still being caught just after the adoption of the moratorium. Let us look briefly at Japan. An agreement between the United States and Japan in 1984 is illustrative for more traditional bilateral power politics and bargaining between the United States and Japan. The United States agreed not to apply the legal provisions to Japan during a three-year period. In turn, Japan undertook to withdraw its objection to the 1981 IWC ban on sperm whaling and its objection to the 1982 moratorium. In the meantime, the United States "allowed" Japan a catch of 400 sperm whales in the two following seasons (Birnie 1986). In this game neither the IWC nor scientific recommendations were left much room. In the case of Japan, however, it is also important to know that key parts of both the business community as well as the central administration were against commercial whaling; they were afraid it might hurt both long-term economic interests as well as political relations with the United States.[20] For the Soviet Union, and later Russia, this is a completely different matter. The USSR/Russia lodged an objection to the moratorium that is still valid. Nevertheless, the USSR declared that it would halt commercial whaling after the 1987 season. This was, however, not a result of U.S. pressure but because its decrepit whaling fleet and the country's economic collapse made whaling unprofitable.

Then there were the small-scale whaling states in Latin America (including Brazil and Chile) and in Asia (including the Philippines and Korea), but these were catching only a few hundred animals a year. Thus, the decision to quit whaling mainly reflected simple cost-benefit analysis. The perceived political and possibly economic costs of continued commercial whaling were much higher than the economic return from continued whaling operations.[21]

Norway and Iceland fall in some kind of a middle category; they both had whaling traditions dating back hundreds of years, and although whaling was no longer a significant industry in either of these countries, whaling was considered both perfectly legitimate as well as economically important in certain local areas (Andresen 1998). Both countries put up a fight against the IWC majority, the United States, and some of the major NGOs. For different reasons, Iceland did not lodge an objection to the 1982 moratorium and was thus put in a difficult position. A large research program was launched, but Iceland decided within the framework of the "comprehensive assessment" to assess the status of the whale stocks relevant to its economy. This strategy was successful in the sense that the IWC SciCom, based on the research conducted, stated in 1990 that the previously unclassified minke whales of the central stock could be classified as an initial management stock, meaning that it could be harvested. However, the IWC refused to let Iceland start whaling before the whole management system was in place. This caused Iceland to exit the IWC in 1992. No whaling has been conducted by Iceland since 1989. Iceland has been active in building up a possible alternative to the IWC, the North Atlantic Marine Mammal Commission (NAMMCO), but has so far not been very successful (Andresen 1997). Norway quit commercial whaling after the 1987 season. Thereafter, it launched a scientific program to assess the status of the Northeast Atlantic minke whale stock. At the 1992 IWC meeting the SciCom endorsed Norway's research and concluded that this whale stock comprised almost 90,000 animals (Cherfas 1992). Consequently, Norway, which had objected to the so-called moratorium, decided to resume commercial whaling from the 1993 season.

Throughout these processes, by far the most important opponent for both Iceland and Norway was the United States, not the IWC.[22] While the IWC was passing critical resolutions, the United States had more powerful means at its disposal. Norway was certified four times, even when catch was down to five whales for scientific purposes. Some of the major NGOs, most notably Greenpeace, in a sense were more important than the IWC, as they launched consumer boycotts in some countries, most notably the United States and Germany. Although the effects of such boycotts are disputed and uncertain, they were certainly taken very seriously by the two countries (Andresen 1998).

Thus, a rather complex pattern emerges when trying to explain why the whaling nations quit whaling. Although the United States played a key role in this process, the significance of the IWC should not be underestimated. This was the arena that made possible the adoption of the so-called moratorium, thereby providing the necessary basis for actions by the United States as well as major NGOs against the whaling nations. The creation of an essentially new IWC cannot be properly understood without considering its institutional aspects. This was made possible partly by taking active use of existing rules and procedures and—when necessary—amending them. As an open organization, any nation could join the IWC irrespective of its interest in the whaling issue. However, from the establishment of the IWC and for approximately the next thirty years, basically only states with some kind of material interest in the issue became members. Membership was stable, fluctuating at around fifteen, but from the end of the 1970s to the early 1980s, membership soared to around forty, consisting mostly of countries with no whaling traditions, many of them coming from the developing world. Thus a three-fourths majority to amend the Schedule and adopt the so-called moratorium was made possible.[23]

Why did these new countries suddenly avail themselves of their right to join the IWC? First, developed nations without whaling traditions (like Switzerland, Finland, and Sweden) joined probably simply to improve their green image at no cost (apart from the membership fee). Some developing countries joined, it has been argued, in connection with the Law of the Sea negotiations. Like deep seabed minerals, whales should belong to all nations and not only to those with relevant capabilities (Hoel 1985). Moreover, there is little doubt that the new members' sudden concern for the whaling issue came as a result of active lobbying by environmental organizations (Gulland 1988, 45). It appears that Greenpeace was particularly important. According to some observers, Greenpeace had a deliberate strategy to pack the IWC with new members: "the whale-savers targeted poor nations plus some newly independent countries . . . and between 1978 and 1982, the operation added at least half a dozen new member countries" (Spencer et al. 1991, 177). A former legislative director of Greenpeace ocean ecology claims that "Using the media and the force of public opinion, environmental and animal welfare groups pushed for an end to commercial whaling. With startling speed they carried out what amounted to a *coup d'état* in the IWC" (Wilkinson 1989, 278).[24] This was possible because IWC was an open organization without entry restrictions. If the IWC had employed the more restrictive membership criteria that characterized, for example, the Antarctic Treaty System, this would not have been an option.

Packing IWC with new members was possible by using the existing open institutional structure of the IWC. However, the antiwhaling nations, with the United States as the leader, also worked actively to open up the organization for subnational actors. The increase in the number of observer international organizations has been even more dramatic; until the early 1970s usually only a handful showed up. Thereafter, the number rose steadily and now approaches almost a hundred. Thus, the yearly IWC has become one of the favorite meeting places for green NGOs. The development has been similar for scientists: from a small and closed group of scientists from the whaling nations, more than a hundred scientists now frequently participate. Although scientists from the member states still dominate, an increasing number of (independent) scientists with various affiliations have participated in the SciCom meetings. From a small and closed club, the IWC has thus been turned more into an open party, and in the early 1990s the participants in various categories just about outnumbered the number of whales that were taken. In general, this more open-access structure also implies that the level of cooperation has increased over time as more and new actors de facto become involved in assessing the effectiveness of the regime.

This development illustrates that leading antiwhaling states, most notably the United States as well as major NGOs, have been clever entrepreneurs—given their goals and perceptions of the issue. Through the use of existing rules as well as by changing the rules, they have made it possible to change the IWC completely, thereby paving the way for the moratorium as well as the virtual cessation of

Table 15.1a
Summary scores: Chapter 15 (IWC)

| Regime | Component or Phase | Effectiveness | | Level of Collaboration |
		Improvement	Functional Optimum	
IWC	1944–1946	NA	NA	Low
IWC	1948–1965	Low	Low	Intermediate
IWC	1965–1979	Low	High	Intermediate
IWC	1979–1995	Significant	Overshoot	Intermediate

commercial whaling—a rare combination of values (NGOs), skills (some scientists), and power (the United States). However, as it is increasingly clear that the politics of the IWC are contrary to a consensual scientific message and that it is not accepted as legitimate by a vocal minority, the situation is unstable (Friedheim 2001).

Conclusions: Determinants of Failure and Success

In the introductory chapter we concluded that effectiveness was by all standards low in the initial phase, increasing beyond doubt in the middle phase, and varied over different indicators in the latter phase. What are the main determinants behind this development? Although weak institutions and a weak knowledge base contributed to the initial failure, by far the most important reason was the malign problem structure (see table 15.1). Excess capacity, fierce competition, and strongly different interests between whaling companies and whaling nations were key features. Although stronger institutions and a significantly improved knowledge basis contributed to the better functioning IWC in the late 1960s and during the 1970s, it is a fact that factors unrelated to IWC institutions were far more important. Reduced interest in whaling due to resource depletion and reduced demand for whale products paved the way for more orderly management as well as better conservation procedures. In short, the problem was no longer malign. This may lead to a rather depressing conclusion based on this case: institutions are effective only when dealing with fairly easy problems. Moving on to the more recent stage, it is uncertain whether this observation holds water. There is no doubt that the problem again became much more difficult due to the infusion of conflicting values

Table 15.1b

Problem Type		Problem-Solving Capacity			
Malignancy	Uncertainty	Institutional Capacity	Power (basic)	Informal Leadership	Total
Strong	High	High	Pushers	Weak	High
Strong	Intermediate	Intermediate	Laggard	Weak	Low
Moderate	Intermediate	Intermediate	Balance	Weak	Intermediate
Mixed	Low	Intermediate	Balance	Weak	Intermediate

and principles. The problem-solving capacity of the IWC, however, is quite strong, not the least due to the power position and dominance of the United States. As much of the U.S. policy is based on bilateral diplomacy and threats, it is difficult to decide how much of the change in IWC policy is based on the regime and how much on U.S. actions outside the regime. As the United States has broad support from the IWC majority and this body serves to legitimize U.S. policy, the link between the two is crucial.

Looking finally to the future, the IWC today seems to be a fragile creature with considerable hostility between its two camps. A proposal from Ireland in 1997 was an attempt to build bridges to secure the future of the IWC. The Irish Proposal would allow limited coastal commercial whaling, but not in the open seas, and would not allow any export of whale products. After some initial movement, there is again a stalemate between the parties on the proposal. Two perspectives may be applied to the future of the IWC: one is that it will collapse as a result of the hostility between the parties; another is that as whaling is such a marginal activity, in practice all parties can live with the IWC as is. So far, the latter interpretation has proved more valid.

Notes

1. Most regimes undergo incremental growth and do not go through distinctive phases. However, the IWC changes are so fundamental regarding the dependent and independent variables that such a tripartite division is necessary. However, the exact phases and years may be debatable. The phases presented here are based on Andresen (1993), Birnie (1985), and Mitchell (1998). The latter has specified the phases somewhat differently.

2. A questionnaire sent to key commissioners of the IWC in the early 1990s confirmed this (Wettestad and Andresen 1991).

3. A main reason for the Norwegian eagerness to conclude agreements was that Norway was on the defense as a whaling nation. The main challenge was the United Kingdom but also the other upcoming whaling nations.

4. One BWU equals one blue whale, two fins, two and a half humpbacks or six sei whales, related to their yield of whale oil.

5. There was suspicion among many whalers that cheating was considerable. In the early 1990s it was confirmed that the catches of the USSR had by far exceeded what was reported (Stoett 1995).

6. The first and more positive evaluation of the NMP given in the quote in the text (Scarff 1977) was made at about the time this new procedure was adopted in 1977. The second evaluation (Gambell 1993) is given some fifteen years later, when it is possible to evaluate how it worked in practice. We judge performance primarily by the standards and knowledge existing at the time regulations were adopted.

7. When we try to *explain* this development in the next section, however, we see that other factors may be more important than the IWC in accounting for the orderly development of the whaling industry.

8. An independent assessment of available research on the Northeast Atlantic minke whales, harvested by Norway, concluded that the scientific basis for the Norwegian whaling had been much weaker than both Norwegian scientists and the authorities had claimed (Andreson et al. 1997).

9. Catch of small cetaceans outside the competence of the IWC is still considerable (Freeman 2001).

10. Some 700 minke whales are hunted yearly (1999). The quota has increased somewhat over time. The quota is set unilaterally as the IWC does not set quotas, but it is well within the estimates for sustainable harvest in relation to the revised management procedure. The stock estimate (Northeast Atlantic minke whales) is some 130,000 animals.

11. St. Vincent and the Grenadines usually catch only one or two humpback whales.

12. According to Schweder (1994), the Dutch scientists in the SciCom represented the most serious obstacle to giving consensual advice in the 1950s.

13. Schweder, a prominent member of the present Scientific Committee, has maintained that these scientists were not necessary more able but that the different institutional (more independent) basis made the difference.

14. I owe this point to Beth DeSombre, a 1997 ISA panel commentator on an earlier draft of this chapter.

15. In principle, however, as whales are common resources, the problem of externalities remains.

16. Such shifts from technical or economic approaches to moral approaches may be rare, but they are not unique. Some of the other cases discussed in this book exhibit similar characteristics. In the concluding chapter we explore what characterizes these issues.

17. Once a country is certified, the president must decide whether to impose sanctions or not (see DeSombre 1995).

18. This provision no longer has practical significance as for quite some time now U.S. fishermen have taken the whole quota.

19. Brazil, Chile, Peru, Korea, Spain, the Philippines, Japan, Norway, Iceland, and the Soviet Union were all conducting commercial whaling. The United States and Denmark (through Greenland and the Faroes) were the only aboriginal whaling nations at the time.

20. For a thorough overview of Japan's positions and negotiation within the IWC, see Friedheim (1996).

21. The United States also leaned heavily on these countries to quit whaling.

22. Both countries were involved in intense bilateral negotiations with the United States. As for Iceland, these also took place long after Iceland quit commercial whaling (see Ivarson 1994).

23. After 1983 membership stabilized at around forty. For a detailed overview of the development regarding membership and attendance by scientists as well as NGOs, see Andresen (1998).

24. As many as two-thirds of the newcomers either left the IWC, failed to show up, or did not pay membership fees (Andresen 1998).

References

Allen, K. T. 1980. *The Conservation and Management of Whales*, (London Butterworth).

Anderson, R., R. Beverton, A. Semb-Johannson, and L. Walløe. 1987. *The State of the Northeast Atlantic Minke Whale Stock* (Norway, Økoforsk).

Andresen, S. 1989. "Science and Politics in the International Management of Whales." *Marine Policy* (April): 99–117.

Andresen, S. 1993. "The Effectiveness of the International Whaling Commission." *Arctic* 46(2): 108–115.

Andresen, S. 1997. "NAMMCO, IWC and the Nordic Countries." In G. Peturdsdottir, ed., *Whaling in the North Atlantic* (pp. 75–89). Reykjavik: Fisheries Research Institute, University of Iceland.

Andresen, S. 1998. "The Making and Implementation of Whaling Policies: Does Participation Make a Difference?" In D. Victor, K. Raustiala, and E. Skolnikoff, eds., *The Implementation and Effectiveness of International Environmental Commitments* (pp. 431–475). Cambridge, MA: MIT Press.

Andresen, S. 2000. "The Whaling Regime." In S. Andresen et al., eds., *Science in International Environmental Regimes: Between Integrity and Involvement*. Manchester: Manchester University Press.

Andresen, S. 2001. "The International Whaling Regime: 'Good' Institutions but 'Bad' Politics." In R. Friedheim, ed., *A Sustainable Whaling Regime*. Seattle: University of Washington Press.

Aron, W. 1988. "The Commons Revisited: Thoughts on Marine Mammals Management." *Coastal Management*. 17(2): 99–109.

Birnie, P. 1983. "Countdown to Zero." *Marine Policy* 7(1): 63–65.

Birnie, P. 1985. "From Conservation of Whaling to Conservation of Whales and Whale-Watching." *International Regulation of Whaling* (Vols. 1–2), (pp. 1–1053). New York: Oceana.

Birnie, P. 1986. "Are Whales Safer Than Ever?" *Marine Policy* 10(1): 63–64.

Burke, W. T. 1996. "Memorandum of Opinion on the Legality of the Designation of the Southern Ocean Sanctuary by the IWC." *Ocean Development and International Law* 27(3): 315–327.

Cherfas, J. 1992. "Key Nations Defy Whaling Commission," *New Scientist* 4(July): 7–11.

DeSombre, B. 1995. "Baptists and Bootleggers for the Environment: The Origins of United States Unilateral Sanction," *Journal of Environment and Development* 4(1): 53–75.

Freeman, M. 2001. "Is Money the Root of the Problem? Cultural Conflicts at the International Whaling Commission." In R. Friedheim, ed., *A Sustainable Whaling Regime*. Seattle: University of Washington Press.

Freeman, M., and S. Kellert. 1992. *Public Attitudes Towards Whales: Results of a Six Country Study*. Edmontons New Haven: University of Alberta/Yale University.

Friedheim, R. 1996. "Moderation in Pursuit of Justice: Explaining Japan's Failure in the International Whaling Negotiations." *Ocean Development and International Law* 27: 349–378.

Friedheim, R., ed. 2001. *A Sustainable Whaling Regime*. Seattle: University of Washington Press.

Gambell, R. 1993. "International Management of Whales and Whaling: An Historical Review of the Regulation of Commercial and Aboriginal Subsistence Whaling." *Arctic* 46(2): 97–107.

Gulland, J. 1988. "The End of Whaling?" *New Scientist* (October 29): 42–48.

Hoel, A. H. 1985. *The International Whaling Commission 1972–84: New Members, New Concerns*. Lysaker, Norway: Fridtjof Nansen Institute.

Hoel, A. H. 1993. "Regionalization of International Whale Management: The North Atlantic Committee for Research on Marine Mammals." *Arctic* 46(2): 115–123.

Holt, S. 1985. "Whale Mining, Whale Saving." *Marine Policy* 9(3): 192–114.

Ivarson, J. 1994. "Science, Sanctions and Cetaceans, Iceland and the Whaling Issue, Centre for International Studies." University of Reykjavik.

McHugh, J. L. 1974. "The Role and History of the Whaling Commission." In W. E. Schevill, ed., *The Whale Problem: A Status Report*. Cambridge: Harvard University Press.

Mitchell, R. 1998. "Forms of Discourse, Norms of Sovereignty: Interests, Science and Morality in the Regulation of Whaling." *Global Governance* 3: 275–283.

Røssum, J. 1984. "Forhandlingene om reduksjon av fangstkvota i Antarktis. Den internasjonale hvalfangstkommisjonen 1960–65" ("The negotiations over reduction of the catch quota in Antarctica. The International Whaling Commission 1960–65"). Master's thesis, Institute of Political Science, University of Oslo.

Scarff, E. 1977. "The International Management of Whales, Dolphins and Porpoises: An Interdisciplinary Assessment," *Ecology Law Quarterly* 6: 323–571.

Schweder, T. 1994. *Intransigence, Incompetence or Political Expediency? Dutch Scientists in the IWC in the 1950s*. (Oslo: University of Oslo.)

Spencer, L., with J. Bollwerk and R. C. Morais. 1991. "The Not So Peaceful World of Greenpeace." *Forbes* (November 11): 174–180.

Stenseth et al., eds. 1993. *The Minke Whale: The Difficult Choice*. Gyldendal, Norway: ad Notam. (In Norwegian).

Stoett, P. 1995. "The International Whaling Commission: From Traditional Concerns to an Expanding Agenda." *Environmental Politics* 14: 130–135.

Tønnesen, J., and A. Johnsen. 1982. *The History of Modern Whaling*. London: Hurst.

Underdal, A. 1980. *The Politics of International Fisheries Management: The Case of the Northeast Atlantic*. Oslo: Norway: Universitetsforlaget.

Wettestad, J., and S. Andresen. 1991. *The Effectiveness of International Environmental and Resource Regimes: Some Preliminary Findings*. Lysaker, Norway: Fridtjof Nansen Institute.

Wilkinsson, D. M. 1989. "The Use of Domestic Measures to Enforce International Whaling Agreements." *Denver Journal of International Law and Policy* 17(2): 271–291.

16

The Convention for the Conservation of Antarctic Marine Living Resources (CCAMLR): Improving Procedures but Lacking Results

Steinar Andresen

Assessing Performance: A Low-Effectiveness Regime

As the history of cooperation within the Convention for the Conservation of Antarctic Marine Living Resources (CCAMLR) is fairly short and incremental, it is not divided into phases. The effectiveness of this subregime within the wider Antarctic Treaty System (ATS) has increased over time in terms of *output* produced by the regime.[1] In terms of *outcome* and even more so in terms of *impact*, the effect of the regime seems to be either modest or uncertain. Consequently, we tend to identify effectiveness—when seen as the ability of the regime to conserve the marine resources in the area—as rather low. CCAMLR faced a severe challenge, however, as it was established almost a decade after the marine resources in the area had been seriously depleted. Thus, it may be argued that CCAMLR never really had a chance of dealing effectively with this problem.

After a very slow start throughout most of the 1980s, some progress was made toward the end of the decade and in the early 1990s knowledge increased, regulations were tightened, and verification improved. Thus, the *conditions* for creating a more effective regime improved. Since the early 1990s we also have witnessed a decrease in catches of marine living resources—fin fish and krill—which helps restore the ecological balance of this fragile marine ecosystem. *If* we could establish that this has happened because of the operation of the regime, effectiveness would seem to be on the increase. To some extent this may have been the case, but it seems equally likely that reduced catches may have been a result of depletion. Moreover, decreased catches were also a result of contextual factors unrelated to the working of the regime. In the 1990s a new challenge faced CCAMLR—the increased unreported catch of Patagonian toothfish within the Convention area. As CCAMLR has not been able to deal effectively with this problem, it strengthens the impression that this is a low-effectiveness regime.

In this chapter we evaluate the effectiveness of CCAMLR in rather narrow terms in relation to the official purpose of the Convention to conserve the marine resources in the area. It may be argued that because CCAMLR is but one small piece in a larger Antarctic regime, other and broader criteria could be applied as well. Some analysts have found that CCAMLR has been quite successful regarding the protection of the wider treaty system (Stokke 1996). More recently, others have argued that the various challenges posed by the increased catch of Patagonian toothfish is a test case of the strength of the entire Antarctic Treaty System (ATS) (Herr 1999). We have chosen a narrow approach when evaluating the regime for reasons that are elaborated on later. Still, the development and working of CCAMLR have to be studied within the context of the ATS.

Background: The Process of Creation, Institutional Framework, and Challenges

Although the *history* of CCAMLR dates back less than twenty years, the history of human exploitation of Antarctica is much older. The Southern Ocean surrounding the Antarctic continent has been an important hunting ground for centuries due to its rich population of marine mammals. In 1778, the first British fur seal hunters started out with South Georgia as their base. The United States and other countries were soon to follow in what was perhaps to become the most effective extinction of an animal that the world has ever seen. Starting at the end of the nineteenth century, a race for the exploitation of the Antarctic whales started. As is commonly known, the history of Antarctic whaling also ended with commercial depletion of the large Antarctic whales.[2] But Antarctic history has been characterized by more than massive resource exploitation. In the first half of the nineteenth century, France, the United Kingdom, and the United States sent scientific expeditions to the Southern Ocean. Scientific curiosity as well as national prestige were motivating forces (Indreeide 1989). In short, the early history of Antarctica and its surrounding waters were characterized by three trends: resource exploitation, scientific investigation, and nationalism. Although the blend of the mixture is different today, these three trends are still the essential components in shaping the development of Antarctica.[3]

The more recent *political history* of Antarctica dates back to the late 1950s. It started out with the International Geophysical Year (IGY) in 1957–1958 in which Antarctica was given special emphasis and a Special Committee on Antarctic Research (SCAR) was established in 1961. Parallel to this scientific work, the United States took a diplomatic initiative in 1958 with the eleven other countries conducting scientific research in Antarctica. In 1959 these twelve *consultative parties* held a conference in Washington, and the *Antarctic Treaty* was available for signatures by the end of 1959 (the Treaty came into force in 1961). The Convention was accepted by all twelve parties involved—not least because the sensitive political issue of sovereignty was frozen. Some of the twelve Antarctic nations had claimed sovereignty over parts of the Antarctic continent, some of these claims overlapped, while other nations rejected national claims altogether.[4] The importance of continued international scientific cooperation is another feature of the Convention. The Antarctic Treaty is still the legal and political cornerstone for international Antarctic cooperation, but it is no longer the only instrument. Today, reference is usually made to the Antarctic Treaty System (ATS). In 1972 a Convention for the Conservation of Antarctic Seals (CCAS) was adopted, and in 1980 CCAMLR, the specific focus of this case study, came into being. In 1988 the Convention on the Regulation of Antarctic Mineral Resource Activities (CRAMRA) was adopted. The most recent institutional arrangement is the 1991 Protocol on Environmental Protection to the Antarctic Treaty.

According to Lagoni (1984), the consultative parties had in mind three major concerns when drawing up CCAMLR: conservation of krill, avoidance of conflict over sovereignty claims between claimants and nonclaimants, and retention of authority over Antarctic affairs in the light of increasing external pressures. Thus the parties were motivated not only by concern for the fragile ecosystem in the area but also by the broader political goal of securing the interests of the consultative parties. Still, the triggering effect was linked to marine resources in the area. It was believed that the harvest of krill would increase dramatically, and the catch of fish in this area also attracted attention. In the late 1960s and early 1970s "it was often suggested that the world's fish catch could be doubled by harvesting the resources of the Southern Ocean" (Heap 1991, 48). Exploitation by long-distance fishing fleets was intensive in the 1970s, but catches declined rapidly. Considering the previous history of resource depletion in Antarctic waters and the delicate and vulnerable balance of the Antarctic ecosystem, this development worried the scientists within SCAR. In 1972, SCAR established a Subcommittee on the Living Resources of the Southern Ocean, upgraded to a Group of Specialists in 1976. The scientists conveyed their concern to the *Consultative Meeting*, the political body of the Antarctic Treaty nations. At the suggestion of the United States, a conference to deal with the conservation of Antarctic marine living resources was organized. A result of this conference was the launching of the BIOMASS program (Biological Investigation of Marine Antarctic Systems and Stocks)—a ten-year program with both scientific and management components. The direct negotiations leading up to CCAMLR started in 1978 and were concluded two years later in Canberra (Indreeide 1989).

The consultative parties wanted to establish an international resource regime that was consistent with their broader ATS interests. First, was not self-evident that the ATS nations should be the founding fathers for the management of marine living resources of the sea. Both UNEP and particularly the FAO had also voiced an interest in the matter (Beck 1986). Second, the Law of the Sea development with increased national jurisdiction on the part of the coastal states triggered the unresolved jurisdictional controversies between the ATS nations.[5] Finally, inspired by the concept of the "common heritage of mankind" introduced in the Law of the Sea negotiations in relation to the deep seabed minerals, some developing countries strongly resented the exclusive nature of the ATS cooperation (Stokke 1996).[6]

These factors could—separately or in combination—destroy the fragile stability of the Antarctic Treaty System, but the consultative parties were able to withstand the external pressure (Hoel 1990, 7–8).[7] Although there was some disagreement regarding the management of the resources, they all had in common the interest to create a subregime to maintain the wider ATS regime.

The goal of CCAMLR is spelled out in its title: *Conservation* of Antarctic Marine Living Resources. The more precise goal of CCAMLR is ambitious. For the first time in an international fisheries agreement, the ecosystem approach was chosen—in contrast to the conventional approach of single-species management.[8] The operationalization of its three-stranded ecosystem standard is spelled out in the Convention.[9] Oversimplified, the goals are to

· Secure maximum net recruitment,
· Secure ecological balance between species, and
· Prevent the nonreversible reduction of any species.

According to article 2, paragraph 2, of the Convention, "the term *conservation* also includes *rational use*."

The organizational structure and the decision-making procedures are fairly traditional. The two main components are the Commission and the Scientific Committee. The main task of the Commission is to

· Facilitate research and comprehensive studies,
· Compile data on the status and changes in populations,
· Ensure acquisition of catch and effort statistics on harvested populations,
· Analyze and publish information,

· Introduce conservation measures, and

· Analyze conservation effectiveness—including designating the quantity of any species that may be harvested in the area.

Thus, apart from the management component, CCAMLR has a high scientific profile, and the Commission shall take "full account" of the advice given by the Scientific Committee (article IX). Fisheries may be regulated by a wide variety of measures such as open and closed seasons, designation of size of quotas and appropriate areas, and protected species.[10] According to the Convention, a system of observation and inspection should also be implemented. The Secretariat is located in Tasmania, Australia.

As to the *decision-making procedures*, the traditional consensus procedure is spelled out, giving the member states a de facto right of veto. In fact, a double veto is ensured by the objection procedure. The *area of application* of the Convention follows the natural boundary area of the Antarctic convergence rather than an artificial delimitation that would have stemmed from political interests" (Cordonnery 1998, 127). The Antarctic convergence is defined as a line varying between 60 degrees South and 45°S, which means the Convention area extends beyond the limits to which the Antarctic Treaty applies. The Convention applies to all marine living resources, defined in article 1(2) as "fin fish . . . and all other species or living organisms, including birds, found South of the Antarctic convergence."[11] As to the *right of membership*, all states have a right to *accede* to CCAMLR, but limitations are made on the right to participate in the decision-making body of CCAMLR as engagement in research or harvesting activity is required. The links to the Antarctic Treaty are evidenced by the fact that parties to CCAMLR, not parties to the Antarctic Treaty, are obliged to accept this treaty and abide by its goals and principles. Also, the harvesting or research requirements do not apply to the twelve consultative states negotiating CCAMLR (including the United States, the Soviet Union/Russia, the United Kingdom, Australia, New Zealand, Argentina, Chile, and Norway, which have played important roles in the regime). The present membership of CCAMLR is limited to twenty-three nations.

The *challenge* facing these members is how to conserve the Antarctic marine living resources in a sustainable manner. Most of these members have no material interests at stake as they are not involved in any harvesting activities, but a distinct minority is involved in harvesting activities. In that sense the problem bears some resemblance with the more recent stage in the IWC. Thus, CCAMLR is a resource-management regime "dominated by the conflict of values and interests between conservationists and fishing states over the implementation of the principles of the ecosystem approach" (Cordonnery 1998, 125).

In sum, the goal of CCAMLR is conservation of marine resources through an ambitious ecosystem approach. The vehicle to achieve this goal is a Commission, with traditional decision-making procedures and specific conditions with regard to membership and rights due to the links to the ATS. The Commission acts on advice from the Scientific Committee. Finally, members with and without harvesting interests are faced with the challenge of managing the resources in the area in a sustainable manner.

CCAMLR: An Effective Ecosystem Regime?

The history of CCAMLR is fairly short, and overall development has been rather incremental. Many observers have pointed to 1987 as one important turning point in its history, and important changes have also taken place in the 1990s. However, we identify these as more normal events and changes that may occur within a regime characterized by a fairly steady development. There are no watershed events like those witnessed within the IWC.

As CCAMLR is partly a subregime or a nested regime, in contrast to the other regimes covered in this book, the problem is how to measure its performance particularly in relation to its ability to conserve the Antarctic marine living resources or in relation to the broader goals of the ATS. Although a broader focus certainly has some merits, these criteria are most relevant in the *initial* phase of the regime. Some of the most difficult questions *within the ATS system* regarding conflict over national jurisdiction were sorted out during the negotiations about CCAMLR (Indreeide 1989). The *external problem* concerning the perceived lack of legitimacy of CCAMLR by actors like the FAO and the United Nations, have also lost most of their relevance (Stokke 1996). Thus, although concern for marine resources was not the only motivation for the creation of CCAMLR (because these other issues have more or less been solved), increased attention has been geared to the real goal of CCAMLR—management of marine living resources.[12]

Has CCAMLR been able to make accessible the ecosystem approach underlying CCAMLR? Doubts have been raised about the extent to which it is possible to give an unambiguous interpretation of this goal—whether an effective operational definition can distinguish between the different levels of the ecosystems (Basson and Beddington 1991). As the knowledge of the ecosystem of the Southern Ocean is still limited, an answer to this question in precise terms is not yet possible. But this is not an exercise conducted by natural scientists, and precise goal achievement is not the ultimate test of whether a regime is effective or not, so this imprecision is not necessarily a major problem in our context.

Knowledge of the ecosystem of the Southern Ocean was almost nonexistent prior to the start of the process leading up to CCAMLR in the early 1970s. Thus, measures taken to increase the *knowledge base* are a necessary—if not sufficient—precondition for effective ecosystem management.[13] Second, considering that many fin-fish stocks were severely depleted prior to the adoption of CCAMLR, *regulatory measures* reducing the catch of *fish* were also necessary to conserve the resources. Considering both the key role of the krill in the ecosystem of the Southern Ocean as well as the fears of the founding fathers that the catch of krill might explode, the role of krill in the regulatory system deserves attention. In this context, it may be asked whether a multispecies ecosystem approach should have been adopted. Finally, what has been achieved regarding provisions for *inspection* and *verification*?[14] These three aspects deal primarily with *outputs* flowing from the regime, and as such they tell us little about whether the regime is effective or not. However, they are usually necessary, if not sufficient, conditions for subsequent behavioral change on the part of its members. First, a factual account of the different types of decisions (outputs) flowing from the regime will be presented.

Outputs: Increasing Knowledge, More Regulations, and Some Verification

As noted, improving the *knowledge base* was given high priority in this Convention. Initially, measures taken to provide relevant information were twofold: scientific working groups and workshops were established on specific data problems, and member states were required to provide data on catch and effort statistics (Hoel 1990, 13). Subsequently, this information has been used by the Committee's long-term program of work. More recently scientific work has become more firmly *institutionalized* through permanent working groups in key areas to secure more sustained research programs in key areas. After a somewhat slow start the data-gathering process has become smoother, and knowledge about the ecosystem has increased significantly, although uncertainties still abound (Stokke 1996).

Regulatory measures are closely linked to the database and providing information; meaningful regulatory measures are not easily adopted without sufficient knowledge. In the period 1982 to 1990, only seventeen regulatory measures were adopted (Hoel 1990). From a slow start in quantitative terms, there has been a gradual increase, especially since the 1987 meeting. From then on CCAMLR also started to use catch limits, a more refined fisheries management measure as compared to the earlier all-or-nothing approach. CCAMLR's choice of management strategy was even initially characterized as "rather conservative and careful" compared to other fisheries bodies (Hoel 1990, 14). Also, the multispecies approach was applied to a limited extent and described as "a real and strong limitation on the rights of the fishing nations" (Indreeide 1989, 82). Since 1991 the Commission has expanded its scope of conservation measures by attempting to minimize the incidental mortality of seabirds through long-line fishing in the area. Measures have also been taken to reduce the considerable incidental catch of fur seals. According to another observer, "As from 1987 the nature of conservation measures has changed dramatically," including "an overall total allowable catch, a reporting system, and a closed season" (Vicuña 1991, 28). Thus, during its first decade CCAMLR made some progress in producing regulations in the area. In the 1990s, the Commission reached agreement on a steadily growing number of conservation measures that have become increasingly stronger over time: "It is indicative that more than 80 percent of the ninety-one conservation measures adopted by May 1995 by the Commission have been made after 1990" (Stokke 1996, 142) This development continued thereafter, and by 1997, 117 conservation measures had been adopted. It should be noted, however, that many conservation measures are designed to have a seasonal influence as they impose catch limitations that are subject to annual review (Cordonnery 1998, 128). Consequently, only forty-three of these measures are recorded as having an ongoing effect. Still, in simple quantitative terms progress has undoubtedly been made.

It took a long time before the krill, at the very heart of the Antarctic marine ecosystem, was included in the regulatory repertoire of CCAMLR. Clearly, this was a major deficiency. A mitigating factor in practice was the fact that the actual catch of krill was moderate compared to the sustainable yield of this stock. Over time, the regulatory progress has also been applied to krill. In 1991, a precautionary TAC was recommended by the Committee and later acted on by the Commission (Scientific Committee 1991, 22). The quota was set at 1.5 million tons for one statistical area. Precautionary catch limits have since been extended to other areas, including the South Orkney Islands and South Georgia (Cordonnery 1998, 129). Although important in principle, this development in practice has not meant much due to the declining interest in this resource.

The same development as outlined above can be seen regarding *reporting procedures*, as nothing happened initially along this dimension. It was not until the 1987 Commission meeting that the issue was addressed, and at the next Commission meeting in 1988, a joint inspection system was adopted, supervised by a Standing Committee on Observation and Inspection (SCOI) (Commission 1988, 29–35). The system came into operation in the 1989 to 1990 season, and inspectors were appointed from the Soviet Union, Argentina, Chile, and the United States. The inspection system was nationally operated as inspectors are appointed by and report to their governments, who in turn report to the Commission. Inspections can also be carried out by inspectors from one country on a fishing vessel of another country. Prosecution (the imposition of sanctions regarding violations) is the responsibility of the *flag state* of the offending vessel and are reported to the Commission. In practice there has tended to be little room for independent monitoring, and few violations are reported: "For example, the Soviet Union during 1990 carried out 118 inspections of its own vessels, one of which resulted in prosecution for violation of the Commission regulation of mesh size" (Cordonnery 1998, 136). Still, the system has been improved significantly over time, and "there is little doubt that CCAMLR is considerably closer to a satisfactory system today than only five years ago" (Stokke 1996, 148).[15] The Commission as well as the SCOI have stated that a system of vessel monitoring (SVM) should be introduced within the Convention area, but progress has been limited. Thus, inspection and independent verification are still real problems, as they are in most international environmental and resource regimes (Victor et al. 1998).

In summary, over time considerable improvements have been made, and institutional growth has occurred. If the focus is exclusively on outputs, the conclusion is that CCAMLR is about to become a smoothly functioning regime. This point has some merits, but it does not necessarily mean that CCAMLR is an effective regime according to our terminology.

Outcome: Modest Behavioral Effect

Has CCAMLR affected the behavior of the Antarctic fishing fleet after it was established? Catch figures prior to and following the establishment of CCAMLR provide some indication of behavior changes. The figures show that fishing carried out in the period *prior* to the establishment of CCAMLR was the main reason for depleted resources. In South Georgia waters 35 percent of the total catch over two decades (1970 to 1990) was taken in the first two years. In South Orkney Islands waters 57 percent was taken in the first year; in Antarctic Peninsula waters 61 percent was taken in the first year; and in South Indian waters 36 percent was taken during the first two years (Heap 1991, 48). CCAMLR can hardly be held responsible for this development. However, other figures indicate that CCAMLR has not done much in the way of reducing the amount of fish taken. For example, the average catch in South Georgia waters in the twelve years before CCAMLR entered into force was 67,139 tons; in the eight years since it entered into force it has been 80,417 tons. Most of the highest-value species were taken *before* CCAMLR came into force, the same pattern that was witnessed with Antarctic whaling (Heap 1991, 49).

It took some time before CCAMLR actually introduced conservation measures. It appears that those measures were too little, too late as up to the end of the decade CCAMLR had yet to bring about the necessary reversal of the previous negative trend: "The pattern of fishing strongly suggests that a number of fish stocks are now in a depleted state and therefore, in terms of the Convention, stand in need of severely restrictive conservation measures" (Heap 1991, 48). Catches have declined strongly in the 1990s, most markedly for krill but also for fin fish. More recently, the harvest of fin-fish stock has varied between 50,000 and 100,000 tons annually. The catch of krill in the early 1990s, when the precautionary quota was set at 1.5 million tons, was only 300,000 to 400,000 tons and has been reduced significantly subsequently. Thus, there is no doubt that krill fishing is conducted on a sustainable level. Without considering *contextual factors*, this would indicate that CCAMLR has had a strong behavioral impact on the level of catch by its members. It may well be that conservation measures have had some effect, but as is shown later the main reason for the strongly reduced catch is unrelated to the working of the regime.

Considering all the attention given to the catch of Patagonian toothfish lately, this point deserves some attention.[16] This Antarctic demersal fish is one of the largest of the Antarctic species, weighing up to 100 kilograms and commanding fairly high prices on the international market. Although the Patagonian toothfish fishery is insignificant in global fishing terms, it is nevertheless significant in Antarctic regional terms. It used to be taken as by-catch and then as a targeted commercial species outside the coasts of Argentina and Chile. As interest increased and the resource was gradually depleted in these areas, the toothfish moved into the CCAMLR Convention area. In the 1990s the catch of toothfish increased strongly, but it is not known by how much due to the considerable amount of unreported fishing. According to FAO and CCAMLR statistics, in 1993 and 1994 only some 20 percent was taken in the Convention area. More recently there are indications, albeit uncertain, that "the balance has shifted strongly to reverse these ratios so that a much greater share of the perceived management responsibility appear to be falling increasingly on CCAMLR and its members" (Herr 1999, 245). Although CCAMLR cannot alone be blamed for the failure of sound management of this resource, it certainly has to bear its share of the responsibility. "When the Scientific Committee (in 1997) estimated that the reported catch from the CCAMLR area may have only comprised 40% of the total taken . . . , it was also an issue of the effectiveness of CCAMLR regulation of this fishery" (Herr 1999, 252).

Uncertain Impact on the Problem

We cannot determine with any confidence the impact of the regime on the problem at hand as uncertainties abound; the causal chain is very long, and there may be many other intervening variables. However, as outlined in chapter 1, one way to bring us a step further in addressing this issue is to examine the extent to which the various types of decisions are in line with the advice given by scientists—provided advice can be considered independent as well as consensual.

The correspondence between scientific advice and political decisions regarding the *goal* and *principles* of the Convention is very high. The ecosystem approach and the initiative to regulate fisheries in the Southern Ocean came from the scientists, channeled through SCAR, and their scientific advice was carefully followed in the drafting of the Convention (Indreeide 1989). Overall, the match between advice and policies has been good regarding *procedures* for data collection and information dissemination. Generally, the scientific recommendations on such matters have been given unanimous support by the Commission. The picture used to be much more gloomy regarding *regulations*. According to most observers, during the first few years after CCAMLR came into force, this indicator was hardly relevant as no consensual scientific advice on thorny issues was produced. According to Vicuña (1991, 30), "The very first meeting of the Scientific Committee ended in a deadlock about the role of this organ, a situation which was to endure throughout the first phase." In the interpretation of scientific uncertainty, a precautionary approach stood against a more optimistic approach.[17] Overall, there has been a tendency throughout the last decade toward a higher degree of consensus in the Committee as well as toward better correspondence between their recommendations and the conservation measures being adopted. After the Committee started to present management advice in the form of quota options in the early 1990s, the Commission tended to adopt the more precautionary approach, and in some examples the Commission has found data insufficient and decided to prohibit harvesting.[18]

According to one observer, "This Committee has extraordinary influence within the regional body both by the high level of expertise . . . and its relative autonomy"

(Herr 1999, 245). An alternative view is that "the Scientific Committee is controlled by the Commission and there is no attempt at insulation from political decisions, and no obligation placed on the Commission to follow its advice" (Cordonnery 1998, 138). At the international level, management bodies are *never* obliged to follow scientific advice, and our impression is clearly that the former view is the most accurate observation. Nevertheless, a basic flaw in the science-policy nexus is that scientific research and monitoring are essentially left to the parties themselves. In addition, in the controversy over toothfish the Scientific Commission was not heard, indicating that the influence of science is reduced as polarization increases. Finally, and important when discussing the impact of the regime on the resources at hand, one reason for the significant attention toward the Patagonian toothfish is that since the 1992 and 1993 season it has been the *only* targeted commercial fin fish "out of the thirteen species of finfish taken in significant quantities in the preceding half decade" (Herr 1999, 248). CCAMLR alone is hardly responsible for the decline of these fish stocks, but it certainly makes the three goals of CCAMLR rather distant ones.

In summary, the development of a more refined and science-based regulatory system, including provisions for inspection, is in itself no small achievement, but output is no real indicator of effectiveness.[19] Although there are significant uncertainties, the score is markedly lower in terms of behavioral change caused by the regime and impact on the problem at hand. CCAMLR seemed to be on the right track, but when the Patagonian toothfish problem emerged, the regime once more showed that it was not capable of dealing with more difficult problems. To account for this, we shift to the explanatory perspective.

Regime Implementation: Institutional Growth but an Inability to Deal with Difficult Problems

The Development of Problem Structure and Contextual Factors

Most of the parties to CCAMLR had no material interests in living marine resources. Consequently, the costs of endorsing a seemingly conservation-oriented ecosystem approach was very small for the majority of parties as the regulatory measures of CCAMLR would have no material impact on them. The difference of interest between the harvesting and the nonharvesting nations and the compromise reached is reflected in the tension between the rational-utilization approach versus the conservation approach, outlined in article 2 of the Convention. The actual balance between these concerns was postponed until CCAMLR started its operations.

Initially, the similarity in problem structure to that of the IWC created tensions between the parties, not least because some of the same key parties (and persons) were involved in both regimes. The harvesting nations feared that CCAMLR might be exposed to the same rather extreme conservationist forces that had taken over the IWC. This explains some of their initial less than cooperative attitudes. As it became clear that this fear was largely unwarranted, this (negative) spillover effect was gradually reduced.

CCAMLR was facing a difficult problem in terms of regulating the exploitation of fish in the area. The fishing companies operating in these waters were long-distance fleets. Short-term economic interests dictated that they should take as much as possible in the shortest possible time and then leave for other (unregulated) fishing grounds—before these were "nationalized" through the Law of the Sea development in the 1970s. The high initial catch confirms this development. However, due to early resource depletion, the malignancy of the problem was reduced as subsequent level of catch was fairly small; for those involved it was a *marginal* economic activity.[20] The total of some 2.7 million tons taken over the twenty-year period 1970 to 1990 averaged some 0.27 percent of the world fish catch (Heap 1991). In the 1990s, the marginalization of the area continued as global catch increased while the Antarctic fishery declined considerably. Moreover, although quite a few nations have been engaged in fisheries in this area, the number of significant actors has generally been very small. Throughout the 1980s, the USSR accounted for some 80 percent of the catch (Indreeide 1989, 61).[21] Apart from a few Eastern Europe countries, most of these states were conducting pilot fishing on a larger scale during the 1980s.

For the most part krill has been a secondary target species, taken when fish are scarce. Over the seven years from 1983 to 1989 a total of some 2.1 million tons have been taken, averaging some 300,000 tons a year, or some 0.6 percent of total world catch in this period (Indreeide 1989). As noted, in the 1990s the catch of krill has been considerably reduced, and in 1996 the krill catch dropped below 100,000 tons. Regarding the number of important actors, the picture has been very much the same as for the fishing activity. Although quite a few nations have been engaged in pilot projects, only the Soviet Union and to some extent Japan have been involved in commercial operations. Considering that the Soviet Union and Japan are among the largest fishing nations of the world, the contribution from the Southern Ocean to their total catch more recently has been next to nothing. Still, this problem structure contributed to explaining the slow start of the regime.

Why has the catch of fin fish and krill been so strongly reduced? The most important change is probably unrelated to the regime but has had a significant effect on the working of CCAMLR. The fall of the iron curtain between East and West Germany contributed to reducing the mutual distrust between some of the key players from the two camps, not the least the United States and Russia. Although hard to measure and not much discussed by other analysts of CCAMLR, it seems likely that some negative issue linkages related to broader security and national interests have been reduced. Second, and more important, the previous socialist countries—now labeled *economies in transition*—have gradually changed their economic systems in a more market-based direction. This has made them lose interest in harvesting marine living resources in these remote areas. Due to the long distance to the fish markets and limited demand for krill, this Antarctic marine harvesting fishery has not been profitable as far as is known (Indreeide 1989). The decision on the part of Russia to quit its harvest of krill after the 1992–1993 season should probably be seen in this light, and the same goes for the reduced fishing activities by both Russia and other Eastern European countries. These two factors, triggered by the same external event, made it easier to reach agreement on the more far-reaching measures of conservation described above, before the Patagonian toothfish controversy surfaced.

Other contextual changes in the international political agenda taking place in the late 1980s and the early 1990s (like the strong greening of public opinion, especially in many Western countries) may also have had an effect on cooperation within CCAMLR. Although the marine ecosystem in Antarctic waters did not receive much specific attention in this period, there was considerable general focus on Antarctic matters as such, not least linked to the issue of a minerals regime in Antarctica as well as the call for making Antarctica into a natural park (Herr and Davis 1996). This development may have had some spillover effect on CCAMLR creating increased transparency as well as strengthening the positions of the parties working for stronger conservation measures. For major countries like the United States, the United Kingdom, and France, with rather varied environmental records, the chance to stand forth as environmentalists, without any (material) costs, in the context of CCAMLR—as well as the IWC—presented a good opportunity for improving their environmental image. Thus, the combined effect of the fall of the iron curtain and the greening of public opinion paved the way for a more benign problem structure and provided the general backdrop for gradually more conservation-oriented policy in the early 1990s.

The fierce controversy over the Patagonian toothfish can also be seen in light of the activity of green NGOs and some key countries. Compared to major fisheries around the world, even if all the anecdotal evidence of massive overfishing as well as illegal fishing is true, "it accounts for a minuscule one percent of global catch" (Herr 1999, 256).[22] Considering the small proportions, it may well be seen as a cheap option for scoring green points when New Zealand found the issue so important that it tried to make it a central issue at the 1997 Antarctic Treaty Consultative meeting. Thus, Herr (1999, 242) is certainly right when he claims that "The explanation for this apparently unusual situation is bound up both with the changing post–cold war priorities for Antarctica as well as the changing global approaches to resource regulation." Still, this is only one side of the coin. The inability of CCAMLR, as well as other relevant international bodies, to deal effectively with this issue represents a credibility problem for CCAMLR and ATS. The most serious problem in this context is that *it appears that members of CCAMLR are engaged in fishing activities under flags of convenience.* The principal providers of flags of convenience appear to be Belize, Panama, and Vanuatu. Countries not members of CCAMLR are not bound by CCAMLR regulations, so their activity is not illegal in a narrow, legal sense. Nevertheless, these flag states, as well as those (CCAMLR states) behind the scenes, severely undermine the effectiveness as well as the legitimacy of the regime.

The fishing that takes place outside the Convention area is beyond the formal authority of CCAMLR, but some members have been eager to use other international vehicles like the FAO and the United Nations 1995 Fish Stock Agreement to deal with the problem, although with very limited success.[23] Again, this is not an illegal fishery, but it clearly runs counter to the spirit of CCAMLR as well as to other international bodies and agreements. So why is CCAMLR not able to deal with fisheries in its own Convention area? It has been reported that "during the 1995/6 season more than a hundred vessels were observed to be engaged in unregulated fishing for toothfish in the CCAMLR area with an estimated catch of more than ten times the harvest reported by CCAMLR members" (Stokke 2001, 11). The fundamental reason for lacking effectiveness is the same as most soft international environmental and resource regimes are facing. They are powerless to halt the parties (and others) if these, due to perceived short-term economic interests, do not see it in their interest to comply with regulations adopted. Thus, problem malignancy has again been increasing. In addition, the sovereignty challenge is very complex in the context of Southern Ocean toothfish in that the CCAMLR area includes not only the high seas but also national waters and areas where national claims are either disputed or not recognized beyond the small group of Antarctic claimants (Stokke 2001, 10).[24] Closely related and adding to the complexity of the problem is the difference in the will and ability of the parties to comply with regime rules. More generally, most of the areas being fished are remote, making surveillance and enforcement difficult. This also relates to the problem-solving capacity of the regime, something we turn to in the final section of this chapter.

Increasing but Insufficient Problem-Solving Capacity

In the 1980s the main fishing nations—the USSR, East Germany, Poland, and Japan—generally played the role of laggards, opposing a number of conservation measures suggested by the more conservation-oriented nations such as the United States, New Zealand, Australia, the United Kingdom, and Norway. The USSR was the main stumbling block within the cooperation in this period because it was *the* major player among the major harvesting nations. The Soviet Union also had the best access to scientific information through its harvesting activities. In consequence, the Soviet Union held unique *issue-specific power*, with Japan in second position. In this initial period, the Soviet Union used its capabilities in line with the law of the least enthusiastic actor, slowing down efforts to move the regime forward (Underdal 1980). The counterforces assembled among them some of the most powerful nations in the world. The issue-specific power of the United States and other major consultative parties among the nonharvesting nations was related to their scientific interests—capabilities and presence in the area. However, some of them were also in possession of considerable *policy-relevant power* both as key Antarctic nations as well as in more general terms, especially the United States. Nevertheless, this was not sufficient to stop the main harvesting nations. The lack of progress within CCAMLR in this period lends support to the saying that "vetoing is far easier than engineering." However, it is also a question of how much energy and the degree of priority given by key consultative parties in this initial phase to challenge the USSR and the other harvesting nations, considering that factors related to the broader ATS were seen as more important.

The fishing nations, with the Soviet Union in the forefront, favored *consensus decisions* throughout the negotiations leading up to CCAMLR, while many of the nonharvesting nations favored different kinds of majority voting. The United States favored the use of a two-thirds majority, enabling the nonharvesting nations to vote down the harvesting nations. However, interests related to the decision-making procedures of CCAMLR should also be seen in light of the ATS. According to Barnes (1982, 250), "The question of a system of voting was involved intimately with the claims situation." For example, Argentina and Chile favored the consensus approach to protect their harvesting interests but especially to protect their national claims in Antarctica. France demanded an objection procedure to give their consent due to the area where CCAMLR overlaps its jurisdiction of Kerguelen and the Crozet islands. The adoption of the consensus procedure should also be seen in light of "twenty years fruitful experience of it [consensus] that had already been acquired by the Antarctic Treaty states" (Edwards and Heap 1981, 358).

The conflict over procedures continued after the creation of CCAMLR. A major conflict arose in the Scientific Committee. A draft rule procedure prepared by Australia suggested that "whenever possible" the Committee would decide by consensus. The Soviet Union, however, insisted on consensus for all recommendations adopted by the Committee, and the USSR got its way (Andresen 1991). The Soviet Union seemed to see the Committee as essentially a political body and not as an independent scientific body. However, a safety clause was provided for through the negotiations as the Committee should inform the Commission of "all the views" voiced concerning the matter under consideration. Later on this was to become an important provision. The malfunctioning of the science-policy dialogue represented an important factor in explaining the low effectiveness in the initial period. Vicuña (1991, 31) maintains that "the situation was evidencing a total lack of dialogue between the two organs." Heap (1988, 23) has pointed to what he considered the main institutional problem between the two organs in this period: "[this lack of dialogue has] allowed the decision-makers to divest themselves of the responsibility for the decisions which should be theirs, and theirs alone, and has pushed that responsibility onto the scientific method." This created a situation where the Commission was paralyzed due to lack of decisive scientific evidence in the 1980s.

CCAMLR is rather exclusive in the sense that it allows state participation in the decision-making machinery for countries with either research or harvesting interests in the area. CCAMLR was also reluctant to allow NGOs as observers, not least due to opposition from the Soviet Union. Thus, compared to some other international resource regimes, including the IWC, CCAMLR and the Antarctic nations have been cautious players along this dimension. Nevertheless, its formal acceptance of NGOs in the Convention "proved to be something of a threshold in the ATS, opening the door of acceptance within the system to the new Antarctic NGOs" (Herr and Davis 1996, 100–101). Initially this did not have much effect, and CCAMLR stood forth as a rather closed regime with low media attention and limited participation. Lack of external scrutiny may have made it easier for the harvesting nations to resist the demand of the majority for more forceful action. Given the basic mistrust between key parties, it seemed that the consensus procedure and the science-policy dialogue were most important in preventing constructive problem solving.

The positive changes taking place in late 1980s and early 1990s cannot be attributed only to the exogenous factors. For one thing, the positive development started prior to the changes in Russia. By the end of the 1980s, "the institution became consolidated" and cooperation "took a positive, technical, and nonpolitical turn" (Vicuña 1991, 32, 34). The new and better procedures seem to have contributed to more conservation-oriented measures being adopted. When the Committee was asked by the Commission to produce more detailed advice, these new obligations demanded more data and contributed to putting more pressure on those reluctant to part with such data, most notably the Soviet Union: "The more detailed the information, the more difficult it is to argue reasonably against regulatory measures when that information, albeit uncertain, suggests that stocks are in jeopardy" (Stokke 1996, 147). Thus, the net spun around the fishing nations was becoming increasingly fine-meshed, making it increasingly difficult for them to escape. The Commission now had to make publicly available all the data presented to it, including "all the views expressed" during the deliberations, probably effective measures in exposing a reluctant minority. This also illustrates that the level of cooperation was gradually increasing somewhat.

It could well be conceived that such exposure would result in increasing polarization, as we have seen in the whaling regime, especially as many of the scientists were operating in both regimes. In contrast to the development in the IWC, however, it appears that the generally consensus-driven ATS cooperation has gotten the upper hand and contributed to a more conciliatory atmosphere of learning and incremental growth.[25] The difference is illustrated by the softer approach within CCAMLR, using shaming as a means to change behavior on the part of reluctant members rather than sanctions as applied by the United States on the whaling issue (De Sombre 1995). This gradual positive change has occurred without changing the decision-making procedures, indicating that there are other factors behind this procedure (like increased trust, fewer stakes, and institutional growth) that are more important. This being said, institutional growth and learning are fairly easy when turbulence is low, as used to be the case in the late 1980s to the early 1990s. When the Patagonian toothfish case exploded sometime later, an institutionally more mature CCAMLR was not able to handle the issue. If this problem should be handled effectively, soft and smoothly functioning procedures were not enough. There was a need for harder procedures related to enforcement, inspection, and control. Although these had improved, they were still inadequate. Also, underlying political conflicts over sovereignty between some key fishing states such as Argentina and Chile, on the one hand, and the United Kingdom, on the other hand, contributed to aggravate the enforcement problems (Herr 1999, 254–255).

The exclusive approach regarding state participation in CCAMLR has continued, but new members have been included if they demonstrate an active interest in the area. South Korea has gained membership in the Commission on these grounds. Admitting some new entrants and thereby including parties that may otherwise harvest outside the Convention but at the same time being rather restrictive, seemed for a long time to be a wise approach.[26] As the conflict over the Patagonian toothfish has shown, the exclusive and cosy ATS approach has not been able to handle the sudden interest in this resource by nonmembers. However, considering new harvesting members, CCAMLR should first clean up its own house by stopping members from harvesting through nonmember flags of convenience. As to the participation by nonstate actors, CCAMLR has opened up somewhat over time. On the one hand, some of the green NGOs have found their way to national delegations of nonharvesting nations like the United States, New Zealand, and Australia (Herr and Davis 1996). In addition, the Antarctic and Southern Ocean Coalition (ASOC), an umbrella organization for the environmental movement, achieved observer status with the plenary of CCAMLR in 1988 and with the Scientific Committee in 1991. Although the activist-oriented Greenpeace has spent a lot of time and money on Antarctic affairs, this organization has not been able to gain any formal position within the ATS, including CCAMLR.[27] This moderately open access structure overall seems to have had a positive effect on the cooperation within CCAMLR. Extreme polarization has been avoided while some NGOs have been included and thereby contributed values, expertise, and adding legitimacy to the regime.

Finally, are there any distinct *leaders* that have contributed to drive the cooperation within CCAMLR forward? In the *regime-creating phase*, the United States seemed to play a dominant role. However, it was quite an easy game as most key parties agreed on the overall unofficial goal of its establishment. Leadership seems to have been less prominent when the cooperation got under way in the 1980s. Detailed studies of the *national* Antarctic polices of some of the key Antarctic nations (including Norway, Australia, the United States, and Chile) underline the significance of the rather *exclusive club of individuals*, both domestically and internationally, dealing with Antarctic matters (Stokke and Vidas 1996). Stability is generally high for the personnel dealing with this issue, domestic attention is very low, and the ties between the key persons at the international level are very close. Another

Table 16.1a
Summary scores: Chapter 16 (CCAMLR)

| Regime | Component or Phase | Effectiveness | | Level of Collaboration |
		Improvement	Functional Optimum	
CCAMLR	1980–1997	Low	Low	Intermediate

pronounced common factor is the close links between the policy-makers and the Antarctic scientists. There is no doubt that this informal, rather consensual and exclusive network, with the common goal of keeping the ATS system in place, has played an important part also in the context of CCAMLR.[28] It may be that over time this network has proven instrumental in toning down the hostility of the mid-1980s and bringing CCAMLR into the more cosy and more consensual ATS family. It is more doubtful, however, whether this had been possible without the positive changes in problem structure and context. Moreover, it is not self-evident that this old network will be able to deal effectively with the new pressure on CCAMLR, but it may resurface when the Patagonian toothfish in the area is depleted.

Conclusions: Key Determinants for the Development of CCAMLR

CCAMLR has made improvements regarding the production of knowledge and regulations as well as verification during its almost two decades of existence (see table 16.1). Still, CCAMLR has not been very important in changing the behavior of the relevant target groups or in restoring the ecological balance in the area; therefore, it stands as a rather ineffective regime. The score might have been higher if the more forceful measures of the late 1980s and early 1990s had been adopted from the very start, but this was not done. This was partly due to lack of knowledge, but the main reasons were resistance from the least enthusiastic actor (the Soviet Union) and inability to counter the laggards from the nonharvesting majority. Had the main nonharvesting forces been more preoccupied with conservation of resources and less with keeping the ATS together, the history of CCAMLR might have been different. However, based on experience from most international cooperative efforts to manage marine living resources, it does not seem very likely this would have helped. Within traditional international regimes, the majority has no means to force the main laggards to change their position, especially if they are dominant actors like

Table 16.1b

Problem Type		Problem-Solving Capacity			
Malignancy	Uncertainty	Institutional Capacity	Power (basic)	Informal Leadership	Total
Moderate	High	Low	Laggard	Weak	Low

the Soviet Union. Moreover, it usually takes time to build effective institutions. Thus, what was needed in the initial phase was not possible to attain within the present structure of international society. The somewhat depressing lesson therefore is that things started to improve only after the problem became much easier to handle and that this change in the problem structure was not primarily brought about by the regime. Unfortunately, this happened too late to provide the basis for effective management.

As long as the problem structure remained essentially benign and considering that CCAMLR had matured significantly, the basis for improved management was present. But when the regime was put under stress by the increased catch of Patagonian toothfish, it revealed that it was still too weak to handle the problem effectively. Short-term economic interests have prevailed over conservation forces: more specifically, procedures for inspection and verification have been too weak.

Notes

1. It is debatable whether output can be used as an indicator of effectiveness. I return to this point later.

2. For an evaluation of international attempts to manage Antarctic whaling, see chapter 15 in this volume.

3. For a comprehensive overview of the legal and political development of the Antarctic regime and its various components, see Stokke and Vidas (1996).

4. Claims to territorial sovereignty over parts of Antarctica were made by several nations in the first half of this century: the United Kingdom (1908), New Zealand (1923), France (1924), Australia (1933), Norway (1939), Chile (1940), and Argentina (1942). Disputes over sovereignty have been particularly acute among the United Kingdom, Chile, and Argentina due to overlapping claims. The United States and the Soviet Union, while reserving their own rights, have expressly refused to recognize claims, and no other countries have given recognition to the claims.

5. Yet another link was established, as according to a member of the U.S. delegation CCAMLR was also seen as a guinea pig for a future minerals convention (Stokke 1996, 124).

6. In the early 1980s, a group of developing countries led by Malaysia launched a critique of the ATS in the United Nations for its exclusiveness (Vidas 1996). Although the issue is raised every year in the United Nations, this should be seen more as a part of some of the routine processes of certain countries in the United Nations rather than a real challenge to the ATS.

7. Special arrangements were made to secure FAO an important position within CCAMLR (Stokke 1996).

8. It was because of the specific characteristics of the Antarctic marine environment—the dependence of other marine living resources on krill—that an ecosystem approach was adopted.

9. See article 2, subparagraphs 3(a–c), for an operationalization ot its three-stranded eco-system standard.

10. Only total quotas are set—no national quotas.

11. Large cetaceans are excluded beeause they are covered by the management responsibil-ities of the IWC.

12. Nevertheless, as is discussed later, jurisdictional conflicts have again surfaced in relation to attempts to regulate the catch of Patagonian toothfish.

13. Research was not done only within the framework of CCAMLR, however. The role of SCAR was important initially. Later on the launching of the BIOMASS program was an important step in increasing knowledge in this area and yet was unrelated to CCAMLR. For further elaboration on the significance of the BIOMASS program, see Stokke (1996).

14. The use of these three output criteria (regulatory measures, inspection, and verification) when evaluating a resource management regime is fairly standard. In earlier assessments of CCAMLR only the first two criteria were applied (Andresen 1991). This was done simply because very little had been done on the last dimension. These three criteria have also been applied by Stokke (1996) in discussing the performance of CCAMLR.

15. An example of the ambiguities of the system is the unilateral enforcement in national economic zones by some of the parties. Moreover, it took a long time (one in 1993 and one in 1994) before the system of inspections was actually carried out in practice (Commission 1993, 20; 1994, 15).

16. For a closer insight and analysis of this problem, see Herr (1999). Herr also has some interesting reflections as to why this rather marginal issue has received so much attention: "And, adding spice to the toothfish controversy has been a liberal sprinkling of 'p' words in the media—piracy, poaching, and plundering" (Herr 1999, 242).

17. A more detailed picture is painted by Stokke (1996) in a detailed account of CCAMLR, pointing out that the Commission followed up advice from the SCICOM on the *marbled nothotenia* in the period 1984 to 1986.

18. This happened in 1991 in the case of mackerel ice fish around South Georgia (see Scientific Committee 1991 and Stokke 1996).

19. In a comparative analysis of the Baltic and the North Sea environmental regimes, the former had a higher score as regards output, while the latter was more effective in a behav-ioral outcome perspective (Andresen 1996).

20. Catch has been concentrated in the South Atlantic around South Georgia and the South Indian Ocean around the Iles Kergulen. The largest catch has been taken around

South Georgia. For an overview of the distribution of catch, see Heap (1991) and Stokke (1996).

21. The following nations used to be engaged in fishing activities: the Soviet Union, Japan, Poland, East Germany, Bulgaria, Chile, Argentina, West Germany, South Korea, Taiwan, and France. New actors have entered the scene more recently.

22. A figure of .05 of 1 percent has been suggested (Herr 1999, 256).

23. Although the Fish Stock Agreement is not yet in force, its provisions may, of course, be applied if the parties so wish.

24. For a thorough elaboration of these and related legal problems, see Herr (1999).

25. According to John Gulland, who participated in this process for a number of years, the general consensus-drive of the ATS has been a major condition for this development to take place (Gulland 1986).

26. For a more general discussion of merits and shortcomings of various participatory approaches in international regimes, as well how this has affected specific regimes, see Wettestad (1999) and Victor et al. (1998).

27. As noted by Herr and Davis (1996), this does not mean that Greenpeace is without influence; it is active within ASOC as well as in delegations of certain countries.

28. It may be that this network does not accord with the Haas (1992) criteria of an *epistemic community*, but it has certainly to a large extent had the same effect as described by Haas regarding other regimes.

References

Andresen, S. 1991. "The Convention on the Conservation of Antarctic Marine Living Resources (CCAMLR)." In J. Wettestad and S. Andresen, eds., *The Effectiveness of International Resource Cooperation: Some Preliminary Findings*. Lysaker, Norway: Fridtjof Nansen Institute.

Andresen, S. 1996. "Implementation of International Environmental Commitments: The Case of the Northern Seas." *Science of the Total Environment* 186: 149–167.

Barnes, J. N. 1982. "The Emerging Convention on the Conservation of Antarctic Marine Living Resources: An Attempt to Meet the New Realities of Resources Exploitation in the Southern Ocean." In J. Charney, ed., *The New Nationalism and the Use of Common Spaces* (pp. 239–286).

Barnes, J. N. 1991. "Protection of the Environment in Antarctica: Are Present Regimes Enough?" In A. Jørgensen-Dahl and W. Østreng, eds., *The Antarctic Treaty in World Politics* (pp. 186–229). New York: MacMillan.

Basson, M. and J. Beddington. 1991. "The Practical Implication of an Eco-System Approach." In A. Jorgensen-Dahl and W. Østreng, eds., *The Antarctic Treaty System in World Politics* (pp. 54–70). London: MacMillan.

Beck, P. J. 1986. *The International Politics of Antarctica*. London: Croom Helm.

Beddington, J. R., R. J. H. Beverton, and D. M. Lavigne, eds. 1985. *Interaction of Fisheries and Marine Mammals*. London: Allen and Unwin.

Commission for the Conservation of Antarctic Marine Living Resources. 1987. *Report of the Sixth Meeting of the Commission.* Hobart: Commission for the Conservation of Antarctic Marine Living Resources.

Commission for the Conservation of Antarctic Marine Living Resources. 1988. *Report of the Seventh Meeting of the Commission.* Hobart: Commission for the Conservation of Antarctic Marine Living Resources.

Commission for the Conservation of Antarctic Marine Living Resources. 1990. *Report of the Ninth Meeting of the Commission.* Hobart: Commission for the Conservation of Antarctic Marine Living Resources.

Commission for the Conservation of Antarctic Marine Living Resources. 1992. *Report of the Eleventh Meeting of the Commission.* Hobart: Commission for the Conservation of Antarctic Marine Living Resources.

Cordonnery, L. 1998. "Environmental Protection in Antarctica: Drawing Lessons from the CCAMLR Model for the Implementation of the Madrid Protocol." *Ocean Development and International Law* 29(2) (April): 125–147.

DeSombre, B. 1995. "Baptists and Bootleggers for the Environment: The Origins of United States Unilateral Sanction," *Journal of Environment and Development*, vol. 4, no. 1, pp. 53–75.

Edwards, D. and J. Heap. 1981. "Convention on the Conservation of Antarctic Marine Living Resources: A Commentary." *Polar Record* 20: 353–362.

Gulland, J. A. 1986. "The Antarctic Treaty System as a Resource Management Mechanism: Living Resources." In *Antarctic Treaty System: An Assessment* (pp. 229–230). Washington, DC: National Academy Press.

Gulland, J. A. 1988. "The Management Regime for Living Resources." In C. C. Joyner and S. K. Chopra, eds., *The Antarctic Legal Regime.* Dordrecht: Martinus Nijhoff.

Haas, P. M., ed. 1992. "Knowledge, Power, and International Policy Coordination." *International Organization* (special issue) 46.

Heap, J. 1988. "The Role of Scientific Advice for the Decision-Making Process in the Antarctic Treaty System." In R. Wolfrum, ed., *Antarctic Challenge III* Berlin: Duncker and Humbloth (pp. 21–28).

Heap, J. 1991. "Has CCAMLR Worked? Management Politics and Ecological Needs." In A. Jørgensen-Dahl and W. Østreng, eds., *W: The Antarctic Treaty in World Politics* (pp. 43–54). London: MacMillan.

Herr, R. 1999. "The International Regulation of Patagonian Toothfish: CCAMLR and High Seas Fisheries Management." In O. Stokke, ed., *Governing High Seas Fisheries: The Interplay of Global and Regional Regimes* (pp. 242–267).

Herr, R. A. and W. B. Davis. 1996. "Decision-Making and Change: The Role of Domestic Politics in Australia." In O. S. Stokke and D. Vidas, eds., *Governing the Antarctic: The Effectiveness and Legitimacy of the Antarctic Treaty System* (pp. 331–360). Cambridge: Cambridge University Press.

Hoel, A. H. 1990. *Managing the Fisheries in Antarctic Waters: Towards Sustainable Management?* Tromsoe: College of Fishery Science, University of Tromsoe.

Indreeide, T. V. 1989. *Forholdet mellom vitenskap og politikk i internasjonal ressursforvaltning: Det internasjonale samarbeidet om forvaltning av levende marine ressurser i Antarktis.* R.007, Lysaker: Fridts of Nausen Institute: FNI.

Infante, M. T. 1996. "Chilean Antarctic Policy: The Influence of Domestic and Foreign Policy." In O. S. Stokke and D. Vidas, eds., *Governing the Antarctic: The Effectiveness and Legitimacy of the Antarctic Treaty System* (pp. 361–383). Cambridge: Cambridge University Press.

Jorgensen-Dahl, A. and W. Østreng. 1991. W: *The Antarctic Treaty in World Politics.* London: MacMillan.

Joyner, C. 1991. "CRAMRA: The Ugly Duckling of the Antarctic Treaty System." In A. Jørgensen-Dahl and W. Østreng, W: *The Antarctic Treaty in World Politics* (pp. 161–186). London: MacMillan.

Joyner, C. 1996. "The Role of Domestic Politics in Making United States Antarctic Policy." In O. S. Stokke and D. Vidas, eds., *Governing the Antarctic: The Effectiveness and Legitimacy of the Antarctic Treaty System* (pp. 409–432). Cambridge: Cambridge University Press.

Lagoni, R. 1984. "Convention on the Conservation of Antarctic Marine Living Resources: A Model for the Use of a Common Good?" In R. Wolfrum, ed., *Antarctic Challenge: Conflicting Interests, Cooperation, Environmental Protection, Economic Development.* Berlin: Duncker & Humblot.

Scientific Committee. 1991. *Report of the Tenth Meeting of the Scientific Committee.*

Støkke, O. S. 1996. "The Effectiveness of CCAMLR." In O. S. Støkke and D. Vidas, eds., *Governing the Antarctic: The Effectiveness and Legitimacy of the Antarctic Treaty System* (pp. 120–152). Cambridge: Cambridge University Press.

Stokke, O., ed. 2001. *Governing High Seas Fisheries: The Interplay of Global and Regional Regimes.* Oxford: Oxford University Press.

Stokke, O. S. and D. Vidas. 1996. *Governing the Antarctic: The Effectiveness and Legitimacy of the Antarctic Treaty System.* Cambridge: Cambridge University Press.

Underdal, A. 1980. *The Politics of International Fisheries Management: The Case of the Northeast Atlantic* (Oslo: Universitetsforlaget).

Victor, David G., Kal Raustiala, and Eugene B. Skolnikoff, eds. 1998. *The Implementation and Effectiveness of International Commitments: Theory and Practice.* Cambridge, MA: MIT Press.

Vicuña, F. O. 1991. "The Effectiveness of the Decisionmaking Machinery of CCAMLR: An Assessment." In A. Jørgensen-Dahl and W. Østreng, eds., W: *The Antarctic Treaty in World Politics* (pp. 25–43). London: MacMillan.

Wettestad, Jørgen. 1999. *Designing Effective Environmental Regimes.* Cheltenham, U.K. and Northampton, MA: Edward Elgar.

VI
Conclusions

17

Conclusions: Patterns of Regime Effectiveness[1]

Arild Underdal

Purpose and Scope

We now shift from an intensive to an extensive mode of analysis. Instead of examining each single case in depth, we now try to pull together evidence from all the fourteen regimes that we have analyzed in part II. Instead of trying to identify the particulars of each case, we now search for factors that seem to shape outcomes in all or at least a large number of cases and try to determine the direction and strength of their impact. The purpose of this chapter is, in other words, to determine the *patterns* of association that exist between regime effectiveness and a set of variables that we assume can affect the formation and performance of regimes.

We proceed in three main steps. First we take a quick look at scores on our *dep*endent variables to get an overall impression of the extent to which the regimes in our sample make a significant difference and achieve optimal solutions. Then we move on to examine *intraregime* changes occurring as a regime spawns a new component (often in the form of a new protocol) or moves from one phase to another. Finally, we examine patterns of variance *across* as well as *within* regimes, drawing on our complete database. The analysis is guided by the hypotheses formulated in chapter 1. In particular, we try to determine how far our core model—conceiving of regime effectiveness as a function of particular notions of problem malignancy and problem-solving capacity—can help us account for the variance actually observed in regime performance. Several of the case studies have attributed causal importance also to other factors, but most of these are variables for which we have not systematically collected data across a larger set of cases.

Before reporting results, a few words are needed to explain the path we have chosen and to point out the most important limitations that pertain to this particular analysis.

The Path and Its Limitations

First, we would like to remind the reader that our unit of analysis in this part of the study is regime *components*, *phases*, or for the intraregime analysis instances of *transition* from one component or phase to another. Unless explicitly stated otherwise, when we speak of cases, we refer to these components, phases, or events rather than to regimes in total.

Second, it bears repeating that the scores we have assigned to our dependent and most of our independent variables are based on complex, qualitative interpretation rather than straightforward observation or measurement. Because of the difficulties involved, for some variables we resort to second-best solutions. For example, ideally we would have liked to base our scores of problem malignancy on actor perceptions rather than on our own interpretations of the objective structure of the problem. Not surprisingly, however, we found it harder to come up with reliable information about actor perceptions than about interdependence mechanisms. We have therefore decided to rely primarily on the latter.[2] In a few instances we have, with the benefit of hindsight, discovered that the conceptualization developed in chapter 1 is incomplete or in other respects inadequate. The most important deficiency pertains to our notion of problem malignancy, which is conceptualized exclusively in terms of conflicts of *interests*—missing the possibility of conflict over *values*. Particularly in one instance (the most recent phase of IWC), this narrow conceptualization leads to severe problems of interpretation. The good news is that by now we are at least aware of this problem and can take appropriate measures.[3]

Third, even though this—to our knowledge—is the most extensive comparative study so far in its field, our database is still too small for anything but the crudest forms of statistical analysis. Our main database includes a total of forty-four cases. At most, thirty-seven of these can be used for studying regime effectiveness.[4] With such a small number of cases we often have to merge categories and combine variables to avoid or at least reduce the empty-cell problem. Merging of categories implies that we cannot fully utilize the analytical distinctions that are made in the codebook and in the individual case studies. Combining variables into composite indexes means that we have to leave some of our hypotheses untested. For example, we cannot adequately distinguish the impact of decision rules (H_5) from that of other elements of institutional capacity. In general, the small number of cases severely constrains the scope for multivariate analysis. Merging categories and combining variables help, but even with dichotomized variables and composite indexes our ability to control for other factors will be very limited. The rule of thumb followed in this chapter is to work with the variables and categories originally used in the coding to the extent that our data files permits, so as to minimize the loss of information. One consequence of this choice is that we sometimes are skating on very thin ice. In particular, the results reported from multivariate analysis should be interpreted with great caution.

Are International Regimes Effective?

The short answer is that most of the regimes included in our sample make a positive difference but fall short of providing functionally optimal solutions.[5] In more than half of our cases improvement measured in terms of behavioral change is considered to be significant or major. On a scale ranging from 0 (negative change) to 4 (major improvement), the average score is 2.5. If we measure effectiveness in terms of accomplishment of functionally optimal solutions on a scale with 1 as the lowest and 3 as the highest value, the average is 1.7.[6] To make scores on these two variables comparable we can convert them to a standardized scale ranging from 0 to 1. The coefficients that come out of this transformation can be seen as a crude measure of the extent to which effectiveness *potentials* are actually tapped. In our data set these coefficients are .51 for behavioral and .35 for functional effectiveness. Nearly 60 percent of our cases score low on the latter dimension. These differences support our argument that the two standards of effectiveness should be clearly distinguished. As expected, the latter is more demanding than the former.

Regimes can have a fair amount of success in dealing also with *malign* problems. The standardized effectiveness coefficients for malign problems are .48 for behavioral change and .34 if measured in relation to optimal solutions. And there is more good news. Where the problem that motivated the establishment of the regime has become less severe, we find that the regime has been as important or more important than other factors in more than half of the cases. In about two out of every three cases the regime seems to have served as an arena for transnational learning about the problem in question and contributed to strengthening the knowledge base for policy decisions. Furthermore, unambiguous *growth* is reported for nine of the fourteen regimes included in this study. Finally, if we look at different generations of regimes, it seems that regulations established during the 1960s and 1970s were on average slightly more effective than those established in earlier periods. No further increase in mean scores can be found for regimes formed or regulations introduced during the 1980s and early 1990s. The latter should not necessarily be interpreted as bad news, though. At least in the early stages, effectiveness is likely to increase with maturity. This implies that the younger regimes may well be more effective than earlier generations were *at the same point in their life cycle.*

Intraregime Variance

To what extent can our core model account for changes in effectiveness *within* regimes—between components or phases of the same regime? To answer this question we constructed a data file with instances of transition from one component or phase to the next as our unit of analysis. We recorded a total of twenty-three such events.[7] Effectiveness measured in terms of behavioral change increased in twelve of these cases, decreased in one, and remained constant in ten.[8] Ten shifts brought us closer to the functional optimum, while one led to a lower score. These figures may plausibly be interpreted as an indication that most of these regimes become more effective over time. Our small data file effectively precludes any attempt at determining the impact of specific components of capacity or aspects of problems. We therefore have to resort to aggregate indexes of problem type and problem-solving capacity. Moreover, we have to merge categories; for each of these composite constructs and for our dependent variables we distinguish between three categories of change: positive, constant or neutral, and negative. The configuration of scores even makes it hard to distinguish the impact of capacity from that of problem type. In this section we therefore concentrate on their *combined* effect.

Table 17.1 offers a first cut. The figures are standardized improvement coefficients ranging from +1.00 (meaning that effectiveness *in*creases in all cases) to −1.00 (meaning that effectiveness *de*creases everywhere). The label *positive* refers to instances where (1) problem type and capacity both changed in a positive direction

Table 17.1
Intraregime change: Standardized improvement coefficients

| | Positive | | Predicted Outcome | | | |
			Constant/Neutral		Negative	
Behavioral effectiveness	1.00	.62	.00	.25	.00	.00
Functional effectiveness	1.00	.60	.22	.30	.00	.00
N	3	13	6	9	1	1

or (2) one of these changed in a positive direction while the other remained constant. The label *negative* is used for instances with the opposite characteristics, while *constant/neutral* refers to cases where (1) neither of the two key independent variables changed or (2) they both changed but in opposite directions. Within each cell, the coefficient to the left refers to the pure case (where both changed in the same direction, or remained constant), while the coefficient to the right includes also cases that are weakly positive (negative) or—for the constant or neutral category— characterized by changes in opposite directions.

The overall pattern supports our basic argument that problem type and response capacity are critical determinants of regime effectiveness. As we can see from the column labeled positive, a shift toward a more tractable problem combined with an increase in problem-solving capacity *always* leads to higher regime effectiveness in our set of cases. Negative shifts and constants yield substantially lower scores, and the order of decline is essentially as predicted for both concepts of effectiveness. Our case studies attribute most of the credit for the increase in regime effectiveness to a strengthening of one or more of the capacity components and to improvement in the state of knowledge.

There is, however, a flip side of the coin: shifts toward higher effectiveness occur also in several instances where our core model would predict no change. As shown in table 17.2, in the least favorable interpretation our core model errs on the pessimistic side in about 25 percent of the cases and on the optimistic side in another 10 to 15 percent. If we consider only transitions that are categorized as strongly positive, negative, or perfectly stable, the model errs on the side of pessimism in one out of ten cases (for behavioral change) and yields overly optimistic predictions in

Table 17.2
Correct and false predictions (in per cent of total)

Dependent var. ↓	Predictions							
	Correct		Too pessimistic		Too optimistic		N	
Behavioral effectiveness	90	65	10	22	0	13	10	23
Functional effectiveness	100	64	0	23	0	14	10	22

Note: A prediction is considered "correct" if, and only if, a positive shift in the aggregate score for the independent variables leads to a shift in the same direction in effectiveness, or if both sides of the equation remain constant.

none. Within each cell, figures to the left are scores for cases that according to our model come out as *strongly* positive, negative, or completely stable.

This record is good but not entirely convincing and suggests that our core model may be deficient in at least one of two respects.[9] One is incompleteness, meaning that there are also other pathways to effectiveness. This is almost certainly the case; we never had the ambition to construct an exhaustive model that would incorporate *all* factors that can influence regime effectiveness. Our case studies have also pointed to several other factors that seem to have facilitated or obstructed regime formation or operation in particular cases. For example, in the South Pacific tuna case, a shift of personnel and a strengthening of coalition cohesion were critical factors. In the ship-generated oil-pollution case, exogenous shocks spurred individual as well as collective action. In the radioactive-waste and dumping cases, actions by NGOs politicized the issues and in a few instances even blocked the actual operations. Several of the issue-specific factors that are emphasized in the case studies can, though, be subsumed into one of our main categories (type of problem and capacity). Examples are the institutional reinforcement that occurred through the establishment of the North Sea Conferences (in the case of PARCOM) and the change in the distribution of power brought about by the turnaround of West Germany in the acid-rain case.

This suggests that another deficiency—inadequate or false specification of causal relationships that are included in the model—may be equally important. It is abundantly clear that the specification used in this analysis is rather poor. In most cases, our crude aggregate indexes give equal weight to all independent variables included. We implicitly assume that causal relationships are all monotonous. Important nuances may well be buried in our crude trichotomies. The format we have chosen for this part of the comparative analysis offers few opportunities for working with a more sophisticated specification of the model. What we can do, however, is explore the impact of one complex exogenous factor for which we have at least some important data: the *political context* within which the regime has been established and operates.

Our core model assumes that each regime or regime component is negotiated and implemented on its own merits only, with no exogenous considerations interfering. We all know that in the real world issues are often linked in functional or political terms. Several of our case studies—including those of LRTAP, CCAMLR, and nuclear nonproliferation—bear this out. This raises two questions: to what extent can such linkages account for the deviance observed in table 17.2, and to what extent can the predictions we have labeled *correct* reflect the impact of context rather than of problem type and capacity?

The short answer to the former question is that if we require that the context be *unambiguously* conducive or obstructive, adding context—as conceptualized here—to the model does not give us much additional mileage. In none of the five cases where our core model—in the weak interpretation—led us to err on the pessimistic side in predicting change in behavioral effectiveness did the regime benefit from a supportive political context or a positive change in context. In the three cases where our predictions were too optimistic, the context is arguably negative in two. With regard to the second question, suffice it here to note that the context is positive in none of the nine cases where the core model correctly predicted increasing behavioral effectiveness. The picture is essentially the same with regard to functional effectiveness. A safe conclusion seems to be that adding context—as conceptualized here—to our core model can, at best, increase the proportion of correct predictions marginally. Overall, we still see a greater increase in effectiveness than our model would predict.

Some of the remaining discrepancies could be accounted for simply by summarizing observations made in the various case studies about other factors at play. There is, however, no compelling reason to assume that the net effect of this set of additional factors will *generally* be positive (although in this particular sample, one of the key factors—exogenous events—most often weighs in on the positive side). The pattern that emerges from this first cut into our database seems to suggest that a more promising strategy would be to search for basic causal mechanisms that work *consistently* to facilitate regime formation and enhance performance over time. At least three partly interrelated mechanisms of this kind can be identified—shifts in *policy priorities*, *learning*, and *institutional maturing*.

At least in the Western world, public demand for and governmental supply of environmental protection have increased significantly over the past three to four decades. Spurred by growing public concern with environmental health and stimulated by a significant buildup of domestic institutional capacity for environmental governance, the environmental policies of most if not all OECD countries (and several other countries as well) were at a more advanced stage by the end of the twentieth century than they were twenty or thirty years earlier (see, e.g., Weale 1992; Jänicke and Weidner 1997). The development of international environmental regimes most likely mirrors this positive shift in basic values, policy priorities, and institutional capacity at the domestic level—although probably with some time lag. Superimposed on the specifics of each particular regime or issue is a *general* growth (or decline) of a policy field.[10] According to this line of reasoning, the overall trend toward increasing regime effectiveness that we find in our sample can be interpreted as a consequence of this general shift in basic values and policy priorities.

This argument has an equally interesting flip side, suggesting that no increase in regime effectiveness can be expected in policy areas or periods where no such positive shifts occur. Since our sample includes mainly environmental regimes, we cannot shed much light on the latter proposition. A careful reading of our case studies does, however, suggest that there are at least two other mechanisms that are generated by the institutions and processes themselves and thus presumably are at work also in stagnant policy fields: learning and institutional maturing.

In several cases—LRTAP being a good example—regimes seem to have benefited significantly from *learning* on the part of the actors involved. Moreover, the regimes themselves have often served as important arenas for the exchange of information and in several cases have initiated and managed research that has contributed significantly to a better understanding of the problem in focus. We can see that the state of knowledge tends to improve over time, and part of this improvement can in most cases be attributed to activities initiated or managed by a regime. Previous experience offers important lessons for the design of future regimes or regulations; solutions that have proved effective in one setting are often adopted in another, while measures that have failed are avoided. Furthermore, a political institution does not work like a piece of technical equipment, producing instant results once it has been installed. Institutions are typically developed and put into operation over a period of several years, in some cases even decades. Whatever results they produce are typically brought about through gradual adjustment of actor behavior and perhaps also modification of actor preferences. This is by no means to suggest a simple law of linear growth; we do know that regimes sometimes move ahead in quantum leaps and sometimes stagnate, decline, or collapse. Moreover, we know that internal features can play an important role in causing or accelerating change. The point we are making is simply this: at least in the early stages of regime formation and implementation, learning and institutional maturing are likely to be mechanisms at work in a large number of cases and also in nonexpanding policy fields. Their combined effect may well be strong enough to bring about an increase in effectiveness even in the absence of growth in capacity or a shift to a more benign problem. Unfortunately, we do not have the data required to determine whether and to what extent these mechanisms can help us account for the deviance observed in our particular sample. What we can say is that there is nothing in the evidence we have that speaks *against* such an interpretation. *All* cases where the predictions derived from our core model proved too pessimistic are instances of institutional growth or improvement in the knowledge base.

Overall Variance

A study of intraregime change can provide important clues but is not sufficient to explain variance in regime effectiveness. The most obvious limitation is that it does not tell us anything about *absolute* scores. Thus, a regime that starts at a very low level of effectiveness may well improve but still fail to reach the level of another regime that started with a high score and remained stable. Conversely, a regime that starts out with a very high score may well decline and still do better than a regime that can point to a record of improved performance. Only by taking a different approach, focusing on absolute scores rather than changes in scores, can we get a reliable basis for determining the merits of the hypotheses we formulated in chapter 1. By shifting gears we will also be in a better position to control for the mechanisms discussed at the end of the preceding section.

In technical terms, the core of this shift of approach is a change in the unit of analysis. In the previous section we studied instances of *intraregime transition* from one component or phase to the next—that is, *events*. In this section, our unit of analysis is *regime components or phases*. We now search for *patterns of variance* across as well as within regimes. We proceed in three main steps. First, we examine bivariate measures of association between effectiveness and our two basic explanatory factors—*type of problem* and *problem-solving capacity*. We then move on to a modest multivariate analysis, eventually including also level of collaboration as an intervening variable. Since type of problem and problem-solving capacity are both complex and multidimensional constructs, a final step is to decompose each of these aggregate constructs into their main constitutive elements. This is particularly important in the case of capacity, since we have hypothesized that different elements serve different functions, meaning that the importance of each element will vary from one kind of problem to another. Throughout this section we combine ordinary statistical techniques with the Ragin Qualitative Comparative Analysis (QCA) approach.[11]

The Overall Pattern

In pursuing this agenda, we use the composite indexes of problem malignancy and problem-solving capacity described in chapter 2. To repeat, the index of political malignancy takes as its basis the distinction between problems of incongruity and problems of coordination. For problems of incongruity and only for this category, we then examine the degree of asymmetry and the presence of cumulative cleavages (see pp. 55–56).[12] In chapter 1 we formulated two hypotheses referring to different

Table 17.3
Impact of problem type and capacity: Bivariate measures (N = 36 or 37)

Independent Variables	Behavioral Change		Functional Optima	
	Spearman	Distance	Spearman	Distance
Problem malignancy	−.10	−.08	−.16	−.06
Malignancy + uncertain	−.39*	−.38	−.34*	−.23
Capacity	.33*	.30	.33*	.21

Notes: * Significant at the .05 level.
** Significant at the .01 level.

notions of type of problem. One (H_1) suggested that effectiveness would tend to decline with increasing political malignancy as defined above, while the other (H_2) focused on the *combined* effect of malignancy and uncertainty about the nature of the problem. Below we report results for both these notions. The index of problem-solving capacity also includes three main components: institutional capacity, the distribution of power, and the supply of instrumental leadership by national governments or delegations or by transnational networks of experts (epistemic communities) (see pp. 13–15 and 55–56).[13]

Table 17.3 shows the bivariate relationships between our two concepts of effectiveness on the one hand and problem type and problem-solving capacity on the other. We use two different statistical measures of impact. One is Spearman's rank-order correlation, measuring the degree of *correspondence* between two ordinal level rankings (range −1 to 1). The other is the difference between the top and bottom categories with regard to standardized effectiveness coefficients. To repeat, these coefficients express actual achievement as a fraction of potential achievement and range from 0 to 1.[14] As the two measures are different in substantive contents—one measuring *correspondence* between two rank orders, the other *distance* in mean scores—there is a priori no reason to expect the two to show a strong pattern of covariance. They should, though, point in the same general direction.

The results can briefly be summarized as follows. Both problem type and problem-solving capacity are statistically associated with both concepts of effectiveness, and the direction of association is as we expected—negative for malignancy and uncertainty and positive for capacity. Somewhat surprisingly, problem malignancy in itself does not seem to make much of a difference. On the other hand, the combination of malignancy and uncertainty seems to be a potent factor. This difference in scores between our two conceptualizations of problem type might suggest either that the state of knowledge is the more important variable or that the two interact so that the consequences of malignancy vary with the state of knowledge (and perhaps vice versa).[15] There is evidence to support both these interpretations. The bivariate correlation between uncertainty and effectiveness is −.40* for behavioral change and −.24 for functional achievement. This indicates that the state of knowledge can be an important factor in its own right.[16] It would, however, be much too early to take political malignancy out of the equation. On closer inspection, we find that the bivariate correlation between malignancy and effectiveness is very weak for problems characterized by low or moderate uncertainty but −.51 (behavioral change) and −.62* (functional achievement) for problems characterized by high uncertainty. In other words, the impact of malignancy tends to increase substantially with uncertainty, as hypothesized in H_2. We may therefore tentatively conclude that both these dimensions are important factors and that they interact as we expected. However, it also seems that the impact of malignancy is more complex and contingent than suggested in H_1.

In chapter 1 we surmised that *problem type* would be the more important determinant of effectiveness. To assess how much of the variance can be attributed to each of the two main factors we have to undertake some kind of multivariate analysis. In table 17.4 we summarize the results from two rounds of logistic regression analysis—one with problem type defined in terms of malignancy, the other defining problem type in terms of the malignancy-uncertainty combination. The results are most easily reviewed column by column. The first column shows effectiveness measured as behavioral change. At this stage we can draw at least two conclusions. First, the trivariate analysis confirms that problem type and capacity both make a difference and that the direction of impact is as predicted. Malignancy still seems to have a very weak impact. The latter is a somewhat puzzling result that calls for further examination (see below). On the other hand, the malignancy + uncertainty combination seems to be at least as important as capacity. Second, taken together, problem type and capacity account for a substantial proportion of the variance observed in this set of cases. The estimated probability of success is .95 when a system with high capacity encounters a nonmalignant problem that is well understood and a dismal .08 when a system with low capacity faces a malign problem clouded by high uncertainty. We predict success more accurately than failure; with the malignancy + uncertainty combination in the equation, we predict correctly 85

Table 17.4a
Problem malignancy and capacity: Logistic regression (N = 36)

	Behavioral Change	Functional Effectiveness
Malignancy	$b = -.22$	$b = -.27$
Capacity	$b = 1.12^*$	$b = .99$
Model chi-square	6.25^* (df = 2)	5.13 (df = 2)
Percent correct predictions	69%	67%

Table 17.4b
The malignancy-uncertainty combination and capacity: Logistic regression (N = 36)

	Behavioral Change	Functional Effectiveness
Malignancy + uncertainty	$B = -1.79^*$	$b = -1.00$
Capacity	$B = .90$	$b = .84$
Model chi-square	12.11^{**} (df = 2)	7.09^* (df = 2)
Percent correct predictions	75%	75%

percent of the high-effectiveness cases but only 63 percent of the cases with no or minor improvement.

The results for *effectiveness* defined in terms of distance to functionally optimal solutions show the same overall pattern, with two exceptions. First, the probability of success under the most favorable circumstances is lower (.80). Second, we tend to err on the optimistic side, predicting more successes than we actually observe. These observations can both be interpreted as indications that functional effectiveness is the more demanding standard. We also see indications of interplay, meaning that the effect of the each of the two main factors seems to vary with the score on the other. The combination of malignancy and uncertainty reduces effectiveness more when capacity is high than when it is low. For its part, capacity makes less of a difference in coping with malign problems wrapped in uncertainty than in dealing with more benign problems that are better understood. For the former type of problems, even high-capacity systems have a modest rate of success ($p = .35$).

The first round of QCA analysis adds two interesting observations. The bad news is that the combination of high malignancy and high uncertainty can be lethal; in a veristic mode analysis it comes out as a *sufficient* condition for low functional effectiveness.[17] The good news is that the opposite combination of a nonmalign problem and low or moderate uncertainty comes out as a sufficient condition for significant or major improvement in actor behavior (but not for high functional effectiveness). Indirectly, these two findings suggest that problem-solving capacity—at least as conceptualized in this study—is most critical when it comes to dealing with moderately malign problems characterized by moderate uncertainty. For easier problems, high capacity does not seem to be necessary, and for very malign problems clouded in profound uncertainty, it does not seem to be sufficient.

In H_3 we hypothesized that regimes dealing with truly malign problems could achieve high effectiveness *only* in the presence of at least one of the following elements: (1) selective incentives for cooperation, (2) linkages to more benign issues, and (3) a system with high problem-solving capacity. The results summarized in table 17.5 support this hypothesis; in no instance do we find high effectiveness in a strongly malign case when all three elements are absent. Had we added uncertainty the picture would have become even more somber; strongly malign problems that are poorly understood *all* have a low score on *both* dimensions of effectiveness. An equally striking result that comes out of table 17.5 is, however, the fairly high rate of success when (at least) two of the three elements are present. And there is more good news. In our sample, this fortunate situation is not a rare exception: it occurs in nearly 40 percent of the strongly malign cases. A safe conclusion seems to be that, at least in combination, the three factors identified make a huge difference.

Before we move on, let us pause for a moment to consider the poor performance of problem malignancy so far in predicting regime effectiveness. The low coefficients in tables 17.3 and 17.4 not only fail to support one of our main propositions (H_1) but also do not square well with the general thrust of argument in our case studies. Moreover, table 17.5 clearly shows that in the absence of high problem-solving capacity or in other fortunate circumstances encounters with strongly malign problems most often fail, as we would expect. These observations strongly suggest that there may be methodological as well as substantive explanations for what appears to be inconsistent results. We have already identified at least one major substantive piece of the puzzle: *interplay*. The effect of malignancy seems to be contingent on the level and scope of uncertainty; malignancy comes out as a highly potent factor *in combination with* uncertainty but not for problems that are fairly well understood. This holds also when we control for capacity. The synergistic relationship works both ways; for its part, uncertainty seems to be more of an obstacle in dealing with the more malignant problems. Moreover, uncertainty may serve to enhance

Table 17.5
The impact of selective incentives, linkages, and capacity for strongly malign problems (N = 18)

Components Present	Behavioral Change		Functional Effectiveness	
	Highest	Higher Two	Highest	Higher Two
0	0%	20%	0%	0%
1	20	53	0	17
2	57	71	57	71
3[a]	—	—	—	—

Note: Percent of cases in each row that reach the highest (or one of the higher two) levels of effectiveness.
a. No data available.

malignancy. In at least three regimes—LRTAP, ozone, and MEDPLAN—new knowledge led to a change in actor perceptions of the political structure of the problem, by and large in the direction of a more *benign* interpretation. This all provides substantial support for H_2, but it also suggests that H_1 may be underspecified and thus fail to capture adequately the impact of problem malignancy.

The inconsistency with the general thrust of the case-study analysis indicates that we also have problems of *measurement error*. A careful reading of some of the case studies (in particular, those of the North Pacific salmon and South Pacific tuna fisheries) brings out at least one: a problem may be strongly malignant for the group of parties as a whole and yet predominantly benign for a subgroup. In such cases, there may be good prospects for coalition-building. And a coalition that can marshal sufficient power will eventually win. Such an outcome does not challenge the basic argument behind our core model, but our database does not grasp this kind of configuration well at all. The reason is simply that our codebook does not provide for differentiation of problem type by subgroup or scale, thereby forcing the coder to make an overall assessment that obscures the point one way or the other.

The analysis of intraregime change offers another important clue.[18] We observed that behavioral effectiveness increased in twelve and functional achievement in ten out of twenty-three (twenty-two) transitions. In only three cases did we find a shift toward a more benign problem. This implies that most of the change in effectiveness that we observed *within* regimes must be accounted for in terms of other factors. A quick look at the case studies is sufficient to confirm this conclusion. This suggests that we may get a better understanding of the impact of malignancy by controlling for stage of regime development. When we do, we find that problem malignancy is a good predictor of success in the *initial* stage[19] but not for further regime development. The small number of valid observations calls for a cautious interpretation, but the implications are sufficiently interesting to be spelled out. First, the finding suggests a reformulation of H_1. The reformulation would differentiate the impact of malignancy by stage. In the preregime phase and in the initial stage of regime formation, problem malignancy seems to be a crucial determinant of outcomes. Once the initial hurdles are cleared, however, the *further* development of the regime seems to depend primarily on other factors—including state of knowledge, problem-solving capacity and exogenous events. If we are right, it follows that the research design that we have adopted for this study fails to grasp the full impact of malignancy. We have studied *regimes*—that is, instances where some cooperative arrangement has been established. To grasp the full impact of malignancy we would, however, have had to focus on *problems*—including those that did *not* generate institutionalized cooperation as well as those that did. Since the perspective we adopted here is shared by almost all of our colleagues, there is a real possibility that *the entire field of regime analysis may be biased in favor of positive findings.*

Our core model includes also one *intervening* variable: *level of collaboration* (see chapter 1). Its relevance in this particular context hinges on two conditions. First, it should not be strongly correlated with either of our *in*dependent variables. This requirement is met. As expected, the level of collaboration tends to decline as malignancy and uncertainty increase and to increase with capacity, but neither of these correlations is strong (−.19 and .24, respectively). Second, our interest in exploring the impact of an intervening variable increases if it is strongly correlated with our *de*pendent variables. This requirement is at least to a large extent met: Spearman's rank-order correlation is .50** for behavioral change and .44** for functional effectiveness.[20]

To get a perspective on relationships among these variables, we have—in this particular case—to switch to a different procedure.[21] In figure 17.1 we report partial correlation coefficients—measures of association in which the other variables in the model are controlled for. Proceeding stepwise, the coefficients shown for the relationship between malignancy + uncertainty (capacity) and level of collaboration on the other are those that obtain when we control for capacity (malignancy + uncertainty).[22] All other coefficients are those that obtain when *both* the other independent or intervening variables have been controlled for.

Two main conclusions can be drawn from these figures. First, introducing level of collaboration as an intervening variable does not upset the conclusions from the

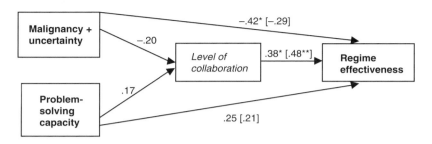

Figure 17.1
Core model: Partial correlations coefficients
Note: Figures in brackets refer to effectiveness measured as functional achievement. We have entered the malignancy + uncertainty combination in this figure on the basis of findings reported above (see table 17.4b, p. 444).

trivariate analysis summarized above. The results in figure 17.1 are consistent also with the observation made in table 17.4b that the impact of both of our two aggregate independent variables seems to be somewhat larger for behavioral change than for functional effectiveness. Second, level of collaboration seems to have an independent positive effect. This strength of this effect depends, however, much on where we draw the distinctions; only at the highest levels do we see significant results. Different analytic techniques give essentially the same result. For the practitioner, this can be interpreted as another piece of good news; the simple take-home message is that more cooperation tends to produce better results. In particular, engaging regime bodies in joint planning and implementation tends to improve the prospects for achieving effective solutions.

Decomposing Problem-Solving Capacity

The time has come to try to decompose the aggregate construct of problem-solving capacity into its main constitutive elements and explore the contribution of institutional setting, power, and informal leadership in enhancing regime effectiveness. In chapter 1, we argued that the relevance of each of these elements is likely to differ from one type of problem or task to another. In this section we therefore specify effect by problem type, using the crude distinction between malign and nonmalign problems as our tool of differentiation.

Since the number of cases in each of these categories is quite small, results from even the crudest multivariate analysis will have to be interpreted with great caution. We

Table 17.6
The impact of different components of capacity, specified by problem type (bivariate analysis, Spearman correlations)

Capacity Components	Behavioral Change		Functional Optima	
	Benign/Mixed	Malign	Benign/Mixed	Malign
Institutional capacity	.25	.24	.31	.30
Informal leadership	−.05	.33	−.14	.23
Power (basic)	.02	.50*	−.29	.49*
Capacity (total)	−.11	.44*	−.22	.42*
N	11–12	24–25	11–12	24–25

Table 17.7
Impact of institutional setting and the distribution of power, specified by problem type (trivariate analysis, partial correlation coefficients)

Capacity Components	Behavioral Change		Functional Optima	
	Benign/Mixed	Malign	Benign/Mixed	Malign
Institutional capacity	.32	.24	.60	.30
	(.32)	(.05)	(.60)	(.37)
Power (basic)	−.21	.51*	−.60	.50*
	(−.21)	(.48*)	(−.60)	(.44*)
N	8	22	8	22

Note: Coefficients in parentheses are those that obtain when institutional capacity is defined in terms of decision rules and role of secretariats and chairs only.

therefore begin by reporting bivariate measures of association (table 17.6) and limit the multivariate analysis to only two of the three capacity components (table 17.7).

Three findings stand out. First, as expected, the role of the various components of capacity varies from one type of problem to another. In this study, power seems to be a critical factor in dealing with malign problems—in particular, *strongly* malign problems—but at best ineffective and at worst counterproductive when it comes to solving problems that are more benign. Since power weighs so heavily in our index of capacity, the same general pattern is found for our aggregate notion of capacity as well. Overall, it seems that our notion of capacity is not of much help in accounting for variance in effectiveness for nonmalignant problems. Of the three main components, only institutional capacity is positively associated with success on both dimensions. This does not come as a total surprise; as we explained in chapter 1, our main interest was to learn more about determinants of success in dealing with malign problems, and our concept of capacity was constructed with that purpose in mind. Nonetheless, it is worthwhile to note that the results do indicate that we would need a different model—including a different specification of capacity—to satisfactorily account for the variance observed for benign and mixed problems.

Second, by comparing the scores for the two dimensions of effectiveness, we find that institutional capacity—particularly when defined in terms of the three organizational elements only—seems to be a more important factor in achieving functionally effective solutions than in bringing about behavioral change. For power it seems to be the other way around.[23] These are intuitively plausible results. The behavioral change brought about by powerful actors may serve themselves better than the group at large. Several of our case studies—including those of North Pacific salmon, radioactive waste, and whaling—seem to suggest that coercion-based solutions may well go *beyond* the technically optimal solution. By contrast, actors in regime roles—notably secretariats and conference chairs—are supposed to work for the common good. We all know that the real world is often less clear-cut than such ideal constructs would indicate. Even so, these findings should at least serve to indicate that the recipe for achieving functionally effective solutions may differ in interesting respects from the recipe for bringing about improvement in behavior.

Third, institutional capacity seems to make greater difference when power is concentrated in the hands of pushers than when it is not. Controlled for malignancy the coefficients are .39 versus .10 for behavioral change and .65* versus .12 for functional effectiveness. By contrast, power is more important in bringing about behavioral change when capacity is low (.48) than when it is high (.26), but contributes more to functional effectiveness when capacity is high (.30 compared to .20). Taken together, these observations suggest that institutions tend to be more dependent on powerful actors than vice versa but also that institutional capacity plays a role in directing the exercise of power towards the pursuit of *common* interests.

In chapter 1 we formulated a set of hypotheses pertaining to specific elements of capacity and different sources of informal instrumental leadership. Our small database does not enable us to distinguish reliably the impact of one of these elements or sources from that of another. It nevertheless seems worthwhile to take one step further in decomposing the notions of institutional capacity and informal leadership to explore whether one particular element or source stands out as substantially more important than others. Table 17.8 offers a first crude cut. The

Table 17.8
Decomposing aggregate notions of institutional capacity and informal leadership (partial correlation coefficients) (N = 34)

Component/Source	Behavioral Change	Functional Effectiveness
Institutional:		
Decision rule (in use)	.09	.16
Role of secretariat	−.01	.19
Conference and committee chairs	.12	.32
Fast-track options	.27	.16
Informal leaders:		
Delegates	.36*	.33
Epistemic communities	.09	.17

coefficients reported are controlled for problem malignancy but not for other capacity components. We are skating on very thin ice here, so these results should be interpreted as suggestions rather than findings.

With this note of caution in mind, at least three observations can be made. First, most coefficients are weak but—with one minor exception—positive. Second, the most promising tools seem to be fast-track procedures and leadership by delegates or chairs.[24] Third, we may note that the scores for most of these factors are somewhat higher for one concept of effectiveness than for the other. Although weak, the pattern is interesting: while fast-track procedures and leadership by delegates seem more effective in changing behavior, the contributions that secretariats and chairs make seem to be more clearly geared to the pursuit of functionally good solutions.

Further scrutiny brings out some interesting observations about interplay. First, active chairs and active secretariats tend to appear on the scene together (Spearman = .70**). This makes it hard to separate the contributions of one from the other, so what we are in fact measuring may well be their combined effect. Second, with the exception noted above, formal leadership by chairs and informal leadership by delegations or delegates seem to a large extent to serve as functional equivalents; combining the two does not seem to generate much additional energy. The relationship seems, though, to be weakly asymmetrical: leaders in formal organizational roles tend to depend more on the support of informal leaders than the other way around.

Third, the presence of an epistemic community seems to enhance the contributions that chairs, secretariats, and delegate leaders can make, at least to promoting functional effectiveness. Thus, where an epistemic community is present, the

correlation between leadership by chairs and regime effectiveness increases to .45*
for functional achievement (controlled for problem malignancy). There is, at best,
a very weak effect the other way. A plausible interpretation seems to be that actors
in formal organizational roles can benefit from the presence of informal transnational networks of experts, while epistemic communities make their influence on
policy primarily through channels to *national* governments, at least in the initial
stages.[25]

Paths to Effectiveness

To get a better grasp on pathways and recipes, we once again have to shift to a different mode of analysis. In this subsection, we search for *combinations* of factors
that seem to lead to high or low effectiveness, using the Ragin QCA technique as
our main tool. Such combinations can be seen as more or less complex clues or
paths leading to a particular outcome. Indirectly, the study of such combinations
can also shed some light on the relative importance of various factors. It clearly
makes sense to say that a factor that appears in *all* combinations or that emerges
as the only sufficient condition is more important than one that appears in none.
Although with less confidence, we may also consider a factor that appears in many
combinations as more important—at least for purposes of political engineering—
than one that appears in few.[26]

We begin by asking which (if any) factors or combinations of factors lead to *high*
effectiveness. We proceed stepwise, first searching for combinations that are not
specific to one particular type of problem and then moving on to differentiate the
answer by distinguishing between malign and nonmalign problems. At each step,
we run models at different levels of specification, much as we did above. Model 1
includes only our two aggregate independent variables (type of problem and capacity). Model 2 decomposes each of these constructs into two or three main components (malignancy and uncertainty; institutional capacity, informal leadership, and
power). Model 3 adds level of collaboration to the equation.

In technical terms, this analysis is a matter of identifying factors or combinations
of factors that emerge as *sufficient* conditions that *invariably* lead to high effectiveness in our sample of cases. The latter five words are important: in reporting
results obtained by using this particular approach, we make no claim that the findings based on this small sample can be generalized to the universe of international
(environmental) regimes.[27] There will almost certainly be some cases where neither
of these combinations will be sufficient to ensure high effectiveness. The results

Table 17.9
Pathways to *high* effectiveness. QCA-Ragin crisp set analysis

	Behavioral Change		Functional Effectiveness	
General	M1: BENIGNKNOW		M1: -	
	M2: INSTCAP KNOW		M2: -	
	M3: LEVCOL + INSTCAP KNOW		M3: -	
Benign/mixed	M2: INSTCAP* INFLEAD* KNOW		---	
	M3: INSTCAP* INFLEAD* KNOW			
Malign	M2: INSTCAP KNOW		M2: POWER INSTCAP	
	M3: LEVCOL + INSTCAP KNOW		M3: POWER INSTCAP	
Overall model fit	M2: .56	(.61)**	M2: .44	(.56)**
	M3: .72	(.78)**	M3: .44	(.67)**

Upper case means presence.
+ means "or", open space between factors means "and".
- means no sufficient condition found.
--- means insufficient data (N ≤ 5).
* indicates necessary condition.
** Model fit coefficients in parentheses accept also factors and combinations that appear only *once*. Model fit coefficients are computed by the "fuzzy set" program.

should therefore by no means be interpreted as identifying foolproof recipes for success. We believe, however, that this kind of analysis can point us in the direction of clues or pathways that are sufficiently *promising* to warrant further study by the researcher and serious attention by the practitioner.

In table 17.9 we have summarized the results of this first cut, in the conventional QCA format. This format itself will probably be unfamiliar to many readers, but with a few clues (see below) the interpretation should be fairly straightforward. Six main conclusions can be drawn at this stage.

First of all, as we can see from the Model 3 results, there is more than one pathway to effectiveness. Had we included more independent variables, we would probably have identified more pathways. Moreover, we may note that in only two instances does the analysis identify a factor that is also *necessary*. In terms of our models, we may consider two or more pathways that appear together as equivalent. It is, however, important to realize that we have no basis for extending that conclusion *beyond* the horizon of these simple models. In fact, sound reasons can be given for hypothesizing that each of these factors or combinations of factors works only under particular circumstances. One important challenge for further research is to help us specify these circumstances more precisely than we are able to do today.

Second, if we were to use this table as a basis for identifying one or two particularly important keys to success, we would point to (1) institutional capacity and (2) knowledge. Admittedly, neither comes out as a sufficient condition in itself. There are, however, few pathways that lead around them—for functional achievements *all* include high institutional capacity. Moreover, for significant and major improvement in behavior with regard to benign and mixed problems, institutional capacity comes out as a necessary condition (and so does informal leadership).[28] Knowledge also appears to be an important key, at least in combination with institutional capacity.

Third, comparing the two notions of effectiveness, we see that there are more paths leading to significant or major behavioral improvement than to functionally optimal solutions. The overall model fit is also higher for the former. This confirms that functional effectiveness is the more demanding standard. Moreover, we have weak indications that at least some factors are more important for one notion of effectiveness than for the other. For example, level of collaboration and informal leadership appear to be more important in bringing about positive behavioral change than in achieving good solutions, while—somewhat surprisingly—power appears only in the functional effectiveness column. It is important to note, though, that it is the *combination* of power and institutional capacity that appears as important in this context, not power in itself. We may also note that had we lowered the threshold, this particular combination would have appeared as a path to behavioral change as well.

Fourth, if we compare the two problem categories, we see that power can be an important asset in dealing with malignant problems but does not enter any of the combinations found to lead to success in dealing with benign and mixed problems. This confirms the conclusion we reached on the basis of tables 17.6 and 17.7.

Fifth, in studying the results of QCA analyses, we have to search also for combinations that occur only or to a large extent as *negative* conjunctures. In table 17.9 there are no negative conjunctures. To settle the question we have to look beyond the crisp-set solutions themselves. In this case, a look behind the scene does not point out any obvious candidates. It does, however, leave us with a few question marks. And a closer look at the case studies confirms that at least one of these questions may be well worth pursuing: do power and knowledge work well together?

At least two of our cases—radioactive waste disposal and whaling—indicate that they sometimes do not. In both of these cases, we see the role of research-based knowledge decline as the dominant coalition flexes its muscles. Habermas would not have been surprised; communicative rationality *requires* the absence of coercion (Habermas 1984–1987). A reasonable interpretation of the scant evidence that we have at this stage would be that institutional capacity combines easily with power as well as with knowledge, while reliance on a knowledge-based strategy does not square well with coercion and vice versa. These are intuitively plausible hypotheses. There is, though, also some evidence to suggest that the relationship between power- and knowledge-based strategies may be contingent on type of problem. More precisely, it seems that when we are dealing with a conflict over *values*, the two interact like oil and water. When the conflict centers on interests, the relationship is likely to be strained but not necessarily characterized by incompatibility.

At this stage, however, these interpretations all remain conjectures. More research is needed to enable us to distinguish with confidence factors or combinations of factors that interact synergistically from those that interact negatively and to distinguish both of these categories from those that do not interact at all. Further advances along that frontier could produce knowledge that would be highly relevant to decision makers. Finally, as we can see from the bottom row, the overall model—meaning here the set of conditions identified—accounts for a fair amount of the variance observed. Again, the model does somewhat better in accounting for behavioral change than for functional effectiveness.

For perfectly understandable reasons, most researchers and decision makers probably are more interested in recipes for success than in clues to failure. It is nevertheless worthwhile to take a quick look also at the latter. In table 17.10 we have used the same three models and the same analytic technique to search for factors or combinations that invariably lead to *low* effectiveness in our sample of cases.

The results can be summarized briefly as follows: First, we see that none of the factors or combinations of factors included in our models comes out as a sufficient condition for failure to bring about significant improvement in actor behavior. This may be interpreted as good news, since a factor that is not a sufficient cause of failure cannot be a necessary condition for success (assuming that success and failure are the only possible outcomes and also mutually exclusive categories). Now, since we require a minimum of *three* valid observations to accept a factor as a sufficient condition, we would need to rerun the analysis with a threshold of *one* to be sure that we have not missed something. Even when we do, we find no single factor that comes out as a definite obstacle to high behavioral effectiveness. Second, as we have

Table 17.10
Pathways to *low* effectiveness. QCA-Ragin crisp set analysis

	Behavioral change	Functional optima
General	M1: -	M1: MALIGNUNCERT
	M2: -	M2: LOWINSTCAP-UNCERT
	M3: -	M3: LOWINSTCAP-UNCERT
Benign/mixed	M2: ---	M2: -
	M3: ---	M3: -
Malign	M2: -	M2: UNCERT
	M3: -	M3: UNCERT
Overall model fit	-	M2: .35 (.35)
		M3: .35 (.55)

Note that in table 17.10 presence and absence refers to *low* scores on the dependent variables. Upper case means presence.
- No sufficient condition found.
--- Insufficient data (N ≤ 5).

noted earlier, the combination of malignancy and uncertainty is an effective obstacle to high functional achievement. By implication, for malign problems poor knowledge appears to be sufficient to block a high score on the more demanding standard of effectiveness. The combination of poor knowledge and low institutional capacity appears to be a sufficient cause of failure when we examine the data file as a whole. This finding extends and supports the conclusion we arrived at earlier about the importance of these two factors as clues to *high* effectiveness.[29]

Concluding Reflections

Now, where does all this leave us? Let us try to review the overall picture from two different perspectives—that of the concerned citizen and that of the student of international regimes.

For the concerned citizen, eager to see environmental problems alleviated, we have a fair amount of good news. First, most environmental regimes do succeed in changing actor behavior in the direction intended. Our study strongly supports Young's conclusion that "regimes do matter in international society" (Young 1999, 249). Processes of regime formation and implementation can have a significant impact even if they do not succeed in producing collective decisions; governments as well as societies quite often make unilateral adjustments in response to new ideas and information or in anticipation of new regulations. Second, even strongly malignant problems *can* be solved effectively. There are exceptions, but the overall progress made in coping with politically malignant problems is higher than we expected at the outset. Third, most regimes tend to grow and become more effective as they develop beyond their initial stages. Moreover, there are some indications that more recent generations of regimes are somewhat more effective than previous generations, compared at the same point in their life cycles. Fourth, the pathways identified include at least some factors that are potential targets of deliberate engineering. Organizational arrangements are political constructs. Knowledge can be improved through investment in research and more effectively transformed into decision premises through the design of institutions and processes for dissemination and consensual interpretation. We need not even consider the distribution of power or the structure of the problem as givens. Our case studies demonstrate that the distribution of power can be changed through multiple maneuvers, including low-cost invasion of an existing organization (International Whaling Convention), entry of new actors into the activity system itself (South Pacific tuna), shift to another arena (radioactive waste), shift to a different regulatory approach (ship-generated oil pollution), and change in macroregimes (Law of the Sea to regional fisheries regimes). We have also seen that new knowledge or ideas can lead to change in actor perceptions of the nature of the problem and thus help transform a malign problem into one that is more benign (LRTAP and MEDPLAN). One important take-home message seems to be "Don't give up!"

That said, there is bad news as well. First, despite a fair amount of success in changing actor behavior, the solutions achieved by and large leave substantial room for improvement. Measured in terms of distance to functionally optimal solutions, the overall score is significantly lower (.35 as compared to .51). Second, some of the overall improvement that we have observed seems to be due to fortunate circumstances for which the regime itself can claim little or no credit. The most fundamental factor intervening is a general growth in public demand for and governmental supply of policies for environmental protection. This warrants a note of caution: the rate of success that we have observed for environmental regimes may not be easily replicated in a stagnant or declining policy field.[30] Moreover, within the field of environmental policy the success may be very hard to sustain in periods with declining public concern. Third, even though malignant problems *can* be solved effectively, the combination of strong malignancy and poor knowledge is a formidable obstacle. And even though it seems that once the initial hurdles are cleared the prospects for further progress are better than we feared, it is abundantly clear that the initial hurdles will often be high. As we have seen, the design of this study—including only cases that have already cleared the first hurdle—leaves us prone to underestimating the impact of malignancy. Fourth, before rejoicing at instances of declining problem malignancy we should realize that the shift is sometimes a consequence of the fact that the situation has deteriorated to the point where there is not much left to fight over (see whaling). Fifth, the optimistic observation that most of the key factors *can* be manipulated one way or the other must be tempered with the warning that political engineering will often be a difficult exercise. Our table-style presentation of QCA results might make things look deceptively simple: all you need is *X* and *Y*. But as any seasoned practitioner knows only too well, building up organizational capacity and a firm base of consensual knowledge sometimes take decades (and, as the IWC case reminds us, in the meantime the patient may well have died). Power *can* be undermined or marshaled through different tactical maneuvers, but—as any battle-experienced general can testify—performing these maneuvers with limited resources and under severe operational constraints is not likely to be easy (and there is no guarantee of victory). Finally, even though soft components such as knowledge and organizational arrangements can make a substantial difference, there is also a darker side of the story: dealing with *strongly* malign problems, hard power—control over the basic game—seems to be the critical asset. In brief, this study provides ample evidence to suggest that we still have a long way to go and formidable obstacles to pass before we can claim that problems are effectively solved. Good (institutional) equipment, skills, and perseverance will most certainly be required. So will, in hard cases, coercive power.

Reviewing the overall performance of the model we developed in chapter 1, we are left with mixed impressions. There is some good news. First and most important, our core model seems able to account (at least in statistical terms) for a substantial proportion of the variance observed, within as well as across regimes. As we have seen, the odds for success measured as significant and major improvement in actor behavior are nineteen to one when a high-capacity system deals with a nonmalign problem that is fairly well understood, compared to one to eight when a low-capacity system encounters a malign problem clouded in high uncertainty. With state of knowledge included in the equation, 75 percent of our predictions were correct (90 to 100 percent for intraregime changes). Taking into account that the core version is a strictly structural and context-free model, its performance is encouraging. Second, our in-depth case studies by and large lend credibility to the key *basic premises* on which the model is based. The general line of reasoning that we pursued in chapter 1 seems to point us toward causal mechanisms that we find at work in most if not all of the regimes included in this study. Combining these two observations, it seems fair to say that our core model at least identifies important factors and mechanisms and yields predictions that are roughly right most of the time.

Having said that, we also have to admit that there is substantial room for improvement. In the analysis of intraregime change, weak predictions derived from the model erred in about 35 percent of the cases, most often on the pessimistic side. And an overall fit rate of 75 percent gives us little reason to bask in glory. Anyone familiar with the technique will realize that with a dichotomous dependent variable we would in the long run be right about 50 percent of the time by pure chance. Although we get the extremes right, the intermediate range is rather fuzzy. Furthermore, our notion of capacity does not take us far in accounting for variance in outcomes for nonmalign problems. Nor does it shed all that much light on effectiveness defined in terms of achieving technically good solutions. Finally, some of the hypotheses we formulated in chapter 1 seem, at best, to be underspecified (a summary overview can be found in table 17.11). H_1 is the most obvious case; our study indicates that malignancy interacts synergistically with uncertainty and is most of an obstacle in the initial stages of regime development. A more sophisticated proposition about the impact of malignancy would thus have to incorporate interplay and differentiate by stage of regime development. Similar things could be said about our notion of capacity. It is abundantly clear that it calls for further differentiation by type of problem, regime stage, and concept of effectiveness.

How, then, can we do better? The easy way out is to make the model more inclusive by adding other independent (and perhaps also intervening) variables. Popular candidates include the role of NGOs and target groups, domestic politics, and the role of shared ideas and discourses (see, e.g., Young 1999, 127–128). Our own case studies indicate that additional mileage can be gained by extending the analysis in at least four directions; context, domestic-level politics, process dynamics, and type of regulation. Yet the temptation to reach out in multiple directions should be tempered with a reminder that adding a new variable or bringing in a new mechanism is useful *only to the extent that we know how it works*.[31] The vague idea that something is important does not take us very far; we have precise and useful knowledge only to the extent that we can specify the *direction*, *strength*, and *form* of impact under different *conditions* and understand *how* this impact is generated. If there is one lesson to be learned from our own problems in getting things right here, it is this: we fail not (primarily) in identifying what is *important* but in specifying *importance*.[32]

In that respect we are in good company. The main challenge for the field at large at this point is not to increase inclusiveness; it is to increase *precision*.

Table 17.11
A summary review of the hypotheses formulated in chapter 1

Hypothesis	Status
H_1: The more politically malign the problem, the less likely that the parties will achieve an effective cooperative solution—particularly in terms of technical optimality.	Underspecified. Malignancy seems to account for a substantial proportion of the variance only in the *early* stages of regime formation and in combination with uncertainty.
H_2: Political malignancy and uncertainty in the knowledge base tend to interact to increase the intractability of problems. Uncertainty in the knowledge base tends to slow down the development of effective responses to benign problems as well, but here the effect on the end result will be less detrimental.	Strong support. Malignancy and uncertainty interact synergistically, and the combination of strong malignancy and high uncertainty can be lethal.
H_3: Regimes dealing with truly malign problems will achieve high effectiveness only if they contain one or more of the following: (1) selective incentives for cooperative behavior, (2) linkages to more benign (and preferably also more important) issues, and (3) a system with high problem-solving capacity. The presence of at least one of these factors is a necessary but not a sufficient condition for high effectiveness.	Strong support.
H_4: The establishment of (negotiation) arenas as formal institutions that exist and are used over an extended period of time tends to facilitate cooperation and enhance effectiveness of international regimes by, inter alia, encouraging actors to adopt extended time horizons and norms of diffuse rather than specific reciprocity and by reducing the transaction costs of specific projects.	Not systematically tested. The general thrust of our case-study analysis is, however, consistent with the basic argument.
H_5: In dealing with malign problems, the decision rules of unanimity and consensus tend, other things being equal, to lead to less effective regimes than rules providing for (qualified) majority voting. More precisely: the use of majority voting tends to lead to more ambitious regulations (output). However, to the extent that this is accomplished at the cost of sacrificing the interests of significant actors, it will do so at the risk of impairing compliance (outcome). Except for strongly malign issues, the former effect will most often be somewhat stronger than the latter.	The available data do not enable us to separate the effect of the decision rules from that of other institutional factors. The decision rule *is* no doubt an important determinant of outcomes but is almost a constant in this sample and therefore cannot account for much of the variance observed here.

Table 17.11 (continued)

Hypothesis	Status
H_6: Actor capacity on the part of the international organization in charge and its subordinate bodies and officials tends, *ceteris paribus*, to enhance regime effectiveness. The impact tends, other things being equal, to be larger in the case of malign than in the case of benign problems and greater for moderately malign than for strongly malign problems.	The first part of the hypothesis supported by some of the case studies and by the statistical analysis. No significant difference found between malign and nonmalignant problems, but the positive effect seems to be stronger for functional achievement than for behavioral change.
H_7: The existence of (informal) networks of experts—*epistemic communities*—contributes to regime effectiveness by strengthening the base of knowledge on which regimes can be designed and operate. The more integrated an epistemic community, and the deeper it penetrates the relevant national decision-making processes, the more effective—*ceteris paribus*—the regime it serves tends to be.	Some support. This does not seem to be a critical factor.
H_8: Concentration of power in the hands of pushers tends by and large to enhance effectiveness, while concentration of power in the hands of laggards has the opposite effect: · In dealing with benign problems, unilateral provision of collective goods by a benevolent hegemon tends, although with some exceptions, to weaken the incentives of others to contribute. · In dealing with malign problems—particularly those characterized by severe asymmetry and cumulative conflict—concentration in the hands of pushers tends to generate fear and withdrawal among laggards. For such problems, the most conducive distribution of power is likely to be one in which the aggregate strength of pushers is roughly balanced by the aggregate strength of intermediaries, with laggards in a weaker position but not completely marginalized.	The first part of this hypothesis is supported but only with regard to malign problems. Overall, power seems to be more important in bringing about behavioral change than in achieving functionally good solutions. We do not have data enabling us to test the two bullet point items.
H_{8a}: The prospects of effective regimes tend to decline if the distribution of power in the decision game differs substantially from the distribution in the basic game.	The available data do not permit adequate testing.
H_9: Instrumental leadership tends to facilitate regime formation and implementation. And the more skill and energy that are available for and actually invested in instrumental leadership, the more likely an effective regime will emerge.	Supported.

Table 17.11 (continued)

Hypothesis	Status
H_{9a}: The need for instrumental leadership tends to increase with problem malignancy. However, supplying leadership tends to become increasingly difficult as malignancy increases. Instrumental leadership thus tends to make most difference in dealing with problems that are *moderately malign*.	Cannot be adequately tested with current database. What we can say is that our notion of capacity *in general* accounts for more of the variance observed for malign than for nonmalign problems. A different specification of capacity seems to be required for benign problems.
H_{10}: Skills in designing effective regimes tend to improve over time as actors learn. This implies (1) that regimes tend to become more effective over the first decade or two of their lifetime and (2) that the latest generation of regimes tend to be more effective than regimes of previous generations.	(1) Supported. (2) Mixed. Regimes established in the 1960s and 1970s were slightly more effective than regimes established earlier. No further increases are seen for later generations, but many of these regimes are still in a period of growth and may eventually do better. Successive generations of regimes should be compared at the same points in their life cycle.

Notes

1. I gratefully acknowledge useful comments from other members of the team and from Gunnar Fermann, Jon Hovi, Olav S. Stokke, Jonas Tallberg, Oran R. Young, and Michael Zürn.

2. The rank-order correlation between objective and perceived malignancy is .74, suggesting that the distinction may be analytically important. We therefore run important parts of the analysis also with the latter variable but report results only if we see significant discrepancies.

3. Where the consequences are most disturbing—especially in the analysis of intraregime change—we simply leave that particular phase out of the comparative analysis.

4. Four cases deal with what we have called the preregime phase—the state of affairs that existed *before* the regime in focus was established. For some of the remaining cases we lack adequate data for one or more important variables.

5. We also set out to measure distance to *politically* optimal solutions. Needless to say, this proved to be a particularly difficult task, so we have decided not to report any results from that exercise.

6. In our codebook, this variable has also a fourth value ("regimes goes beyond functional optimum"). In this analysis, however, we have merged this category with category 2 ("regime does not meet these requirements fully, but the gap is not very large either"). The reason is simply that we do not consider overshooting to be a good thing; category 3 ("regime meets or comes close to meeting the requirements of a technically optimal solution") thus indicates better performance than category 4.

7. The analysis reported below includes only twenty-two of these cases for functional effectiveness. One phase of the IWC regime has been left out because of conflict of values over the ideal solution (see note 3).

8. A prediction is considered correct if and only if a positive shift in the aggregate score for the independent variables leads to a shift in the same direction in effectiveness or if both sides of the equation remain constant.

9. An alternative line of explanation would focus on measurement error.

10. This is certainly an explanation focusing on *context*, but these aspects are not captured in our database, which highlights issue- or regime-specific links or motives.

11. This is a particular approach to qualitative comparative analysis, relying on Boolean algebra (a particular format of logical expressions) and designed to come up with a list of *configurations* of circumstances associated with a given outcome. For a good introduction, see Ragin (1987).

12. Technically, this produces an index ranging from 1 (predominantly benign problems) to 6 (strongly malign problems, characterized by severe incongruity + asymmetry + cumulative cleavages), weighted so that the incongruity and cost-effectiveness dimension becomes the most important (range 1 to 4). Categories are then merged pairwise, leaving us with a truncated scale with only three levels. In this particular sample of cases, this composite index produces an ordinal ranking that is almost identical with that of the incongruity variable (Spearman = .95**). It therefore makes no real difference which scale we use in this particular analysis. For convenience we have used the incongruity variable in the analysis reported below.

13. This gives an index ranging from 2 to 5, with power weighted as the most important component (range 1 to 3). To reduce the empty-cell problem, the index has been truncated to a three-level scale.

14. Since we have only two truly benign cases, we have merged the benign and mixed categories throughout this part of the analysis.

15. There is, of course, also the possibility of measurement error (see below).

16. This observation is consistent with the main conclusion from a comparative study of the use of inputs from scientific research as premises for international environmental policy decisions. In that study the state of knowledge was found to be—by far—the most important determinant of adoption of inputs from science (see Underdal 2000).

17. Note that this is a more pessimistic conclusion than that produced by the logistic regression analysis. This discrepancy reflects a basic difference between the two techniques. The veristic mode of the QCA program considers any factor that brings about a particular

outcome in all cases *included in the data file* as a sufficient condition. By contrast, the notions of necessary and sufficient conditions are alien to logistic regression analysis.

18. Careful comparison with the case study indicates that we have also another measurement problem; interesting nuances are lost as we compute indexes and merge categories. This is, however, a problem that pertains not merely to the malignancy variable.

19. The Spearman correlation with effectiveness measured as behavioral change is −.73 (for functional achievement one of the variables has a constant value). Malignancy also seems to be a good predictor of (noncooperative) outcomes in the preregime phase.

20. To avoid or at least reduce the empty-cell problem, we have truncated the level-of-collaboration scale by merging of the initial six categories into three. Only the two highest categories have sufficient data to be included in the analysis. The dependent variables have both been dichotomized.

21. In principle, logistic regression could also have been used for this purpose. With such a small number of cases, however, it becomes very sensitive to outliers, and in this particular case it produces some results that cannot be given a substantively meaningful interpretation ("exaggerating" the positive impact of level of collaboration). The partial correlation technique that we resort to here assumes linearity and is therefore not ideal for our purposes. On the other hand, any bias is likely to be on the conservative side.

22. Note that on the basis of findings reported above we have entered the malignancy + uncertainty combination in this figure.

23. This is, though, a less robust finding, due in large part to differences with regard to benign and mixed problems. For the entire data set, coefficients are .33 for behavioral change and .19 for functional effectiveness, when problem malignancy and institutional capacity have been controlled for.

24. This finding must be interpreted with great caution since the range of variance is higher for some components than for others.

25. The latter suggestion seems to be consistent with the emphasis in Haas (1992).

26. As indicated, this is not a truly compelling argument. The number of combinations in which a factor appears need not correlate perfectly with the proportion of variance that it accounts for.

27. Note that the results reported contain no information as to *how frequently* a particular factor or combination of factors is associated with success. The default option of the QCA program is to report all conditions that coincide with a positive outcome at least *once*. In this analysis, we require at least *three* valid observations. Note also that with such a small number of observations, none of the paths identified by the veristic mode of analysis would pass the default criteria of the probabilistic mode.

28. In this file, institutional capacity and informal leadership are so highly correlated that we cannot reliably separate the effect of one of these variables from the effect of the other.

29. This finding is not merely an *implication* of the former. Two different data sets (truth tables) are used—one sorting cases on the basis of positive scores, the other on the basis of negative scores. The residual categories of nonpositive and nonnegative include also intermediate values.

30. The study of the nuclear nonproliferation case does, though, indicate that the mechanisms at work are quite similar.

31. Besides, a complete model would also be useless, except for crude heuristic purposes. A good model is a *simplified* representation of an object.

32. Arguably, much of the debate about the relative merits of the power-based, interest-based, and knowledge-based approaches could have benefited from a clearer distinction between these two questions. The fact that a particular hypothesis fails need not tell us much about the importance of a composite factor like power or knowledge. For example, the observation that the hegemonic stability hypothesis has an extremely disappointing record in recent comparative studies (Hasenclever, Mayer, and Rittberger 1997, 103) by no means warrants the conclusion that power has no place in the equation. Some of our findings in this study may in fact be interpreted as strong support for power-based theory—but for a different proposition, referring to a different concept of power. More generally, our analysis suggests that although useful in highlighting different sets of independent variables, the distinction between power-based, interest-based, and knowledge-based approaches should not be seen as describing self-contained islands of theory. Some of the most interesting insights obtain by *combining*, for example, power and interests.

References

Haas, P. M., ed. 1992. "Knowledge, Power, and International Policy Coordination." *International Organization* (special issue) 46.

Habermas, J. 1984–1987. *Theory of Communicative Action*. Boston: Beacon Press.

Hasenclever, A., P. Mayer, and V. Rittberger. 1997. *Theories of International Regimes*. Cambridge: Cambridge University Press.

Jänicke, M., and H. Weidner, eds. 1997. *National Environmental Policies: A Comparative Study of Capacity-Building*. Berlin: Springer.

Ragin, C. 1987. *The Comparative Method*. Berkeley: University of California Press.

Underdal, A. 2000. "Comparative Conclusions." In S. Andresen, T. Skodvin, A. Underdal, and J. Wettestad, eds., *Science and Politics in International Environmental Regimes*. Manchester: Manchester University Press.

Weale, A. 1992. *The New Politics of Pollution*. Manchester: Manchester University Press.

Young, O. R. 1999. *Governance in World Affairs*. Ithaca: Cornell University Press.

18

Epilogue

Edward L. Miles and Arild Underdal, with Steinar Andresen, Elaine Carlin, Jon Birger Skjærseth, and Jørgen Wettestad

Chapter 17 states our findings in technical terms. In this Epilogue we emphasize policy implications that may be of interest to a wider group of readers beyond the scholarly community. We begin with a look at the overall balance sheet, asking how well the fourteen regimes analyzed in this study perform in terms of our standards of effectiveness. Our next step is to examine patterns of variance, the basic question being why some regimes or regime components are more effective than others. We conclude with a few reflections about the basic model that has guided our research and the methodological approach we have adopted in this study. Knowing what we know now, what would we have done differently?

The Overall Balance Sheet

1.   Regimes do make a significant difference.
Most of the regimes studied in this volume have succeeded in changing actor behavior in the directions intended. There is powerful support for the counterfactual argument that, in the absence of regimes, things would have been worse. In more than half of our cases the improvement brought about by the regime is considered significant or even major. Particularly encouraging is the finding that regimes have a fair amount of success also in coping with *malign* problems. Regimes are, in other words, not merely fair-weather constructions. Moreover, our comparative analysis indicates that more cooperation leads to better average results. More precisely, we find that regimes engaging in activities beyond standard-setting—in particular, functions such as planning and implementation—tend to be more effective than those that do not. And there is more good news: in about two out of every three cases regimes seem to have served as arenas facilitating or fostering transnational processes of learning and contributed to strengthening the knowledge base for policy

decisions. These results are, of course, not real substitutes for concerted action. In a dynamic perspective they are nevertheless important in that they can help *transform* intractable problems into more manageable tasks and stimulate actors to reconsider present policies and positions. Regimes work not merely by establishing formal rules and regulations but also by generating and fostering social processes that can gradually transform their task environments.

2. Instant success is a rare thing.

Most regimes—even those that eventually achieve significant results—have modest beginnings. The initial stages of regime formation are often prolonged uphill battles. "Aller Anfang ist schwer" ("Every beginning is difficult"), as the German saying goes. Complex issues need much cultivation to become ripe for resolution, and new organizations need time and practice to find their mode of operation. The good news is that there is evidence to suggest that *if* one can pass the initial hurdles—the most critical being those of achieving a common understanding of the essential characteristics of "the problem," a commitment to a long-term objective, and the establishment of a procedural or organizational framework for cooperation—chances are reasonably good that more substantial results can be achieved later. Unambiguous *growth* is reported for nine of the fourteen regimes included in this study. Furthermore, younger generations of regimes seem to be somewhat more effective than previous generations at the same stages in their life cycles. Admittedly, these developments are to some extent due to exogenous drivers for which no single regime can claim direct credit. In particular, international environmental regimes have benefited from the general growth in public demand for and governmental supply of policies for environmental protection. Yet our study indicates that this is only part of the explanation. Regimes often mature and grow also by the force of endogenous mechanisms, such as learning and the momentum often generated by political processes. There seems to be a cumulative impact of knowledge and experience in the international community with regime design and implementation, enabling parties to draw on lessons learned from previous instances of success and failure.

3. Most regimes fail to achieve optimal solutions.

Even though the regimes we have studied by and large make a positive difference, most fail to solve the problems they were designed to solve and often fail by a wide margin. Nearly 60 percent of our cases score low in terms of what we have called functional or problem-solving effectiveness. For environmental problems this is an important limitation because nature keeps adding up the bill. Moreover, enthusiasm over fair achievements with regard to behavioral change must be tempered with a reminder that some results may come late and at a high price. In at least one case (whaling), the patient almost died in the meantime. In this study we have made no systematic attempt to calculate the costs of regime formation and operation. Nor have we tried to estimate the time taken to develop and implement the various measures. In rating a regime as *effective* we do not claim that it is also *efficient*—that it achieves its results at low or minimum costs—nor do we claim that it increases *net* social welfare. All we say is that it contributes to alleviating the problem it was designed to cope with. That is by no means a trivial accomplishment. But effectiveness thus defined is only a *necessary*—not a *sufficient*—condition for efficiency and net welfare gains.

Explaining Variance: Why Are Some Regimes More Effective Than Others?

Behind these broad generalizations, we find a considerable range of variance in effectiveness within as well as among regimes. In our search for explanations we have pursued two paths. One focuses on the nature of the problem. The basic assumption was that some problems are intellectually less well understood or politically more malign than others and hence more difficult to solve. The other path started out from the elusive and complex notion of problem-solving capacity, the basic argument being that some systems have at their disposal more powerful institutional tools and can mobilize more skill and energy to tackle the problems they encounter. Our main findings may be summarized as follows:

4. As stand-alone problems, both uncertainty and malignancy can—up to a point—be overcome, but the combination can be lethal.

Our basic assumption is supported: uncertainty about the seriousness and causes of the problem and a politically malign configuration of interests are both very real hurdles. It takes more resources to deal with a malign problem or one that is not well understood. Success in coping with a truly malign problem was achieved *only* in instances where malignancy was tempered by links to a more benign problem or by selective incentives or where the problem was attacked by a system with high problem-solving capacity. This does not, however, warrant the conclusion that voluntary cooperation will fail whenever the parties face such problems. As long as they have to cope with only *one* of the two challenges, they quite often succeed in mobilizing what it takes to make significant progress. We were in fact surprised—

and encouraged—by the success rate for malign problems. We were, however, also struck by the degree to which uncertainty and malignancy interact synergistically to make problems intractable. The success rate for malign problems declines sharply as uncertainty increases, and uncertainty becomes more of an obstacle as problems get more malignant. In dealing with a malign problem, substantial uncertainty about its seriousness and causes serves as a functional equivalent of live ammunition in the hands of laggards. The combination of high uncertainty and strong malignancy is most often lethal.

The main policy implication of this finding is straightforward: when faced with a truly malignant problem, begin by trying to build a base of consensual knowledge about its essential characteristics. In some cases, more knowledge will lead actors to redefine their interests in a direction that makes the problem itself appear more benign (as we have seen in the case of LRTAP). Even if such a transformation does not occur, new knowledge can at least provide a basis for real negotiations. But consensual knowledge is no panacea leading automatically to agreement on effective policy measures. In fact, there are at least two types of situations in which this recipe will not work well at all. One is where new knowledge removes a veil of ignorance that leaves parties unable to determine with confidence who will win and who will lose from particular measures. The other is where the problem is characterized by conflict over basic values (whaling being a good example). Yet with these exceptions, we may for all practical purposes consider a base of consensual knowledge about the basic characteristics of the problem to be a necessary, though by no means sufficient, condition for achieving effective solutions to truly malign problems.

5.   If you cannot make significant progress in dealing with the substantive problem (yet), focus first on building capacity to act.

In our analysis, problem-solving capacity comes out as a very important determinant of regime effectiveness. Now, that may be neither surprising nor clarifying since our notion of capacity includes such a wide range of components, from decision rules to the distribution of power. We have more to say about each of the main components below, but let us pause for a minute to reflect on the implications of this initial and very general observation. In media reports and activist assessments we are often told that regime negotiations fail if they do not produce binding agreements committing the parties to significant substantive measures. If we see a particular meeting or conference as the beginning and the end, it is often easy to sympathize with such somber verdicts. Regime development is, however, better understood as a *series* of events, where each step—particularly those taken in the initial stages—must be evaluated not merely in terms of whatever substantive commitments it produces but also in terms of the contribution it makes to building capacity to act and a platform for subsequent moves. In a long-term perspective, the latter may well be more important than the former. This is, of course, not to say that organizational and procedural arrangements can substitute for the real thing. The point is simply that capacity-building measures can be important *means* of achieving substantive results further down the road. When institutional capacity is weak and progress on substantive measures slow and meager, decision makers may be well advised to focus much of their energy in the initial stages on capacity-building measures.

6. There is no simple cure-all treatment.

As it stands, this brief sentence merely states the obvious: a cure must somehow match the disease. It becomes interesting only to the extent that we can specify which cures work where and when. Our study suggests at least some interesting clues. For example, while particular organizational arrangements (see item 7 below) and the presence of transnational networks of experts seem to help in dealing with malign as well as benign problems, basic game power seems largely ineffective and sometimes even counterproductive in tackling benign problems. Furthermore, while some capacity components (such as secretariats and epistemic communities) seem to interact synergistically, others appear to serve largely as functional equivalents (such as formal and informal leadership), and some seem to be mutually repellent in large doses (such as knowledge-building strategies and coercion). A decision maker would want to combine components of the first category, concentrate scarce resources on only one from the second, and be cautious about mixing elements of the third. There is also another aspect to the opening sentence: a particular result is rarely if ever the product of one single factor. In several instances we can point to one particular move or event that tipped the scales. But that does not imply that this move or event is also in a more fundamental sense the most important factor. It might have been just the straw that broke the camel's back. The general lesson is that this is an area in which a search for a quick fix is likely to yield disappointing results. The same applies to simple copying of a solution that has worked well in some other instance. Rather, we should think of capacity building as a matter of manipulating causal complexes and of problem solving as a question of developing and pursuing conditional, composite strategies.

7. Institutional capacity is an asset, particularly in promoting functionally good solutions.

Organizational capacity to integrate and aggregate actor preferences—expressed in decision rules and active roles for secretariats and conference and committee chairs in providing inputs—makes a positive difference in dealing with malign as well as benign problems. Moreover, as demonstrated by, for example, the North Sea pollution cases, so-called fast-track options (provisions enabling a subset of parties to move faster or further than others) will sometimes enable front-runners to escape the law of the least ambitious program. We also find that creating a forum where government ministers themselves meet can, at least for moderately malign issues, help inject political energy into a process that is about to get stuck in discussions over technicalities. That said, we should recognize that there are important limitations to all these tools. Our study clearly suggests that organizations tend to be more dependent on powerful actors than vice versa. We should also recognize that some institutional arrangements (such as decision rules) are not easily manipulated in the "right" direction. Others are likely to have a bill attached; for example, strengthening of secretariats normally increases the costs of running an organization. These are important caveats. Yet the evidence we have does suggest that institutional capacity is an asset in dealing with almost any type of problem and that even moderate increments *can* make a positive difference, particularly in combination with other elements. Institutional capacity contributes to upgrading the common good in situations where actor interests diverge.

8. In a highly decentralized international system, entrepreneurial leadership is a critical supplement to and often a substitute for IGO capacity.

Our analysis clearly points to entrepreneurial leadership—by governments, delegations, transnational networks, or even small groups of individuals—as an important factor in developing and implementing international regimes. Informal leadership by delegates or delegations seems at least as important as leadership provided by secretariats or conference and committee chairs. There is even weak evidence to suggest that conference and committee chairs tend to be more dependent on the support of informal leaders than vice versa.

9. When the going gets tough, power is the ultimate tool.

In previous research, the so-called power-based theory has most often been equated with the proposition that the presence of a single hegemon is a necessary condition for regime establishment and implementation. This hegemonic stability hypothesis has "an extremely disappointing" record in recent empirical research (Hasenclever, Mayer, and Rittberger 1997, 103). It would, however, be a mistake to interpret the lack of support for this particular hypothesis as suggesting that power in a more generic sense has no place in the equation. Our study clearly indicates that basic game power—defined as control over the system of activities to be regulated—is a critical factor in coping with politically *malign* problems. Power seems to be a more potent factor in bringing about behavioral change than in promoting the common good and at worst seems counterproductive in dealing with benign problems. These observations suggest that the behavioral change that powerful actors can bring about will sometimes serve themselves better than their partners. The general implication thus has a distinct realist flavor to it: for malign problems the chances that a particular option will be adopted and implemented tend to increase when that option caters to the interests and values of powerful actors.

Reflections on Our Basic Model and Methodological Approach

Looking back at what we done, we have no regrets about the methodological strategy we adopted at the outset. Although many details leave significant scope for improvement, we confidently recommend the kind of dual-track research strategy that we have pursued here—one combining in-depth case studies with more extensive comparative and statistical analysis and combining the analysis of intra-regime change with comparisons across regimes. The analysis has also confirmed that it makes sense to distinguish between effectiveness as improvement of behavior and effectiveness as the achievement of an optimal solution. Regimes should be evaluated not merely in terms of whether they matter but also in terms of whether they solve the problem they were designed to solve. The question that most worries us is whether our choice of cases introduces an inherent and systematic bias in favor of findings inflating the importance of regimes.

A study of regime effectiveness, of course, has to focus on *regimes*. The results tell us something about how well these institutions have performed. What this approach cannot provide is a basis for determining the effectiveness of *regime building* as problem-solving strategy. For the latter purpose, we would have had to focus on *problems*—including those that did *not* generate institutionalized cooperation as well as those that did. This distinction is not merely a matter of academic hair splitting. If the former category is large or regime building is a slow and costly process, the finding that most regimes make a positive difference *once in place* does not warrant the conclusion that regime building is an effective *strategy* of problem solving. Since the perspective that we have adopted here is shared by almost all of our colleagues, there is a real risk that the entire field of regime analysis may be biased in favor of findings inflating the importance of international regimes as tools for solving collective problems.

Assessing the basic model we developed in chapter 1, there is some good news and some bad news. The good news is that it identified a set of important factors and accounted for a fair amount of the variance observed in regime effectiveness. The odds for success measured as significant or major improvement in actor behavior are nineteen to one when a high-capacity system deals with a nonmalignant problem that is well understood, compared to one to twelve when a low-capacity system encounters a malign problem clouded in high uncertainty. The bad news is that it has become equally clear that the model is incomplete and in some respects not well specified. We expect some of our reviewers to revel in pointing out missing factors. To us, specification failure is the more serious concern. Suffice it here to point to two of our core variables where we think important lessons can be learned. First, we can now see that we need a more complex concept of problem malignancy. The most obvious deficiency of the construct we developed in chapter 1 is that it does not capture conflict over *values*. As indicated by cases such as whaling and radioactive waste disposal, conflict over values will sometimes be a salient feature of the problem and an important determinant of processes and outcomes. We can also see a need to differentiate by subsets of parties; a problem may be more malignant within one subset than in another. Second, much work remains to be done before we have a sufficiently sophisticated concept of problem-solving capacity. The crude notion that we developed in chapter 1 needs first of all to be further *differentiated* by type of problem or function. We have already seen that it gives us little mileage in accounting for success and failure in dealing with relatively benign problems. There is also much to indicate that it needs to be differentiated by stage (agenda setting, formation, implementation) and by concept of effectiveness. Furthermore, we need a notion capturing the *interplay* between or among components. As we have shown above (item 6), some components interact synergistically, some serve largely as functional equivalents, while others tend to be mutually exclusive. It is abundantly clear that a simple additive notion of capacity cannot help us detect nor understand such interplay.

These are by no means trivial challenges. Yet we are confident that by building systematically on previous research and employing the full repertoire of relevant methodological tools, significant progress can be made in addressing all of them.

Appendixes

Appendix A
Incongruity Problems: A More Technical Description

We have briefly describe the generic problem of incongruity in more formal terms (see figure A.1). Let q_a be the fraction of total group benefits from a certain measure taken by actor A that enters A's own cost-benefit calculations and k_a be the corresponding fraction of total costs. From figure 4.1 we can easily see that if $q_a > k_a$ (meaning that actual costs are underrepresented in A's decision calculus), it will tend to pursue that option too far. Conversely, if $q_a < k_a$ (meaning that actual benefits are underrepresented), a particular course of action will appear as less attractive to A than it is to the group as a whole. Other things being equal, the larger the discrepancy between q_a and k_a, the more A's behavior will deviate from what would

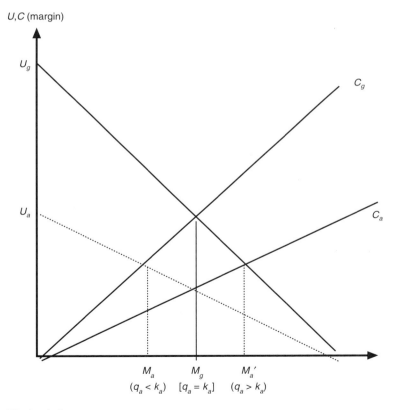

Figure A.1
The problem of incongruity

be the collective optimum. This raises the question of which actor one should look at in determining the degree of incongruity of a problem. The theory of collective action suggests that we probably would want to look first at the problem as seen from the perspective of the largest actor (see the notion of "privileged group," Olson 1965, 48f) and then perhaps view it from the perspective of the smallest. The average level of incongruity seems a less interesting notion. By implication, for the group optimum to be secured through individual utility-maximizing behavior under conditions of interdependence, we must require that $q_i \cong k_i$ (assuming $0 \le q_i, k_i \le 1$ for all group members (Olson 1965, 30–31). The more the options involved are discrete and far apart in utility terms, the more this requirement can be relaxed.

Reference

Olson, Mancur. 1965. *The Logic of Collective Action*. Cambridge, MA: Harvard University Press.

Appendix B
Seattle/Oslo Project Codebook: Selected Variables

Case Identification

Var01. Regime name
Var04. Component number
Var05. Phases
      0. Not applicable
      1. Preregime
      2. Regime formation
      3. Regime implementation
Var06. From (year)
Var07: To (year)

Independent Variables

Var15. *Type of problem*, considered on its own substantive merits only
      1. Predominantly benign (relationship of synergy/contingencies)
      2. Mixed (balanced or close to balanced)
      3. Predominantly moderately malign (mainly externalities)
      4. Predominantly strongly malign (significant element of competition)
Var16. Does some hidden agenda significantly affect the overall character of the problem as perceived by the parties involved?
      0. No, or only to a minor extent or for a few unimportant parties
      1. Yes, adds benign elements
      2. Yes, adds malign elements
Var17. Type of problem, as *perceived* by the parties involved at the time
      1. Predominantly benign
      2. Mixed
      3. Predominantly moderately malign
      4. Predominantly strongly malign
Var18. Is the structure of the *system of activities* being regulated
      1. largely symmetrical?
      2. moderately asymmetrical?
      3. strongly asymmetrical?

Var19. Is the structure of the *problem* itself (in terms of, for example, exchange of externalities or impact)

> 1. largely symmetrical?
>
> 2. moderately asymmetrical?
>
> 3. strongly asymmetrical?
>
> 4. indeterminate?

Var20. Was this how the parties perceived the problem at the time?

> 1. Yes, essentially
>
> 2. No, at least some parties perceived it as being more symmetrical.
>
> 3. No, at least some parties saw it as being less symmetrical.
>
> 4. Both 2 and 3

Var21. Are there significant *functional (substantive) linkage(s)* to other problems beyond the regime's domain? If so, is (are) the problem(s) to which the regime (or regime component) is linked more or less malign?

> 1. No significant functional or substantive linkages
>
> 2. Linkage(s) mainly or exclusively to more benign problem(s)
>
> 3. Linkage(s) to problem(s) of similar character or to more benign as well as to more malign problems
>
> 4. Linkage(s) mainly or exclusively to more malign problem(s)

Var22. If substantive linkages existed, did they influence actor behavior?

> 1. No, or only in rare cases or to a truly minor extent
>
> 2. Yes, in at least several cases or to a significant extent

Var23. Did at least some parties have significant *ulterior motives* in promoting or designing the regime (see the regime as an instrument for achieving other purposes, beyond the problem ostensibly addressed)?

> 1. No
>
> 2. Yes

Var24. Are there significant selective incentives (positive side benefits of rules and norms) involved for some actors (such as institutionalized rewards for compliance or indirect effects flowing from regime provisions, such as improving the competitive edge of particular industries or companies)?

> 0. No, or only to a minor extent
>
> 1. Yes, the regime itself explicitly include such provisions.
>
> 2. Yes, the regime *indirectly* provides selective incentives.
>
> 3. Yes, both 1 and 2

Var28. Is the *configuration of interests* for different components of the regime characterized predominantly by (applies only when a regime has two or more components)

1. crosscutting cleavages?

2. overall crosscutting cleavages but with one or a few major actors standing out as exceptions?

3. cumulative conflict (pitting the same actors against each other on different issues)?

4. overall cumulative conflict but with one or a few major actors standing out as exceptions?

5. some (balanced) mix of crosscutting and cumulative cleavages?

Var40. Institutional setting: *formal decision rule* for substantive decisions

1. Consensus or unanimity explicitly required

2. (Qualified) majority decisions permitted, with right of reservation

3. (Qualified) majority decisions permitted, no right of reservation

4. Other, please specify:

Var40a: Institutional setting: *actual decision rule* for substantive decisions ("rule in use")

1. Consensus or unanimity

2. (Qualified) majority, with right of reservation

3. (Qualified) majority, without right of reservation

4. Other, please specify:

Var43. Does the regime include specific provisions for any of the following fast-track options, or did the parties in fact resort to one or more of these (circle as appropriate)?

0. None

1. Explicit recognition of more ambitious (voluntary) efforts by a subset of actors

2. Provisional treaty application (before treaty or protocol formally enters into force)

3. Resort to less formal instruments (soft law) than originally envisaged

4. Both 1 and 2

5. Both 1 and 3

6. Both 2 and 3

7. All three

Var44. Institutional setting: role of *secretariat* (treat as a cumulative scale; score the *highest* response category applicable)

0. No secretariat (of its own)

1. Confined to office and record-keeping functions only

2. Provides some independent inputs into negotiation processes but only of a descriptive or informational nature (includes also monitoring functions)

3. Provides (also) some political inputs but of a low-key nature; its political role essentially that of a mediator or go-between

4. Provides (also) "political" inputs; acts as advocate promoting own ideas and solutions

Var45. Institutional setting: role of *conference presidents and committee chairs* (treat as a cumulative scale; score the *highest* category applicable)

1. Acts essentially as a process manager; no significant political role

2. Provides (also) some independent political inputs but of a low-key nature and essentially as mediator or go-between

3. Acts (also) as a political advocate promoting own ideas and solutions

Var46. Did some delegates or delegations play particularly important roles with regard to providing entrepreneurial leadership in the negotiations?

1. No, or only at the margins

2. Yes, provided some/weak leadership

3. Active, strong leadership

Intervening Variables

Var58. The *knowledge base*: level and scope of uncertainty

1. Low uncertainty; the basic causal mechanisms and relationships are known, and descriptive knowledge is, by comparative standards, solid.

2. Intermediate; includes high score on one dimension (theoretical or empirical) and low on the other, as well as intermediate on both

3. High uncertainty; applies to theoretical understanding of cause-effect relationships as well as to descriptive knowledge

Var59. Did the knowledge base improve significantly over time (if the score refers to a specific phase: during this phase)?

1. No

2. Yes, particularly with regard to estimates of inputs (pollution) or harvest (biological resources)

3. Yes, particularly with regard to the state of the recipient, stock or ecosystem

4. Yes, particularly with regard to the causal relationship between 2 and 3

5. Generally, along all three dimensions (2 to 4)

Var60. How much of the improvement in knowledge base (if any) can be attributed to the functioning of the regime itself?

0. Not applicable (no significant improvement did occur)

1. Only a small amount (well below 50 percent).

2. The regime itself did not contribute much to knowledge building, but indirectly it stimulated national or NGO activities. Direct and indirect contributions taken together seem to have been about as important as exogenous factors.

3. The contributions of the regime itself (not including indirect effects) seem to have been about equally important as exogenous factors.

4. When direct contributions and indirect effects are taken together, most or all can be attributed to the regime.

5. Most or all can be attributed directly to contributions by the regime itself.

6. Cannot be determined

Var61. Presence and role of *transnational epistemic communities* (use Peter Haas's criteria)

1. No transnational epistemic community can be seen operating in this case

2. Yes, but it seems rather loosely integrated (by no means a coherent actor), did not penetrate deeply into national governments or administrations, and did not play an active or influential role in regime formation or implementation processes.

3. Yes, it seems fairly well integrated, penetrated national governments or administrations to a significant extent, and played an active and influential role in regime formation or implementation processes.

Var62. *Level of collaboration*

0. Joint deliberation, but no joint action

1. Coordination of action on the basis of tacit understanding

2. Coordination of action on the basis of explicitly formulated standards, but with national action being implemented solely on a unilateral basis; no centralized appraisal of effectiveness

3. Same as level 2, but including centralized appraisal of effectiveness

4. Coordinated planning combined with unilateral implementation; includes centralized appraisal of effectiveness

5. Coordinated through fully integrated planning *and* implementation; includes also centralized appraisal of effectivenes

Dependent Variables

Var63. *Regime effectiveness*: behavioral change in relation to the hypothetical state of affairs that would have existed in its absence (refers to change in behavior regulated by the regime)

0. Negative (net) improvement (behavior changed in the wrong direction)

1. Situation unchanged, or some negative and some positive effects with no clear *net* impact on behavior either way

2. Small (marginal, slow) improvement

3. Significant, but not truly major improvement

4. Major improvement

Var66. *Regime effectiveness*: distance to collective optimum—*functional* (technical) judgment

1. The regime falls short, by a large margin, of meeting the requirements of a functionally optimal solution.

2. The regime does not meet these requirements fully, but the gap is not large.

3. The regime meets or comes close to meeting the requirements of a technically optimal solution.

4. The regime rules and regulations go *beyond* what is considered to be functionally optimal.

Var68. Has the regime substantially changed the contents or priorities of the international political agenda or the overall relationship among the participating states?

1. No (neither)

2. Yes, affected international political agenda

3. Yes, affected the overall relationship among at least some of the participating states

4. Yes, both (2 and 3)

Var69. Has the regime itself served as an important arena for transnational *learning*?

1. No, or only to a very modest extent

2. Yes, to a significant extent

Var70. *Regime development path*: Looking back at the time period covered in the case study, how would you characterize the overall path of regime development?

1. Stability no significant growth or decline

2. Fairly stable trend of incremental growth

3. Overall growth, but not stable and incremental

4. Decline

5. Other, please specify:

Var 113r. Power skew, basic game

1. Skewed in favor of laggards

2. Balanced, or in favor of intermediates

3. Skewed in favor of pushers

Indexes

Problem structure:
Var15 (1–4)
If Var15 $\geq$ 3, then add
Var18 (1 = 0) (2,3 = 1)
+ Var28 (1,2,5 = 0) (3,4 = 1)
Recoded (1,2 = 1) (3,4 = 2) (5,6 = 3)

Malignancy + uncertainty:
Var15
+ Var58
Recoded (1–3 = 1) (4,5 = 2) (6,7 = 3)

Institutional capacity:
Var40A (1 = 1) (else = 2)
+ Var44 (0–2 = 1) (else = 2)
+ Var45 (1 = 1) (else = 2)
+ Var43 (0 = 1) (else = 2)
Recoded (4,5 = 1) (6–8 = 2)

Informal leadership:
Var46 (0,1 = 1) (2 = 2)
+ Var61 (1 = 1) (else = 2)
Recoded (2 = 1) (3,4 = 2)

Problem-solving capacity:
Institutional capacity
+ informal leadership
+ Var113r
Recoded (1 = 1) (2,3 = 2) (4,5 = 3)

Appendix C
Data File: Key Variables

VAR01	VAR02	VAR04	VAR05	VAR06	VAR07
TELCOM	1	1	2	1958	1971
TELCOM	1	1	3	1958	1971
TELCOM	1	2	3	1972	1992
INPFC	14	1	2	1907	1952
INPFC	14	2	3	1952	1976
INPFC	14	3	3	1977	1984
INPFC	14	4	3	1985	1992
RADWASTE	3	0	1	1964	1972
RADWASTE	3	1	2	1972	1977
RADWASTE	3	2	3	1972	1982
CCAMLR	15	0	3	1980	1997
NPT	2	1	1	1948	1953
NPT	2	2	2	1953	1970
NPT	2	2	3	1953	1970
NPT	2	4	2	1970	1995
NPT	2	4	3	1970	1995
OSCOM	5	0	2	1972	1987
OSCOM	5	0	3	1987	1992
LRTAP	9	0	2	1977	1979
LRTAP	9	1	3	1983	1993
LRTAP	9	2	3	1985	1994
LRTAP	9	3	3	1988	1996
LRTAP	9	4	3	1991	1996
IWC	13	0	2	1944	1946
IWC	13	0	3	1948	1965
IWC	13	0	3	1965	1979
IWC	13	0	3	1979	1995
CITES	11		1	1900	1960
CITES	11		2	1960	1973
CITES	11		3	1973	1996
FFA	16	1	2	1976	1979
FFA	16	1	3	1980	1995
OILPOL	4	1	1	1920	1954
OILPOL	4	1	2	1954	1969
OILPOL	4	1	3	1969	1996
OILPOL	4	2	2	1969	1978
OILPOL	4	2	3	1978	1996
OZONE	10	0	2	1982	1985
OZONE	10	1	3	1985	1987
OZONE	10	2	3	1987	1996
PARCOM	6	0	3	1974	1987
PARCOM	6	0	3	1987	1995
MEDPLAN	7		3	1976	1985
MEDPLAN	7		3	1985	1995

VAR15	VAR16	VAR17	VAR18	VAR19	VAR20
1	2	1	1	1	1
2	2	2	1	1	1
2	2	2	3	1	3
3	2		3	3	1
4	0	4	3	3	1
4	2	4	3	3	1
4	2	4	3	3	1
1	0	1	3	3	2
3	2	3	2	3	1
4	2	3	2	3	1
3	1	3	2	2	1
4	2	4	3	3	1
2	2	2	3	3	1
2	2	2	3	3	1
4	2	4	3	3	1
4	2	4	3	3	1
3	0		3	2	1
3	0	3	3	2	1
4	1	2	2	3	2
4	0		2	3	1
4	0		2	2	1
4	0	3	2	2	1
4	0		2	3	1
4	0	3		3	2
4	0	3	3	3	2
3	0	3		2	1
2	2	4	2	2	
1	0	1	1	1	1
1	0	2	1	1	3
2	0	3	1	1	3
2		2	3	3	1
2	1	2	3	3	1
4	2	4	3	3	1
4	2	4	3	3	1
4	2	4	3	3	1
4	2	4	3	3	1
4	2	4	3	3	1
3	0	3	2	1	1
2	0	3	2	1	1
2	0	2	2	1	1
4	0	2	2	3	
4	0	4	2	3	1
4	2	4	2	2	4
3	2	3	2	2	1

VAR01	VAR21	VAR22	VAR23	VAR24	VAR28
TELCOM	4	1	2	1	1
TELCOM	4	1	2	1	5
TELCOM	4	1	2	1	1
INPFC	4	2	2	0	3
INPFC	4	2	1	1	3
INPFC	4	2	1	1	3
INPFC	1		2	0	3
RADWASTE	1		1	1	
RADWASTE	4	2	2	1	3
RADWASTE	4	2	2	1	3
CCAMLR	4	1	2	0	
NPT	1		2	0	5
NPT	4		2	1	5
NPT	4		2	1	5
NPT	4	1	1	1	5
NPT	4	1	1	1	5
OSCOM	4	2	1	0	
OSCOM	4	2	1	0	
LRTAP	3	2	2		4
LRTAP	1		2	0	4
LRTAP	1		2	0	4
LRTAP	1		2	0	4
LRTAP	4	1	1	0	
IWC	1		1	0	
IWC	1		1	0	
IWC	1		1	0	
IWC	1		1	0	
CITES	1		1	0	
CITES	4	1	1	0	
CITES	4	2	1	0	
FFA	1		2	1	5
FFA	3	2	2	1	3
OILPOL	3	2	2		
OILPOL	3	1	2	2	1
OILPOL	3	2		2	1
OILPOL	3	2	2	2	1
OILPOL	3	2		2	1
OZONE	1	1	1		3
OZONE	1	1	1	1	3
OZONE	4	1	1	1	3
PARCOM	2	1	1	0	
PARCOM	2	1	1	2	
MEDPLAN	1		2	3	
MEDPLAN	1		2	3	

VAR40	VAR40A	VAR43	VAR44	VAR45	VAR46
2	2	7	3	2	2
2	2	7	3	2	2
3	2	0	3	3	2
1	1	0	1	1	2
1	1	0	1	1	2
1	1	0	1	1	2
3	3	0	1	1	2
2	1	1	4	2	2
2	2	3	2	2	2
2	2	3	2	2	2
1	1	0	1	1	1
4	4	0	0		0
2	2	0	4	3	1
2	2	0	4	3	1
2	2	0	4	3	2
2	2	0	4	3	2
1	1	0	2	1	1
1	1	5	2	1	1
1	1	0	2	1	1
1	1	0	2	1	1
1	1	0	2	1	1
1	1	0	2	2	1
2	2	0	0	1	1
2	2	0	0	2	0
2	2	0	1	1	0
2	2	0	2	2	0
					1
2	2	0	3	3	1
2	2	0	3		1
2	1	0	0	3	2
1	1	6	3	3	
3	3	0	2	1	0
3	3	0	2	1	0
3	3	2	2	1	0
3	3	2	2	1	0
				3	1
2	1	0	4	3	2
2	1	0	4	3	2
2	1	0	3	2	1
2	2	5	3	2	1
2	2	1	3	3	2
2	2	1	3	3	2

VAR01	VAR58	VAR59	VAR60	VAR61	VAR62
TELCOM	1	1	0	3	2
TELCOM	1	1	0	3	2
TELCOM	1	2	5	1	5
INPFC	3	5	5	1	2
INPFC	1	5	2	1	2
INPFC	1	1	5	1	2
INPFC	1	3	5	1	4
RADWASTE	2	5	5	2	5
RADWASTE	1	5	5	3	5
RADWASTE	1	5	5	3	5
CCAMLR	3	3	2	1	2
NPT	3	1	0	1	0
NPT	3	6	4	2	2
NPT	3	6	4	2	2
NPT	1	5	5	3	2
NPT	1	5	5	3	2
OSCOM	2	2	3	1	3
OSCOM	2	3	3	1	3
LRTAP	3	2	1		3
LRTAP	2	4	3	2	3
LRTAP	3	4	3	2	3
LRTAP	3	2	4	2	3
LRTAP	1	5		2	3
IWC	3	1	0	1	0
IWC	2	3	2	1	2
IWC	2	4	2	1	2
IWC	1	5	3	1	3
CITES	3	1		1	1
CITES	3	3	0	3	2
CITES	2	3	2	3	2
FFA	3	1	0	1	2
FFA	3	5	2	3	5
OILPOL	1	1		1	0
OILPOL	1	1	0	1	2
OILPOL	1	1	0	1	2
OILPOL	1	1	0	1	2
OILPOL	1	1	0	1	2
OZONE	2	1	0	2	0
OZONE	2	1	2	3	3
OZONE	1	5	4	3	3
PARCOM	3	1	0	1	3
PARCOM	2	2	4	1	4
MEDPLAN	3	2	3	3	3
MEDPLAN	2	2	3	3	3

VAR63	VAR66	VAR68	VAR69	VAR70	VAR113R
3	1	1	1	2	3
3	1	1	1	2	3
4	2	1	2	2	3
3	1	1	2	1	3
1	1	2	1	4	3
4	3	1	1	3	3
4	4	1	1	3	3
		2	2	1	3
4	3	2	2		3
4	3	2	2		3
2	1	1	1	2	1
		4	2	3	2
3	2	2	2	3	3
3	2	2	2	3	3
4	3	4	2	3	3
4	3	4	2	3	3
1	1	1	2	2	1
4	3	1	2	2	1
				2	1
2	1	1	2	2	2
2	1	1	2	2	2
2	1	1	2	1	2
3	1	1	2	2	2
					3
1	1	1	1		1
2	3	1	2	2	2
3	4	3	2	5	2
				2	3
1	1	1	1	2	3
2	2	2	2	3	2
2	1	3	1	1	1
4	3	3	2	5	1
					2
1	1	1	1	2	1
1	1	1	1	2	1
3	1	1	1	2	1
3	1	1	1	2	1
					2
1	1	1	2	3	2
4	3	1	2	3	2
1	1	1	2	1	1
3	2	1	2	1	1
1	1	1	2	2	2
1	2	1	2	2	2

Contributors

Steinar Andresen, Senior Research Fellow, Fridtjof Nansen Institute, Lysaker, Norway

Elaine M. Carlin, Research Scientist, Joint U.S./Norwegian Research Team, School of Marine Affairs, University of Washington

Maaria Curlier, Vancouver, B.C., Canada.

Edward L. Miles, Virginia and Prentice Bloedel Professor of Marine and Public Affairs and Senior Fellow at the Joint Institute for the Study of Atmosphere and Oceans, University of Washington

Jon Birger Skjaerseth, Research Scientist, Fridtjof Nansen Institute, Lysaker, Norway

Arild Underdal, Professor of Political Science, University of Oslo and the Center for International Climate and Environmental Research (CICERO)

Jørgen Wettestad, Senior Research Fellow, Fridtjof Nansen Institute, Lysaker, Norway

Index